GLOBAL TELECOMMUNICATIONS STRATEGIES AND TECHNOLOGICAL CHANGES

GLOBAL TELECOMMUNICATIONS STRATEGIES AND TECHNOLOGICAL CHANGES

Edited by

Gérard Pogorel

School of Business Administration
National Institute of Telecommunications
Evry, France

1994

NORTH-HOLLAND

AMSTERDAM - LONDON - NEW YORK - TOKYO

ELSEVIER SCIENCE B.V.
Sara Burgerhartstraat 25
P.O. Box 211, 1000 AE Amsterdam, The Netherlands

Library of Congress Cataloging-in-Publication Data

Global telecommunications strategies and technological changes /
edited by Gérard Pogorel.
p. cm.
Papers presented to the Ninth International Telecommunications Society (ITS) International Conference, held in Sophia Antipolis, France, in June 1993.
Includes bibliographical references and index.
ISBN 0-444-89960-X (alk. paper)
1. Telecommunication policy--Congresses. 2. Telecommunication--Congresses. 3. Technological innovations--Congresses.
I. Pogorel. Gérard. II. International Telecommunications Society. International Conference (9th:1993:Sophia Antipolis, France)
HE7645.G583 1994
384' .068--dc20 93-44144
CIP

ISBN 0-444-89960-X

This book is printed on acid-free paper.

Printed in The Netherlands.

Preface

The papers collected in this volume were all presented to the Ninth International Telecommunications Society (ITS) International Conference, held in Sophia Antipolis, France, in June 1993. ITS is a Learned Society composed of telecommunications professionals from all countries. Although many participants in this conference are personally engaged in telecommunications management or policy, they spoke there on their own behalf, with no immediate operational purpose. They met as both actors and analysts, to freely exchange the results of their research and their thoughts in the stimulating areas of communication. A community of companies, universities and research centers, consulting firms, telecom users, and international organizations, supports its activities as an open forum for the exchange of ideas in the areas of telecom economics and management.

The choice from the 160 papers presented at the Conference, made by an editorial committee composed of Cristiano Antonelli, François Bar, Laurence Caby, and myself, is based on their common relevance to a theme exposed in the title of this book: Global Communications Strategies and Technological Changes. Each co-editor has taken care of one section.

As Conference Chairman, I would like to thank the authors and the co-editors for the work which has been involved in revising and completing the papers for this publication. Our special gratitude goes to *Digital Equipment*, and to its Sophia Antipolis-based *Worldwide Telecom Group*, which has kindly accepted to sponsor the publication of this volume.

Since the last ITS International Conference, held in Venice in 1990, many changes have occurred. In Eastern and Central Europe, an economic and ideological system has collapsed. There is still no bread to be seen in many bakeries there, but telecom networks are blooming. Telecommunications will build the infrastructure of the expected technical leapfrogging. In Asia, and especially in South-East Asia, the economic development is based on a swift implementation of technical changes. This means in most cases a communication-based development. In those countries which paved the ways of telecommunications development, a new wave of technologies is booming. But the debates on market structures and regulation are protracted, and opposing views flourish. This holds in the U.S.A., where the existing regulation is still being debated. It holds in Europe too, where basic, pro-competitive changes in the regulatory framework will be implemented in 1998. In developing countries, it is to be questioned whether it will be possible to meet the communications needs without increasing the gap between the communication haves and have-nots.

ITS in general, and the contributors to this volume, intend to contribute to finding appropriate ways and means to take positive advantage of the formidable opportunities presented by telecommunications technologies.

Gérard Pogorel
Ninth ITS International Conference Chair

Contents

Global Telecommunications Strategies
and Technological Changes
Edited by G. Pogorel

Introduction

Telecommunications networks in transition: a 21st century perspective

Gérard Pogorel

School of Business Administration, National Telecommunications Institute, F-91011 Evry Cedex, France

For more than a century, the production, sale and exchange of telecommunications services in the major world markets have had to be aligned with national, regional or international regulatory systems which combined different types of monopoly and competition. This co-existence has created asymmetry and tension between contradictory forces. On the one hand, the monopolies, linked by standardization or harmonization, tended toward the establishment of unified or interconnected networks with the aim of providing what it is agreed can be called a market place. On the other hand, competition, diversification of products and services, less specificity and a greater degree of complementarity in telecommunications services, together with other activities, are favouring the recognizedly contagious game of competitive advantages.

The provision of communication services has become technically accessible to suppliers of a more varied type or in accordance with increasingly varied procedures. As newcomers and different forms of complementarity appear on the scene, synergy is established with other activities. Networks which are distinct from the public network are created. Some utilize the infrastructure supplied by the latter (leased lines), others are partly or entirely built on their own infrastructures (satellites, radio transmitters, cable networks, etc.). There is a danger of fragmentation which, in certain countries, is actually already occurring. Industrial type restructuring is taking place and is leading to a growing resemblance between the existing supply of communication services and those offered by other categories of service suppliers, which is helping to create a degree of banality in the sector. The questions to which this gives rise are related to the resultant of these forces in the foreseeable future, their effects on the development of the industrial structures of the sector and on the formulation of competitive network strategies. We will put special emphasis on future developments in Europe, where in-depth evolutions are to take place in the forthcoming years.

1. Telecommunications in Europe: the road to competition

Telecommunications within the Community have long seemed to be in a metastable state. Contagion from British, American and Japanese deregulation has first been contained on the Continent. Each of the public telecommunications operators has largely preserved its field intact. In regard to network interconnection, the operators have hitherto favoured the forms best known to them, from long-standing co-operation to a number of alliances, thus preserving the gentlemen's agreements to which they have been accustomed. This brings them into contact with modern strategies, which have developed spectacularly in all sectors in recent years. However, although it is possible to be captivated by this modern tale of prince meets princess which can only result in a happy ending, there must be no hiding away from the dominating reality of the great industrial and service activities practised around the world, closer to competitive sports where there are only few places reserved on the podium: the creation of a large world oligopoly structure.

The European Community and European Governments have now engaged telecommunications activities in a process of changeover. The conjunction of technological and competitive forces brings about such complex effects that even the best informed admit that they are incapable of foreseeing what the world of communications will be like in ten or twenty years. Until late 1992, the principal task has been to introduce into the GATT framework procedures similar to those now being adopted within the Community. The contents of the telecommunications annex connected to the draft 1992 GATS (General Agreement on Trade in Services) were indeed expressed in a very European language. Representatives of the Community had succeeded in having telecommunications services, which are both services in themselves as well as supports for other services, recognized as a specificity. From this specificity, two factors have emerged: firstly, the need for ONP type systems and, secondly, the recognition of the necessity for developing countries to create their own infrastructures. However, the combination of reserved services plus ONP on the one hand, and open services on the other, could not eliminate the question of an offensive European telecommunications strategy. In this connection, two issues stand out: to start with, for reasons mentioned below relating to the functioning of international institutions, the achievement of the ONP could resemble a long-distance race with the development and ever more rapidly growing diversification of technologies and services.

Secondly, in view of the globalization taking shape in the realm of services, this policy appeared too cautious, conservative and stripped of any strategic vision contributing to the opening-up of world markets. The risk of seeing certain weakened European operators surviving only as suppliers of commonplace services at falling prices could not be overlooked. To give consideration only to the regulatory and structural status quo hypothesis therefore seemed quite inadequate. Different scenarios must now be imagined which are more likely to energize the operators and ensure a competitive strategy embracing services and equipment. What must be done, therefore, is to explore the connections that may exist between the formulation of a European telecommunications policy and the European attitude in the relevant international forums concerned. Institutional and industrial issues have to be raised.

2. *International negotiations: an overlap with industrial strategies*

At the institutional level, it is the respective roles of the national governments of Europe, of the Community organs (and the different Directorates of the Commission), of the CEPT, ITU and CCITT and of GATT which are at stake. International negotiations are becoming increasingly complex and the results more uncertain because of the existence of several dimensions of complexity.

2.1. The ITUs vocational crisis (1)

Up to the present time, the ITU and the technical bodies and conferences which are working towards the establishment of a world telecommunications network have built up one of the most noteworthy examples of successful multilateral co-operation. Operators apply the technical and regulatory standards drawn up by the ITU, and have been doing so for more than a century. The ITU has thus escaped the crises of legitimacy from which the UNO and various specialist organizations have suffered at different times.

Changes in the regulatory environment, however, are putting the functioning of multinational co-operation to the test.

The factors which have facilitated multilateral co-operation within the field of telecommunications are extremely powerful. As long as there were only state monopolies, international communications could only be ensured by means of joint agreements. Obviously, a multilateral standardization of the clauses of these agreements makes them easier to follow and respect. Moreover, this co-operation is further facilitated when the number of participants is limited (one for each country), when each is direct control of his environment, when technical change occurs gradually and when there is consensus on the technique and nature of the public service.

These are the characteristics which have given stability to the international system governing telecommunications and which are now being questioned. Technological developments and the increase in the number of players participating in national markets now make it economically feasible to provide services without recourse to joint agreements. In addition, an acceleration in the rhythm of technical progress, the uncertainties and risks which service markets of the future will be facing, and the growing competition for the conquest of these markets are affecting the spirit of co-operation which has long characterized the world of telecommunications.

The standardization process has already felt the effects. While it is still enjoying some success (GSM in Europe), it has been overtaken in other sectors by isolated or regional initiatives (ISDN). The role of the ITU has been called into question by both established operators and newcomers who claim to be able to ensure the provision of international services defined by themselves, without following the path of multilateral standardization.

The specificity of telecommunications has for long retained its strength and has allowed this sector to steer clear of conflicts of ideology and interests. Today, idealogical differences and competition are becoming more pronounced. True, in the case of the ITU, it is even more so where INTELSAT and CEPT are concerned.

2.2. The break-up of historical structures

The second dimension of complexity, which is instilling a sense of disquiet in those whose world is telecommunications, is the inclusion of their activity, considered as specific for a century, in the 'potlatch' of the GATT negotiations on goods and services, resulting inevitably in a process of bargaining in which different sectors, e.g., agriculture, broadcasting and telecommunications, are mixed together.

The third is the end of the systematically direct representation of telecommunications operators by the government with the introduction of varying forms of denationalization, autonomization, liberalization of competition or privatization, and the appearance of the corresponding players on the scene. Although firms (equipment suppliers and operators) were already represented, de facto, in the ITU standardization committees, this multiplicity has created a new element in the functioning of GATT and transformed the CEPT into a regulatory organization, while European operators have created their own structure (ETNO). As to the role of the different Community instances and of the Services of the Commission, the only possibility is for it to grow. The general implementation of the Maastricht decisions, some of which will have institutional consequences yet to be experienced, will be combined with specific Community choices regarding the role of Brussels in the telecommunications sector and the legal means to back it.

3. Telecommunications: toward a world oligopoly?

Looking now at the subject of competitive strategies within telecommunications, the following question arises: will this sector become an oligopoly tomorrow like the others? To what degree can a comparison be drawn with the international development of other sectors of activity in recent decades? The first elements of a reply to this question may lie in an elementary analysis of the internationalization or multinationalization process of the major companies and then in examining whether these factors are applicable to telecommunications.

The search for low-cost resources (raw materials, manpower) is only of minor relevance here. Nonetheless, some operators can be seen accompanying equipment suppliers in the creation of R&D centres (software, optoelectronics) in Central and Eastern Europe in order to take advantage of the access to cheap, qualified manpower. With regard to the search for new markets, this goes hand in hand with a circumvention of current protectionist barriers to give operators a place beyond their own borders. The exploitation of competitive advantages (product cycle) is a further powerful factor the major operators are counting on for ensuring a future place for themselves, with the simultaneous aim of creating technological and financial strike forces (critical mass). Lastly, the provision of global services for global customers is an important factor in the restructuring process. In an endeavour to adapt themselves to the size and structure of the market, operators are involving themselves in a mimetic development: to supply the major world companies, it is necessary to resemble them.

On the world-embracing telecommunications chessboard, the game has started and it is instructive to observe the moves in progress. It is estimated that there are between 200 and 300 operators in the world at present. They work in a situation of dynamic competition, conforming to classical processes on the one hand (newcomers, establishment of companies, a high mortality rate, mergers, takeovers and already the beginning of concentration into large groups), and on the other to new processes such as agreements and the creation of consortia and networks of diversified alliances.

These different international actions follow varying paths:

- in the case of traditional monopolies, a reliance on interconnected international lines in which they are strongly rooted,

- the beginning of globalization in certain markets on a technological basis (use of satellites) or in an open market segment (value-added services, EDI),
- participation in the activities of foreign national operators (Argentina, Mexico, Chile, Australia, Thailand)
- the provision of new services in some countries where there is relatively little know-how (mobiles),
- otherwise, independently of the acceptance of competition in the value-added services market, competition is also being introduced into the rapidly expanding market sectors comprising mobiles, cable and satellites.

Among these forms of action, only one possesses a truly global characteristic, tending towards the provision of one or more services on a worldwide scale, while the others are of a multi-local nature conducted in juxtaposition with similar activities in several countries but each within an individual framework. It should be noted, however, that for an activity with a great connectivity potential this distinction is weaker than in other sectors: accompanied by technical and regulatory development, the addition of these multi-local operations could result in a global activity.

A process of this nature will tomorrow give birth to major, multi-activity international operators, with others occupying niche markets and the rest being confined to stagnant markets or left by the wayside.

In all the principal sectors (petroleum, automobiles, chemicals, electronics, information processing and telecommunications equipment), there are 12 to 20 companies which shape activities on the world plane in a significant fashion, following the paradigm of competition among the few. Although alliances certainly play a role in this context, it is internal and external growth that predominates. This perspective is not altogether incompatible with the establishment of a market place, but this would have to be done according to particular methods and it would affect the modes of operation of the European Union in formulating industrial and political competition strategies.

4. Towards a European competitive strategy

The required definition of a European attitude on the international telecommunications scene must meet the following legitimate objectives:

- to provide private persons and firms in Europe with access to the best of services at the best of prices,
- to maintain the growth of its strong equipment industry by permitting a satisfactory number of European companies to retain a place in the world equipment oligopoly (as is the case at present),

- to improve the competitiveness of operators in the telephone and network sectors and in the many services which have yet to prove their validity or which are still to be defined,
- to make sure that European telecommunications operators are among the 12 to 20 of the future oligopoly, which is the case today and which should remain so.

Time is running short now, and an insufficient degree of competitiveness of European communication firms would result in an invasion of European conduits by non-Community information suppliers. At the same time, however, it is not certain that keeping to the juxtapositioning of state ex-monopolies is an acceptable, competitive solution. Whatever their past or present technological performance, these monopolies, or at least some of them, remain exposed to the dangers which traditionally threaten monopolistic organizations, whatever their past or present technical excellence.

The ISDN is a good example. For a decade, it was considered that this was the perfect vehicle for the creation of a market place. Monopoly operators in Europe had time to utilize their powers of decision and good means, especially those of a Community nature, to spread a European network endowed with a common standard. Despite unduly late efforts, this has not been achieved and the Europeans are benefitting only gradually from the continental ISDN which might have given them a competitive advantage. It has to be noted, therefore, that the present European structure, comprising juxtaposed monopolies linked bilaterally or multilaterally, is no guarantee for the creation of a market place. Now, for future purposes, a certain number of services should be put on offer on a European scale: broadband leased lines, broadband ISDN, intelligent networks, broadband networks, etc. The question is: who will offer this Global European Network? If it is a consortium of historical multi-headed ex-monopolies, how autonomous will it be and, consequently, how commercially effective? Or will there be newcomers penetrating the market: Hermes from the railways or other RBOCs which can already be seen weaving a web in the mobiles sector, multi-local today, perhaps global tomorrow?

These considerations cannot but be reinforced by the fact that, as the work of the Commission has shown, there is a strong interaction between the industrial success of prosperous equipment manufacturers and the financial capacity of the operators who constitute their principal clients. Now, the particular structure of the American and Japanese markets favours vertical integration there, de jure with AT&T and AT&T Technologies in the United States, de facto with the NTT family in Japan. It is thus all the more positive that the structuring and international extension of services and therefore of operators come closer to that found in the industrial sector. The development of the one and the other, within Europe and outside Europe, will have to go hand in hand.

In this respect, the essential question is thus one of date and tempo which must combine the following:

- continued development within the communication and transmission equipment industries and the strengthening of terminal equipment manufacturers (the same manufacturers or others);
- at the same time, an increase in the strength of European operators, both on the continental and international plane.

5. Competition, public service, transborder operators: the three elements of a European strategy

To be competitive, a European strategy must comprise three inseparable elements: public service, transborder operators and competition. As competition is now under way, the two remaining elements have to be considered seriously as well.

By reason of its culture and the historical tradition of its operators, Europe has a prime role to play in defining and promoting within the international organizations the notion of public service in telecommunications.

5.1. The future of the public service around the world: the freedom of communication?

In the industrialized countries, the notion of a universal public telephone service has remained a (relatively) simple notion only where there is essentially one single supplier. The burgeoning of new techniques complicates the issue. In the developing countries, the problem is how to remedy the deficiencies in communications other than by widening the gap between the haves and the have-nots of communication. Communication and exchange needs increase exponentially, but do so under different conditions of classical telecommunications development in earlier industrialized countries which have witnessed the slow spread of an almost uniform, general-purpose product.

The present technological proliferation on the one hand, and the combination of delays and technological leaps on the other, are thus now creating widely varying conditions. Should the notion of a public telephone service henceforth be regarded as outdated? The recent historic experience in Central and Eastern Europe prompts us to think otherwise. The archaic tyrannies established there, of which there unfortunately remain many examples in the world, used mass communication as a tool and made it a principle to limit interpersonal and communal communication. One cannot believe that a general availability of telephones is sufficient to create a democracy and make it work, although it must be admitted that it helps. It is difficult to imagine todays citizens without unrestricted means of remote communication.

As emphasized by the Secretary-General of the ITU in his introductory speech to this conference, the freedom of communication is a corollary to the freedom of expression and information and, as such, should be included in the texts of the United Nations. Admittedly, these rights remain quasi-rights, written in both national and international texts, but still not always practised. It is necessary to define the contents of this freedom of communication, to include in it the universal public service principle in both advanced and developing countries, to give it substance. This will give rise to legal, political and technical problems. Europe must encourage the examination of these problems-an examination which should be conducted by the specialized international and regional organizations (ITU, CEPT), which would represent a contribution within trade organizations (GATT) and which could involve the drawing up of codes of good conduct by governments or between operators. It would be a positive factor if communication could find a framework within GATT that would give it an improved role in world exchanges. This does in no way, however, obviate the need to carry on discussions in other United Nations organizations, and especially within the ITU, leading to technical and legal advances (standards and a definition of public service).

5.2. Transborder operators

Eventually, however, it is the structure of relations between operators in Europe which will have to suffer a change. It would be a far better bet for Europe and European operators to envisage a recomposition of their landscape, close alliances, a pooling of means, the provision of suitably scaled common services in well-defined sectors.

These advances oblige operators to develop their managerial culture. Their growth has long been based on a product (the voice), which still accounts for 90% of their revenue. They endeavour to maintain this growth by conquering markets and by selling new services. For a large number of traditional operators, however, this has to be achieved by means of a difficult organizational transition. They are compelled to develop a competitive culture, but it is a complex operation to change from an administrative organism into a company confronted by international competition and it does not proceed without painful consequences.

Moreover, operators exercise a profession, or rather a set of competences ranging from the sale of low-priced products and services for the general public to the supply of capital goods worth billions. Tomorrows operator will have to address mass-consumer markets immediately understable messages, while possessing the creativity of consumer-electronics manufacturers, for instance, and the capacity for devising complex avant-garde systems like NASA or Arianespace. This exercise of exceptional scale, scope and speed implies absolutely top level strategic and managerial capacities. Human resource management, an activity often neglected in an environment with a highly technical culture, must reconcile hardly reconcilable demands:

- to become more competitive,
- link technical culture and research with marketable applications,
- to ensure internal and external flexibility,
- to maintain the confidence of personnel in the management and their leaders.

They will have to prove this know-how vis-à-vis customers, competitors, shareholders and suppliers of capital for activities whose needs are among the greatest. It is not surprising to discover a race in the search for managerial structures capable of meeting these objectives, a race which is accompanied by a trend towards a break-up of organization and market structures. It is now urgent for European firms to implement the creation of large European telecommunications firms with world ambitions.

Part 1: New Telecommunication Services

Coordinated by Laurence Caby, *National Institute of Telecommunications*

Chapter 1

Business demand for value added services in Europe

Matthias-Wolfgang Stoetzer

Wissenschaftliches Institut für Kommunikationsdienste, Rathausplatz 3–4, D-53604 Bad Honnef, Germany

1. Introduction

This paper takes a first look at the business demand for New Telecommunication Services (NTS) — also called Value Added Services (VAS) — in Europe. There exists no consistent and uniform definition of the term VAS, but they include online database access, electronic document interchange, electronic mail and bulletin boards, travel reservation systems, telebanking, videoconferencing and so on. From a technical point of view, their development is due to the progress of the data processing and transmission capabilities and the integration of computer and telecommunication technologies. It has been argued that NTS are at the leading edge of the moves towards the 'information society'. Growing business demand for these services is seen as reflecting the fact that NTS are vital for the competitive performance of European industries. This is especially important in circumstances characterized by rapid innovation, internationalization of competition, etc. Here the fast access to information and rapid communication flows become strategically central.

With regard to business demand for VAS in Europe very little is known. Many hypotheses, allegations, confirmations and common sense considerations exist, but only very rarely are they based on firm empirical facts and statistical tests. Therefore, section 2 summarizes very briefly the theories of business demand for telecommunication services and presents the basic hypotheses and empirical findings as to the factors influencing the demand for VAS. The following part (section 3) describes the data used and provides a short descriptive analysis of the demand for VAS in the European Community. Univariate and multivariate analyses of the data set are the subject of section 4 in order to test some basic hypotheses.[1] Finally, section 5 summarizes the main findings and shortcomings and points out future directions of research.

2. *Business demand for VAS*

Despite the fact that revenues generated by business telecom services are a substantial part of the carriers income, business demand has not received much study. A few empirical studies exist but there is still no theoretical model of business telecommunication demand (Brink and Nieswiadomy, 1992, p. 11; Taylor, 1980). There are several reasons for this lack of attention. Firstly, business customers purchase much more complex services in comparison to private households. These services comprise leased lines, centrex, ISDN, VSAT, Value Added Services and so on. Residential demand in general is limited to the simple plain old telephone service. Secondly, theories of private demand can rely on the theory of consumer choice and maximization of utility. In contrast, business demand should be derived from production theory. Demand for telecommunication services is not an end in itself. Telecommunications is an input to the production process and the business-demand function therefore depends upon the firms production function. Thirdly, because households do not vary much as to their tastes it is rather easy to model their behaviour. Businesses show a great heterogeneity as to their size, corporate structure, number of locations, communication needs and so on. A theory of business demand for telecommunication services would have to incorporate all these diversities. Thus business-demand models are becoming rather disaggregated and complex theories.

With regard to demand for telecommunication services in general, it turns out that as to (voice) telephony a lot of empirical research exists. These studies rely on sometimes large data sets, examine price and income elasticities

---

[1] From a macroeconomic point of view, the European market of information and telecommunication industries is supposed to have an increasing importance. A critical examination of the forecasts with regard to the development of the market volumes for VAS in Europe from 1980 to the year 2000 reveals that the methodological foundations of these estimates and forecasts are weak or even non-existent. These shortcomings as to the volume of the market for new telecommunication services are due to the fact that research on the microeconomic level is lacking or only restricted to examinations of the supply side. These supply-side market estimates are in general based only on a few case studies or personal interviews (see Stoetzer, 1992a, b). Demand forecasting based on market surveys is proposed by the International Telecommunications Union (see Engvall, 1991).

and become more and more sophisticated.[2] Compared with this situation, only a limited number of empirical demand-side analyses of VAS are at hand. Most of them concentrate on special types and applications of telecommunication services and do not try to be comprehensive.[3]

At a theoretical level the relevant literature argues that all these NTS are supposed to offer new or enhanced functionalities for the private and business user. Therefore both types of users should exhibit a great amount of potential demand. In economic terms this means the consumer and the enterprises should be very willing to pay for VAS. Concerning the private household this argument is not further developed. Some examples such as, e.g., picturephone and videotex are mentioned to prove the market potential of new telecommunication services.[4] As to the business demand the argument is simply that these telecommunication services are necessary in order to gain a competitive edge or at least not to loose relative competitive strength. A sophisticated version of this reasoning says that the worldwide economic development in general and the European Single Market of 1992, in particular, lead to an internationalization of the markets. This implies an increasing competition. The emergence of more competitive markets in many sectors results in an elevation of the transaction costs of the firms. The transaction costs are the costs of using the market as a device for coordinating demand and supply. They include the costs of looking out for suppliers and customers, negotiating a deal and subsequently monitoring and enforcing contracts. All these parts of transactional costs are to a great extent linked to problems of communication and information. Therefore, new telecommunication devices offer a possibility to communicate, gather and spread information in a more rapid and reliable way. Thus, they are reducing transaction costs and turn out to be a key factor of competitiveness.

Summarizing the literature, it follows that in the first place, business demand is a neglected field of research and that in the second place, the lack

---

[2] Nearly all of the existing studies concentrate on the use of the telephone network because of the availability of valid and reliable data (see Bewley and Fiebig, 1988; Breslaw and Pizante, 1989/1990; de Fontenay et al., 1990; Nijdam, 1990; Zona and Jacob, 1990; Lang and Lundgren, 1991; Appelbe et al., 1992; Hakim, 1992; Walter, 1992). In general, these analyses do not make a difference between varying types of calls, e.g., voice telephony, fax and file transfer by means of a modem. The demand for the telex service in Portugal is estimated by Cabral and Leite (1992). For LANs and WANs see Linhart et al. (1992).

[3] For Germany see Fritsch (1987), Ewers et al. (1990) and Cantzler (1991), as to the special E-mail service Teletex see Köhler (1987), for Switzerland see Müdespacher (1987), for telecommunication services in general in Sweden see Taymaz (1990). As to the use of E-mail and other electronic networks by individuals inside companies in the USA, see Bishop (1992).

[4] Empirical studies of residential demand for the adoption of High Definition Television, cable-TV and satellite dishes exist (Dupagne and Agustino, 1991; Litman et al., 1991).

of knowledge is particularly important as to Value Added Services. In particular, there is still no theoretical model of business demand for Value Added Services despite the evolving literature in the field of business telecommunications demand in general (Taylor, 1993).

Therefore, hypotheses concerning the use of Valued Added Services have to be developed more or less ad hoc. They are part of a common-sense knowledge gained from many case studies and several representative empirical investigations. Five influencing factors are identified as being important for the use of these new telecommunication services. [5]

Firstly, all the studies conclude that the dissemination of VAS varies with the countries at hand. It is well accepted that the United Kingdom leads the development of VAS in Europe, followed by France and that Germany and in particular the countries in the south of Europe show a certain lag as to their use thereof (OECD, 1989, pp. 85–112; PACE 90, 1990, p. 40). This dissimilarity may be simply due to different industrial structures and stages of economic development. Other explanations are restrictive regulatory regimes limiting the supply of new telecommunication services, different technical quality of the telecommunication infrastructure or different telecommunication tariffs (Antonelli, 1989, p. 257). Also a government policy subsidizing certain VAS may be a determining factor (e.g., Videotex in France).

Secondly, even if the knowledge as to the role of telecommunications in the production process is very limited, it is clear that this role will vary from business to business. It seems obvious that certain products and production processes are very information intensive in comparison to others. Therefore the demand for VAS should be high in business sectors that are information intensive and rather low in other sectors of the economy (Taylor, 1980, p. 61; Porter and Millar, 1985; Antonelli, 1989, p. 257; Antonelli, 1989/1990, p. 50). For instance, finance and transport are claimed to depend on a high level of information and communication flows (Thomas and Miles, 1989, p. 11; Cantzler, 1991, p. 188; Benzoni et al., 1992, p. 454).

Thirdly, the relevant literature points out that establishment size has an important influence on the use of VAS (Stoneman, 1983, pp. 97–100; Fritsch, 1987, p. 67; Miles and Thomas, 1990, p. 46; Taymaz, 1990, p. 167; Witte and Dowling, 1991, p. 451). Two arguments support this allegation. In the first place, new telecom services tend to be regarded as both innovative and complex. Smaller enterprises have problems in using these services because of

---

[5] Antonelli (1989) discusses the importance of prices in comparison to other factors determining the demand for NTS. Also meaningful independent variables of a demand equation are income and number of users. Cronin et. al. (1993) analyse the interrelationships between telecommunications and other factor inputs and the influence of relative factor prices on the demand for telecommunications in general across US industries.

a lack of awareness and a lack of know-how. In addition, it seems obvious that an increase in firm size engenders the necessity of internal coordination and leads to a growth in the volume of internal and external communications, thus emphazing also the use of new and more efficient forms of telecommunication services.

Fourthly, several studies put forward the idea that there exists a strong link between the adoption of NTS and the use of data-processing, computers and information technology in general (Fritsch, 1987, p. 68; Antonelli, 1989, pp. 259, 263; Cantzler, 1991, p. 189). Information technology is closely linked to the availability and internal supply of specific technical knowledge that can be instrumental for sophisticated process innovations like VAS.

A fifth hypothesis concludes that the degree of internationalization of an enterprise, e.g. concerning the marketing of services and products offered, favours the adoption of new telecommunication services (Antonelli, 1985; Preißl, 1988, p. 7). A company selling its outputs on the world markets is likely to use the latest information technologies, e.g., E-mail services and online databanks.

3. *Description of the sample*

The data set was collected by Scicon Networks, London, as part of a study of the European market of Value Added Services for the Commission of the European Community. It consists of a telephone survey of over fifteen hundred establishments of all sizes structured across industry sectors and all member states of the European Community. In every organization the people most responsible for telecommunications and computing were interviewed. The telephone interviews are based on a questionnaire that was designed to obtain information on the use of VAS by businesses. The questionnaire covers aspects such as: are enterprises users or non-users of Value Added Services ?; which services were being used?; what are the reasons of using or not using particular services?; the size of the enterprises and their industry sectors, and so on.

The sample of the survey tries to give good statistical coverage of the main industry sectors. These include manufacturing, distribution, retail, finance, transport and tourism. They contribute nearly two thirds of the Community's total Gross Domestic Product. All twelve member states in the European Community are included. The number of interviews conducted in each country was chosen to be broadly comparable to the country's size in terms of number of business establishments and total GDP. The sample survey design is claimed to be representative of the population of interest. It is based on a random sampling from business directories of each industry sector. Within this sampling scheme, quota sampling is subsequently used to obtain a repre-

Table 1
Number of establishments and types of VAS

Type of service	Establishments
Online Database Access	326
E-mail	146
Electronic Data Interchange	111
Electronic Funds Transfer	112
Managed Network Services	46
Telemetry	24
Videoconferencing	10
Other	24

Source: Data provided by Scicon Networks (1989).

sentative coverage as to the size of the enterprises. The unit of reference of the data collected is an 'establishment'.[6] In total, a number of 5726 telephone calls were made and more than 1550 interviews were completed. The study identifies seven types of VAS: Online Database Access, Electronic Messaging (E-mail), Electronic Data Interchange, Electronic Funds Transfer, Managed Network Services, Telemetry and Videoconferencing. To be exhaustive, a category 'other services' was added for the telephone interviews. The study concentrates on business demand. It does not include demand for services from the residential sector. The questionnaires are limited to VAS provided by an external supplier, which means that in-house use of VAS is not considered. The interviews were made during the first half of the year 1989.

The database is available from the Commission DG XIII for scientific purposes. It includes the collected raw data of the interviews. This raw data material was used for the following analyses. It provides a number of 1563 cases that can be included for the analyses.[7] Table 1 describes the diffusion of new telecommunication services in the European Community. In total, 43% of the enterprises use externally supplied VAS and 57% are not adopting them. Most of the demand is currently for Online Database Access services (326 users). These services include Videotex and all kinds of online information services. Other relevant NTS are E-mail, Electronic Data Interchange and Electronic Funds Transfer. The remaining services all have relatively low levels of use. Only ten out of the more than fifteen hundred enterprises interviewed are adopting videoconferencing.

---

[6] An establishment is a site where an organization carries out business. In many cases, especially small organizations, it is equivalent to the company or enterprise. Thus, in the following both expressions are used as synonyms.

[7] With regard to several questions, certain enterprises refused to answer. Therefore a varying number of interviews are missing cases.

These findings are in line with the results of other empirical investigations and especially private market studies. They confirm the fact that database services represent the most well established and well known of all VAS, while other services like telemetry and videoconferencing still play a minor role.

4. *Testing hypotheses*

4.1. Cross-tabulations

To test the hypotheses developed in section 2, a simple approach is to use a cross-tabulation. Table 2 shows the variable use (adopters and non-adopters) of VAS on the one hand and the member states of the EC on the other (variable country). The number of establishments for each combination of values of the two variables is displayed in a cell of the table. The cells also provide information about percentages and are used to test for relationships between the variables. The hypothesis that the use of VAS does not vary with the country of origin of an enterprise, which means that the two variables are independent of each other, is tested with the Pearson chi-square statistic. Its value is 89.37. With eleven degrees of freedom and choosing a significance level of 0.05 [8], the hypothesis that the two variables are independent is rejected. [9] The residual in each cell points out whether there is a deviation between the actual number of users/non-users and the expected number, which means the number of adopters/non-adopters in the case that the variable country has no influence at all. In Germany 93 enterprises are using VAS compared with an expected number of 148, and thus the hypothesis of a lag concerning the use of externally supplied VAS in Germany is corroborated. On the contrary, in France more enterprises than expected adopt new telecommunication services. This is of course due to the widespread use of the French Videotex system Télétel. The allegation of the leading position of enterprises in the United Kingdom as to the demand for VAS is not confirmed. The UK does not exhibit a greater use of externally supplied VAS. This finding contradicts a widely held view but is not implausible. In the UK leased lines are cheap and for that reason internal supply of VAS may be a cost-effective solution. As to the countries in southern Europe (Italy, Spain, Greece, Portugal) only Spain deviates from the expected distribution under the hypothesis that the use does not vary with the countries. The results of Table 2 show that only

---

[8] The significance level of 0.05 defines the probability level that is to be considered too low to warrant support of the hypothesis being tested.

[9] Assuming that the data are random samples from multinomial distributions and with a minimum expected frequency of 11.2, the conditions for using the chi-square statistic are met.

Table 2
Cross-tabulation: Use by country

	Country												
	Germany	United Kingdom	France	Italy	Spain	Greece	Nether-lands	Ireland	Belgium	Denmark	Portugal	Luxem-burg	Row Total
Non-Adoptors													
Count	252	140	106	126	83	14	33	37	25	35	16	24	891
Expected Value	196.9	141.6	154.1	129.6	65.1	14.8	38.2	39.4	32.0	36.0	16.0	27.4	
Column Percent	73.0%	56.5%	39.3%	55.5%	72.8%	53.8%	49.3%	53.6%	44.6%	55.6%	57.1%	50.0%	57.1%
Residual	55.1	−1.6	−48.1	−3.6	17.9	−8	−5.2	−2.4	−7.0	−1.0	0.0	−3.4	
Adoptors													
Count	93	108	164	101	31	12	34	32	31	28	12	24	670
Expected Value	148.1	106.4	115.9	97.4	48.9	11.2	28.8	29.6	24.0	27.0	12.0	20.6	
Column Percent	27.0%	43.5%	60.7%	44.5%	27.2%	46.2%	50.7%	46.4%	55.4%	44.4%	42.9%	50.0%	42.9%
Residual	−55.1	1.6	48.1	3.6	−17.9	0.8	5.2	2.4	7.0	1.0	0.0	3.4	
Column Total	345 22.1%	248 15.9%	270 17.3%	227 14.5%	114 7.3%	26 1.7%	67 4.3%	69 4.4%	56 3.6%	63 4.0%	28 1.8%	48 3.1%	1561 100.0%

	Value	*DF*	*Observed significance*
Pearson Chi-square	89.37	11	0.00
Minimum Expected Frequency	11.16		

31 Spanish enterprises are adopters compared with an expected number of 49 users. But the outcomes for the smaller European countries are probably not very robust, taking into account the limited number of interviews in these countries (Greece: 26, Portugal: 28).

Table 3 examines the influence of the business sectors on the demand for new telecommunication services. In general, the use of VAS is not independent from the business sector (chi-square 63.51, significant on the 1% level). Tourism (travel agencies) and finance (banking and insurance) show a great number of enterprises using VAS, thus confirming hypotheses found in the literature. It is interesting that, using a univariate method of analysis, transportation does not seem to be a sector with great demand for new telecommunication services, such as, e.g., EDI. The manufacturing sector has only 160 users of VAS. If the sector had no influence we would expect 217 enterprises to adopt them. Therefore, the use is less than expected. This finding contradicts the idea that the manufacturing sector is a leader as to the adoption of NTS (Antonelli, 1989/1990, p. 47).

The size of an enterprise as an influencing factor is examined in Tables 4 and 5. As an indicator of the size of an enterprise, the turnover and the number of employees are used. Applying the 5% significance rule, the hypothesis that the use of VAS does not vary with the size of an enterprise cannot be rejected in the case of the variable turnover indicator. More big enterprises are adopting VAS but this is only because more of these major firms are in the sample. A different result can be obtained using the number of employees as an indicator of the size of an enterprise, thus supporting the hypothesis of the influence of the pure size of an enterprise (Table 5). Hence, the results are ambiguous and do not allow a firm conclusion. A different explanation may be that with regard to the use of new telecommunication services these two variables do not necessarily indicate the same fact. In some business sectors (e.g., finance) turnover is only a poor indicator of the size of an enterprise.

There is however, a strong dependence of the adoption of VAS on the amount of money an enterprise spends on information technology, such as computers, data-processing, telecommunications and so on. Table 6 points out that the spending on information explains the use of VAS (chi-square 40.9, significant on the 1% level). As far as the expenditure on information includes the spending on new telecommunication services, this is simply a tautology and the causal relationship goes from the use to the spending and not vice versa. But inasmuch as the spending also covers the whole range of computers, data-processing and telecommunications equipment it is also an indicator of the innovation orientation and the Information Technology expertise of an enterprise. Thus, the results of Table 6 back the idea that the use of Information Technology in general is associated with the use of VAS.

Table 3
Cross-tabulation: Use by business sector

	Business sector							
	Manufact-uring	Distribu-tion	Retail	Finance	Transport	Tourism	Other	Row Total
Non-Adoptors								
Count	342	151	96	108	66	51	62	76
Expected Value	284.4	148.5	106.0	142.2	71.4	69.7	53.9	
Column Percent	68.1%	57.6%	51.3%	43.0%	52.4%	41.5%	65.3%	56.3%
Residual	57.2	2.5	−10.0	−34.2	−5.4	−18.7	8.2	
Adoptors								
Count	160	111	91	143	60	72	33	670
Expected Value	217.6	113.5	81.0	108.8	54.6	53.3	41.2	
Column Percent	31.9%	42.4%	48.7%	57.0%	47.6%	58.5%	34.7%	43.3%
Residual	−57.6	−2.5	10.0	34.2	5.4	18.7	−8.2	
Column	502	262	187	251	126	123	95	1546
Total	32.5%	16.9%	12.1%	16.2%	8.2%	8.0%	6.1%	100.0%

	Value	*DF*	*Observed significance*
Pearson Chi-square	63.51	6	0.00
Minimum Expected Frequency	41.2		

Table 4
Cross-tabulation: Use by turnover

	Turnover				
	<5 Mill. $	5–25 Mill. $	25–100 Mill. $	>100 Mill. $	Row Total
Non-Adoptors					
Count	56	96	131	140	423
Expected Value	63.0	90.0	118.0	152.0	
Column Percent	45.2%	54.2%	56.5%	46.8%	50.8%
Residual	−7.0	6.0	13.0	−12.0	
Adoptors					
Count	68	81	101	159	409
Expected Value	61.0	87.0	114.0	147.0	
Column Percent	54.8%	45.8%	43.5%	53.2%	49.2%
Residual	7.0	−6.0	−13.0	12.0	
Column	124	177	232	299	832
Total	14.9%	21.3%	27.9%	35.9%	100.0%

	Value	*DF*	*Observed significance*
Pearson Chi-square	7.29	3	0.06
Minimum Expected Frequency	60.96		

4.2. Logistic regression

This isolated hypotheses testing does not allow of examining whether the influence of a certain variable is only spurious, considering the other variables simultaneously. Therefore discrete choice models are used to analyse the influences of the different variables together. A logistic regression model is estimated with the dependent variable USE and the independent variables COUNTRY, BUSINESS, SPEND (spending for information technology) and two alternative indicators for size of an enterprise, TURNOV (turnover) and EMPLOY (number of employees). [10] Two independent variables serve as indicators for the degree of internationalization of business activities. ECVOICE and INTVOICE are the percentage of telephone calls with countries inside the EC and all other international calls. The logistic regression function directly estimates the probability of adoption or non-adoption depending on these variables.

Table 7 provides the relevant regression results. Unfortunately, due to the

[10] A logistic regression is used instead of a discriminant analysis because the main focus of the paper is on the influence of the independent variables and not on the classification of a particular enterprise in the group of adopters or non-adopters. Furthermore, the logistic regression requires fewer assumptions than dicriminant analysis (multivariate normality of the independent variables, equal variance–covariance matrices in the two groups).

Table 5
Cross-tabulation: Use by number of employees

	Number of employees						
	<10	10–25	26–50	51–100	101–500	>500	Row Total
Non-Adoptors							
Count	74	112	164	128	281	115	874
Expected Value	67.0	97.7	159.6	149.9	286.2	113.6	
Column Percent	62.7%	65.1%	58.4%	48.5%	55.8%	57.5%	56.8%
Residual	7.0	14.3	4.4	−21.9	−5.2–5.4	1.4	
Adoptors							
Count	44	60	117	136	223	85	665
Expected Value	51.0	74.3	121.4	114.1	217.8	86.4	
Column Percent	37.3%	34.9%	41.6%	51.5%	44.2%	42.5%	43.2%
Residual	−7.0	−14.3	−4.4	21.9	5.2	−1.4	
Column	118	172	281	264	504	200	1539
Total	7.7%	11.2%	18.3%	17.2%	32.7%	13.0%	100.0%

	Value	*DF*	*Observed significance*
Pearson Chi-square	14.51	5	0.013
Minimum Expected Frequency	50.99		

cumulation of incomplete interviews and therefore missing cases, the number of observations is limited to 499. Nevertheless, this sample size remains large and the desirable asymptotic properties of the maximum likelihood estimations can be assumed. The observed dependent variable USE has two realizations. It amounts to 1 when an enterprise is not using VAS and 0 when it is. The estimations of the explanatory variables show their influence on the probability of refraining from adopting NTS.

Several possibilities exist to see whether the model fits the data. Overall, the number of correctly classified cases is about 76%. The model chi-square tests the null hypothesis that the coefficients for all of the terms in the current model, except the constant, are 0. In this logistic regression the model chi-square is 169.3 and the hypothesis of a missing influence of all variables has to be rejected. The test that a particular coefficient is zero can be based on the Wald statistic. Using a significance level of 5% only the coefficients for country (COUNTRY), business sector (BUSINESS) and spending for information technology (SPEND) appear to be significantly different from zero.

The variables COUNTRY and BUSINESS are categorical variables. The regression coefficients of Table 7 tell us how much each category influences the probability of non-adoption compared to the average effect of all categories. As in the simple cross-tabulation in Germany and Spain the probability of an enterprise belonging to the group of non-users increases significantly.

Table 6
Cross-tabulation: Use by spending for IT

	Spending for IT				
	<50 Tsd. $	50–250 Tsd. $	250 Tsd.– 1 Mill. $	>1 Mill. $	Row Total
Non-Adoptors					
Count	188	103	61	23	375
Expected Value	162.4	94.9	67.5	50.2	
Column Percent	58.8%	55.1%	45.9%	23.2%	50.7%
Residual	25.6	8.1	−6.5	−27.2	
Adoptors					
Count	132	84	72	76	364
Expected Value	157.6	92.1	65.5	48.8	
Column Percent	41.3%	44.9%	54.1%	76.8%	49.3%
Residual	−25.6	−8.1	6.5	27.2	
Column	320	186	133	99	739
Total	43.3%	25.3%	18.0%	13.4%	100.0%

	Value	*DF*	*Observed significance*
Pearson Chi-square	40.86	3	0.00
Minimum Expected Frequency	48.8		

Also the cross-tabulation result with regard to the missing influence of Great Britain is confirmed.[11]

The categorical variable BUSINESS reveals that belonging to the manufacturing sector increases the probability of being part of the group of non-adopters of VAS. The same is true for the distribution sector. These coefficients are significant — on the 1% level — together with the coefficient for other business sectors. The multivariate logistic regression verifies the fact that enterprises belonging to the finance sector are likely to use VAS only on the 10% significance level. It also shows that this is the case for the transport sector, contrary to the findings of the cross-tabulation.

The regression coefficient SPEND points out that with an increase in the amount of money spent on information technology the probability of not using VAS decreases. Interestingly, the two indicators of the internationalization of an enterprise's activities — the percentage of calls in the EC and other countries ECVOICE and INTVOICE — turn out to be not significant.

The same is true with regard to the coefficients of the two indicators of the size of an enterprise: turnover (TURNOV) and number of employees (EM-

[11] In contrast to the cross-tabulation, the logistic regression reveals a significantly decreasing probability of non-use for Belgium. Due to the limited number of cases, the estimation results for the small countries should be left out of consideration.

Table 7
Logistic Regression: Dependent variable USE

Independent Variables	Estimation results			
	B	S.E.	Wald	Observed Significance
COUNTRY			63.5	0.00
Spain	2.3	0.45	24.9	0.00
Netherlands	−0.4	0.45	0.9	0.33
Denmark	−0.2	0.45	0.2	0.65
United Kingdom	−0.3	0.31	0.9	0.34
Ireland	−0.5	0.40	1.7	0.20
France	−0.6	0.36	2.6	0.11
Belgium	−1.9	0.76	6.4	0.01
Luxemburg	1.4	0.76	3.6	0.06
Italy	0.2	0.30	0.7	0.40
Portugal	0.6	0.63	0.8	0.38
Germany	1.3	0.26	23.0	0.00
BUSINESS			56.4	0.00
Manufacturing	1.4	0.24	33.2	0.00
Retail	−0.5	0.28	2.7	0.10
Distribution	0.7	0.25	8.7	0.00
Finance	−0.7	0.37	3.3	0.07
Transport	−0.8	0.42	3.4	0.07
Other	1.4	0.35	16.9	0.00
EMPLOY	−0.01	0.11	0.02	0.89
TURNOV	0.04	0.14	0.08	0.78
SPEND	−0.62	0.13	22.5	0.00
ECVOICE	−0.02	0.02	2.1	0.15
INTVOICE	0.01	0.02	0.2	0.63
Constant	0.61	0.45	1.8	0.18
Number of cases:			499	
Percentage of correctly classified observations (overall):			75.8%	
Model Chi-square:			169.3	DF: 22
Observed significance:			0.00	
−2 Log Likelihood:			522.4	DF: 476
Observed significance:			0.07	

PLOY).[12] Hence, the logistic regression comes to a clear result in comparison to the mixed outcomes of the cross-tabulations. In this respect the approach backs the idea that univariate analyses may confound existing correlations between variables.

[12] This results holds in the case of a separate inclusion of only turnover and only employees, which means that it is not contaminated due to collinearity of both variables. Also the treatment of both variables as categorical variables and a forward selection method of inclusion of variables lead to the same conclusion.

This finding is not very unlikely bearing in mind the fact that the size of an enterprise only serves as an indicator variable for many possible influencing factors. For instance, spending for information technology may be a better indicator of most of these factors, thus eliminating the influence of the pure size. Another interpretation may be that huge enterprises are only using more internal VAS and thus do not have to buy telecommunication services on the market.

5. *Conclusions*

New telecommunication services are claimed to be one of the fastest growing sectors of the European economy. There is a general thrust towards the use of VAS for competitive advantage. This paper is a first attempt to evaluate the status and determinants of their adoption in the EC, relying on a representative sample.[13] The analysis of the main factors influencing the demand for new telecommunication services shows that some of the generally accepted hypotheses as to the use of externally supplied VAS can not be confirmed, e.g. the empirical results indicate that the size of an enterprise does not play a significant role. The study points out the need of disaggregated empirical research on the dependent and independent variables. Firstly, a disaggregated analysis focusing on the demand for specific VAS is desirable. Such evaluations could reveal differences and similarities as to the use of online databases, EDI, Videoconferences and so on. Secondly, it is obvious that explanatory variables like the country of origin of an enterprise or its size are only very poor indicator variables. They are correlated with a lot of other very different variables, thus disguising relevant relationships.

It is necessary to develop more specific hypotheses linking the use of certain VAS with, e.g., characteristics of the production process, organizational and marketing aspects, human skills, internationalization of markets and so on

---

[13] Unfortunately the study has a number of important shortcomings. The dataset does not allow any conclusions as to the intensity of demand, e.g., the amount of money spent for VAS. It is limited to a 'yes' or 'no' information as to the use of certain types of VAS. In so far it shares the limitations of other adoption studies. In addition, only externally supplied VAS are taken into consideration. Also as to its reliability and validity, the data set is not beyond doubt. The number of enterprises adopting VAS seems too great. The pretest of a current research project at the WIK shows that in the German retail and manufacturing sectors only a tiny fraction of the enterprises knows anything about, e.g., E-mail or online databases. Looking at the questionnaire used in the Scicon Network study also reveals that the questions are sometimes confusing, e.g., the difference between in-house provision of VAS and of externally supplied VAS is not very clear. Bearing these problems in mind, the results have to be considered tentative outcomes which need further exploration.

(see Fritsch, 1987, pp. 71–78; Taymaz, 1990, p. 164). Hence, it would be feasible to discriminate between several possibilities as to the meaning of the influence of the independent variables country and business sector. [14] For instance, as to the country of origin of an enterprise it is difficult to detect the causes of the observable differences. Varying institutional arrangements (privatization and regulation of the telecom sector), economic factors (prices, quantity and quality of the telecommunication networks) and social factors mingle. This development of specific hypotheses could be based on a variety of reasoning like decision theory, production models and regulatory analyses.

References

Appelbe, T.W., Dineen, C.R., Solvason, D.L. and Hsiao, C., 1992. Econometric Modelling of Canadian Long Distance Calling: A Comparison of Aggregate Time Series Versus Point-to Point Panel Data Approaches. *Empirical Economics* 17, 125–140.

Antonelli, C., 1985. The diffusion of an organizational innovation, International data telecommunications and multinational industrial firms. *International Journal of Industrial Organization* 3, 109–118.

Antonelli, C., 1989. The diffusion of information technology and the demand for telecommunication services. *Telecommunications Policy* 13, 255–264.

Antonelli, C., 1989/1990. Information Technology and the Derived Demand for Telecommunication Services in the Manufacturing Industry. *Information Economics and Policy* 4, 45–55.

Benzoni, L., Lebart, L. and Rowe, F., 1992. Market dynamics and technological segmentation: The case of the French professional telecommunication market. In: C. Antonelli (ed.), *The Economics of Information Networks*, pp. 451–468 (North-Holland, Amsterdam).

Bewley, R. and Fiebig, D.G., 1988. Estimation of Price Elasticities for an International Telephone Demand Model. *Journal of Industrial Economics* 36, 393–409.

Bishop, A.P., 1992. Electronic Networking for Engineers: Research from a User Perspective. *International Journal of Computer and Telecommunications Networking* 25, 344–350.

Breslaw, J. and Pizante, G., 1989/1990. Lag structure in telecommunications demand analysis. *Information Economics and Policy* 4, 325–345.

Brink, S. and Nieswiadomy, M.L., 1992. *Private Line Demand: A Point-to-Point Analysis, Irving.* University of North Texas, Denton, Texas (unpublished paper).

Cabral, L.M.C. and Leite, A.P.N., 1992. Network Consumption Externalities: The Case of Portugese Telex. In: C. Antonelli (ed.), *The Economics of Information Networks*, pp. 129–140 (North-Holland, Amsterdam).

Cantzler, F., 1991. *Quantitative und qualitative Beschäftigungswirkungen neuer Technologien* (Minerva, München).

Cronin, F.J., Gold, M.A., Hebert, P.L. and Lewitzky, S., 1993. Factor Prices, Factor Substitution, and the relative Demand for Telecommunications across US Industries. *Information Economics and Policy* 5, 73–85.

---

[14] Currently, a large-scale empirical research project at the WIK is trying to analyse some of these questions with regard to the demand for VAS in Germany. First results are published in Stoetzer (1993).

de Fontenay, A., Shugard, M. and Sibley, D., 1990. Telecommunications Demand Modeling: An Integrated View. *Contributions to Economic Analysis* (Amsterdam).

Dupagne, M. and Agostino, D.E., 1991. High-Definition Television: A survey of potential adopters in Belgium. *Telematics and Informatics* 8, 9–30.

Engvall, L., 1991. Using market surveys for demand forecasting of various teleservices. *Centennial Scientific Days of PKI, 20–22 November* (Budapest).

Ewers, H.-J., Becker, C. and Fritsch, M., 1990. *Wirkungen des Einsatzes computergestützter Techniken in Industriebetrieben* (Berlin).

Fritsch, M., 1987. Frühe Nutzer der Telematik, Charakteristika von Adoptoren im Vergleich zu (noch) Nicht-Adoptoren neuer Telekommunikationstechniken. In: B. Hotz-Hart and W.A. Schmid (eds.), *Neue Informationstechnologien und Regionalentwicklung*, pp. 65–80 (Zürich).

Hakim, S.R., 1992. The Determinants of Traffic and Duration in International Telephone Communications. *ICTM Research Paper Series 10, International Center for Telecommunications Management* (Omaha).

Köhler, S., 1987. *Der Diffusionsprozeß von Teletex in der Bundesrepublik Deutschland* (München).

Lang, H. and Lundgren, S., 1991. Price Elasticities for Residential Demand for Telephone Calling Time. *Economic Letters* 35, 85–88.

Linhart, P.B., Radner, R. and Tewari, R., 1992. On the Markets for Data Networking Products. In: C. Antonelli (ed.), *The Economics of Information Networks*, pp. 141–156 (North-Holland, Amsterdam).

Litman, B., Chan-Olmsted, S. and Thomas, L., 1991. Estimating the demand for backyard satellite dishes: The U.S. experience. *Telematics and Informatics* 8, 59–69.

Miles, I. and Thomas, G., 1990. The development of New Telematics Services. *STI Review* 7, 35–63.

Müdespacher, A., 1987. Diffusionsprozesse der Neuerungen der Telematik: Die Verbreitung von EDV, Tele-Datenfernverarbeitung und Telefax in der Schweiz. In: B. Hotz-Hart and W.A. Schmid (eds.), *Neue Informationstechnologien und Regionalentwicklung*, pp. 81–97 (Zürich).

Nijdam, J., 1990. Forecasting Telecommunications Services using Box-Jenkins (ARIMA) Models. *Telecommunication Journal of Australia* 40, 31–37.

OECD, 1989. *Telecommunication Network-based Services: Policy Implications* (Paris).

PACE 90, 1990. *Perspectives for Advanced Communications in Europe: 1990, Vol. VI, IDS 3: Business Data Communication Services.* (Commission of the European Communities, DG XIII, Brussels).

Porter, M.E. and Millar, V., 1985. How Information gives you competitive advantage. *Harvard Business Review*, July–August, pp. 149–160.

Preißl, B., 1988. Telekommunikation und das Auslandsgeschäft von Banken und Versicherungen. *WIK-Diskussionsbeiträge* Nr. 36 (Bad Honnef).

Scicon Networks, 1989. *The Market for Value Added Services in Europe, Vol. I–VII* (London).

Stoetzer, M.-W., 1992a. Value Added Services: Problems of Definitions and Data. *Telecommunications Policy* 16, 388–400.

Stoetzer, M.-W., 1992b. New telecommunication services in Europe: Which factors influence their adoption? *Communication and Strategies* 8, 13–32.

Stoetzer, M.-W., 1993. *Der Einsatz von Mehrwertdiensten in bundesdeutschen Unternehmen: Eine empirische Bestandsaufnahme*. WJK Diskussionsbeitrag Nr. 116.

Stoneman, P., 1983. *The Economic Analysis of Technological Change* (London).

Taylor, L.D., 1980. *Telecommunications Demand: A Survey and Critique*, 1st ed. (Cambridge, Mass.).

Taylor, L.D., 1993. *Telecommunications Demand: In Theory and Practice*, 2nd ed. (in press).

Taymaz, E., 1990. A Micro-simulation Analysis of Manufacturing Firm's Demand for Telecommunications Services. In: G. Eliasson, S. Fölster, T. Lindenberg, T. Pousette and E. Taymaz (eds.), *The Knowledge Based Information Economy*, pp. 157–182 (Stockholm).

Thomas, G. and Miles, I., 1989. *Telematics in Transition* (Harlow).

Walter, J.L., 1992. The Demand for Residential Telephone Service: Who are the Non-Subscribers? *Bellcore Discussion Paper* (Livingston).

Witte, E. and Dowling, M., 1991. Value-added services: Regulation and reality in the Federal Republic of Germany. *Telecommunications Policy* 15, 437–452.

Zona, J.D. and Jacob, B., 1990. The total Bill Concept: Defining and Testing Alternative Views. *Bellcore and Bell Canada's Industry Forum, Telecommunications Demand Analysis with Dynamic Regulation*, Hilton Head, South Carolina, April 22–25.

Global Telecommunications Strategies
and Technological Changes
Edited by G. Pogorel

Chapter 2

Sharing in telematic network organizations: opportunities and entry barriers

Ettore Bolisani, Giorgio Gottardi and Enrico Scarso

Dipartimento di Innovazione Meccanica e Gestionale, University of Padova, Padova, Italy

1. Introduction

Despite the fact that the literature generally stresses the wide advantages of telematics in facing problems due to market globalization and to the needs of innovative and flexible organization solutions, the real effects of the adoption of telematics seem to be ambiguous. At present, many technological, strategic and organizational studies offer ample and generic considerations on the advantages of telematics, but the works on their effects provide contrasting interpretations (Etheridge and Simon, 1992; Cash et al., 1988; Antonelli, 1988). In the industrial organization approach, many studies have tried to make clear the effects of the diffusion of telematics and the growth of electronic network organizations on the economic systems. Telematics can make the beginning and development of organizational networks easier, from both structural and functional points of view. There is no doubt that the use of telematics tends to reduce both coordination and transaction costs; but this is just what makes it difficult to decide whether the future situation will favour either the market or the hierarchy.

In Williamson's perspective, the reduction of transaction costs allowed by telematics is strongly in favour of a growth of activities managed by the market. It is generally admitted that the diffusion of information technologies determines: an increase in the number of possibly connected subjects; a reduction in uncertainty; and in general a growth of market efficiency. This might be convincing if there were no barriers to its use. At present, however, the most evident effects seem to be induced by the strategies of certain actors (mainly large or multinational firms) with particular capabilities or specific aims in developing appropriate uses of information technologies. Moreover, if on the one side telematics makes for more intensive interfirm relations and communication flows, on the other it is not clear whether cooperative rather than competitive opportunities are generated.

According to the strategic perspective, firms aim to develop their ability to use information technology in a competitive manner. Those requiring a high volume of transactions usually take the lead in the development of a competitive use of telematics (Johnston and Vitale, 1988). It is commonly agreed that better opportunities for differentiation and simplifications in coordinating production and marketing functions should be combined with: a wider use of multi-sourcing; risk reduction; and finally with improved business opportunities.

The fact that a strategically effective use of telematics may require special abilities derives from the extreme flexibility offered by this technology. Its output is in no way prescribed or defined *ex-ante*. Particular criteria in using this innovation for specific aims are widely determined by the user, and depend on the specificity of objectives, knowledge, intangible assets and learning capacity. In substance, innovations in using telematics are mainly developed by users according to their individual characteristics (Gottardi and Quaglio, 1991).

As innovations in use are firm-specific, they are not easily transferable to contexts which differ from the original ones. This effect of *localized* innovation (Stiglitz, 1988) penalizes a number of subjects in their use of telematics. Therefore, the diffusion of this technology cannot be symmetric, and in reality it spreads in a highly selective way; consequently, new sources of structural heterogeneity are generated in the market. Moreover, when telematics is introduced in an organizational network, the differences in the roles played by firms may be stressed; then some operators can achieve the benefits of information technology better than others. In a cooperative organization this may unbalance the distribution of advantages and power, and it may generate competitive forces.

An evaluation of corporate opportunities and effects resulting from the adoption of information technology could be carried out following Porter's well known framework (1985). According to Porter's perspective, firms can use information technologies in order to obtain new business opportunities and competitive advantages; these latter may be achieved by means of differentiation and cost leadership strategies. Compared to the transaction costs approach, Porter's analysis implies that the transition to telematics probably leads to market forms different from competition, because it stresses the use of information technology in order to create corporate advantages and barriers. These barriers are related with *information* itself, thus they lead to new sources of cognitive heterogeneity between firms, consequently reducing the market efficiency.

The influence of telematic technologies on interorganizational networks is discussed in the economic literature. While differential advantages in sharing a network have been widely stressed, the specific mechanisms generating both entry/exit barriers and asymmetries in the roles played by various operators have not been fully studied (Johnston and Vitale, 1988; Jarrillo, 1988; Rockart and Short, 1991). When trying to evaluate the sources of these barriers due to the use of telematic systems, the tendency to organizational solutions, like interfirm networks, and the characteristics of these networks should be briefly recalled.

2. *Cooperative network organization and information technologies*

The present trend to establishing cooperative networks seems to derive from the increased complexity and operational risks that have emerged in recent years (Conctractor and Lorange, 1988). According to Rockart and Short (1991), shared goals, work and decision making (through the enhanced access to critical information across the network) characterize such an organization. The network as a whole benefits from: expertise and knowledge, timing and issue prioritization for critical issues, responsibilities and risks which are distributed within the network itself.

There are several reasons for participating in a cooperative network: the sharing of entrepreneurial risks, the subdivision of costs, the pooling of scarce resources, the complementing of mutual technical and market knowledge. Generally speaking, the basis of network sharing is the complementarity of components; the network must be seen as a whole, where each firm's distinctive competence (Teece, 1986) is enhanced by its participation therein.

Thanks to the network, each partner can focus on his core activities and technologies thus obtaining reduced time-to-market, improved customer service, higher product quality and lower production costs. Hence, the network may enhance operational efficiency, market positioning and strategic capabilities. Furthermore, when the tendency to a reduced product life-cycle is considered, the sharing in interorganizational networks makes it possible to speed up innovation processes, above all when it may take too long to find and manage all the physical and intangible assets needed to develop new products (Bolisani et al., 1992).

When firms share in network organizations they can take advantage of synergic and complementary effects which finally multiply the resources of various subjects. Meaningful relationships result in cross-fertilization processes which add value to the partners' capabilities and knowledge, thus forming the basis of their competitive advantages. Following this perspective, Jarrillo (1988) uses the expression 'strategic network' and also affirms that essential to it is the concept of a 'hub' or focal firm, i.e. the firm taking a pro-active role in setting up and managing the network.

The industrial network concept is used in general to identify a number of entities which are interconnected (Axelsson and Easton, 1992). In the case of industrial networks, the linkages concern economic transactions among different firms, whatever technology is employed to link them. A key aspect of network analysis is that these structures are essentially heterogeneous in nature. On the grounds of strategic considerations and accessible resources, firms can form various kinds of networks which differ from structural and functional points of view. The structural parameters concern the networks global dimension and its topological configuration (i.e. total number of nodes, their spatial distance and localization). The functional parameters mainly concern the role played by nodes, the intensity and the contents of information flows.

In this sense, the organizational and technical solutions adopted by networks (depending on members' goals, knowledge and capability) are largely firm-specific and difficult to transfer to other contexts. The resulting configuration tends to make the network considerably stiff, thus limiting the number of subjects that can profit from networking. Hence in entering or leaving a network a firm has to support costs which strongly depend on its distinctive features, and in particular on (Pontiggia, 1991):

- *structuredness*: the degree of interdependence among firms;
- *homogeneity*: the degree of similarity;
- *hierarchical position*, indicating the existence of dominance relations;
- *exclusiveness*: regarding the possibility of participating in different networks;
- *degree of dependence from the resources inside the network*, concerning the possible exit of firms from the network.

All these factors concern the specific nature of relations linking all participants. Depending on the idiosyncratic nature of the exchange and its *criticality* (Williamson, 1975), partners may in fact be *loosely coupled* or *tightly coupled*, and preferential directions (centre-periphery) of information flows may exist.

Some studies suggest that, if the globalization of competition is inducing firms to adopt innovative and flexible organizational solutions and to increase interdependencies with other companies, telematics is a useful tool for a more effective achievement of this goal. By allowing more complex information flows, telematics can speed up the rising of new relationships among economic agents. So network organizations and telematic networks lead to the exploitation of strategic and operational benefits given by the new kind of intra- and inter-firm, and market relations.

In spite of the previously mentioned advantages, telematics seems to have various and complex effects on the firms and networks that they tie. A deeper study of these effects draws more attention to the entry/exit barriers which must be overcome when a firm enters or leaves a network, and to the costs reducing the benefits of belonging to such an organization. Ways and sources of these barriers/costs are more easily established by analyzing the nature and the structure of electronic networks.

3. Electronic networks

In investigating electronic networks whose nodes are connected by means of telematic technologies (Kambil and Venkatraman, 1990), their specific technical solutions must be considered. On the basis of the technical devices employed, the nodes can assume a common or a specialized role; for example a unique node elaborates and sends information to various common receivers which are connected with it. In considering the nature of telematic networks, four classes of factors may be distinguished:

(1) the logical structure of the network and the logical content of its information flows;

(2) the physical connection between nodes;

(3) the hardware devices that support information exchange;

(4) the communication language.

3.1. Logical structure and content

The *logical structure* depends on the frequency, duration, complexity and specificity of inter-firms transactions (Williamson, 1975). The higher the value of these variables, the greater the opportunity to integrate the transactions into a hierarchical organizational form.

In relation with the *logical content* of information flows, four distinct (and hierarchically related) roles can be performed by firms (Venkatraman, 1991):

(a) *transaction*, when they exchange structured data using a mutually accepted format;

(b) *inventory*, when a firm's inventory is made 'available and visible' from one party to another without excessive time delay, thus making it possible to trigger the movements of goods;

(c) *process*, when partners integrate their value chain activities (Porter, 1985) in order to enhance common benefits, even by sharing strategic information;

(d) *expertise*, when partners exchange knowledge (i.e., unstructured information) thus creating a 'virtual intellectual network'.

3.2. Physical connection

Four options can be considered (Etheridge and Simon, 1992).

Local Area Networks (LAN). These are used for data communications in a local area such as a building or a group of neighbouring buildings. Essentially, a LAN is privately owned and operated; the user organization is entirely responsible both for the operations and for the physical equipment, and so specific skills and equipment must be provided.

Private Wide Area Networks (Private WAN). These are employed for connections spatially wider than a LAN. The proprietary organization is responsible for the operations and management of communications, while the physical linking circuits are generally rented from a Public Telecommunications Operator (PTO) for exclusive use of the renting organization.

Public WAN. The user is responsible for the end systems that access the public network, for the local data computations and for the organization of the information flows among the linked nodes; the national PTO manages the physical linking circuits and the data routing inside them. The user pays a fixed fee and a variable sum, generally proportional to the resource use. Public WAN are usually *packet-switched* networks, with access points sited in the main cities of a country, and with bridges toward public WAN of other countries. Examples are ITAPAC for Italy, TRANSPAC for France, and so on. Many networked organizations can use the same public WAN to support their communication: the WAN manager provides access protection to the various co-using organizations, and a correct evaluation of the fee charges attributed to each node.

Third Party Networks. These are services provided by PTOs or by private operators, variously referred to as Value Added Networks (VANs), Managed Data Networks (MDNs) and so on. Services can consist of data computation, storing and routing operations, and the managing of the physical linking; these operations are performed by the service supplier for the user's sake. For these services a relatively high fee is charged to users. Electronic mailing, data banks, electronic fund transfer, and EDI networks management are examples of supplied services. Users are responsible only for the management of local devices and of some local data computations.

3.3. Hardware devices

Three basic types of networks can be identified (Brousseau, 1990):

- *computer to computer* networks;
- *computer to terminal* networks;
- *terminal to terminal* networks.

We distinguish between *computer* and *terminal* on the basis of their capability of processing data autonomously (see OSI classification). So here 'terminal' means a simple 'transmit-receive device', which can perform only coding–decoding operations; 'computer' means a more complete and 'intelligent' device which performs sophisticated computations.

Computer to computer. These systems are used to operate repetitive transactions of structured information among partners agreeing on communication standards. Applications range from simple transmission of administrative documents to more sophisticated uses that support production processes. An example are EDI systems: data coding is very complex and implies high level processing enabling users to interpret and utilize exchanged information. EDI networks are generally 'private', in the sense that they are implemented and managed on the basis of private agreements. They are used for structured information interchange among several subjects such as: distributed manufacturing plants, sale points, clients or sub-suppliers.

Computer to terminal. In these networks there is a central processing unit (a 'host computer') and a number of local 'transmit–receive' devices. Data banks and electronic mailing systems are examples of such networks. We can also consider the case of a broadcast manufacturing system with a central processing unit directly controlling many remote production units; another case is represented by a number of customers that send orders to a central 'sale point' (e.g. an airline booking system). In these examples the networks assume a *star* configuration, with a unique central governance and a number of peripheral units.

Terminal to terminal. These systems are used to perform connections without complex and predefined standard codes. A fax network is a typical application: the content of a paper document is simply scanned and transmitted in numerical form. The processing level of each local unit is clearly low. In this case a 'head' node is not necessary.

3.4. Communication language

The communication language used in the network involves several issues: transmission protocols, data format, data coding and encrypting procedures. Network nodes must agree on these languages before performing any communication. To this aim two aspects must be considered. The first is the *information content* of the transmission: the higher it is, the more complex the 'language' must be. In fact a fax transmission or a simple ASCII file transfer requires a simple coding; on the contrary, an EDI transmission needs a highly structured and complex data format. The second aspect is related to the *communication standard* adopted in the network, and in particular to how much a network is open to the external world. We will discuss this question more widely in section 4.3.

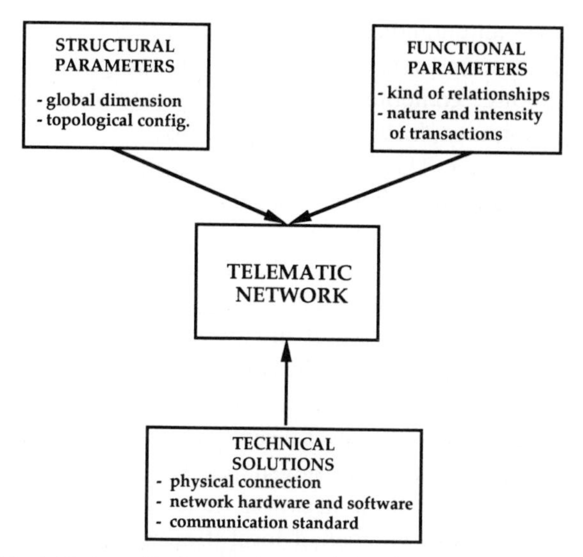

Fig. 1. Factors affecting the characteristics of telematic networks.

Factors affecting the characteristics of telematic networks are summarized in Fig. 1. Moreover, there exists a bi-directional relation between *technical and logical variables* (and also between the network's general features and the adopted telematic solution). On one hand, the fulfillment of a specific logical relationship calls for a particular technical solution; on the other, the availability of technical facilities offers the opportunity to exploit different inter-firm organizational forms. Note that at present in implementation of telematic networks, strategic and organizational dimensions rather than technical problems seem to increase the advantages and the risks of investment.

4. Network entry and exit barriers

Industrial organization theory analyzes in detail how entry and exit barriers originate in a market. Basically, the size of these barriers is tied to the dimension of capital stock which is requested (or decided) by firms when they operate in particular markets: only those having sufficient investment capacity can enter. By taking an appropriate decision as to its own production capacity (and hence as to sunk costs), the first-comer can slow down or block the entry of potential new-comers.

Capital may be embodied in various forms. The development of 'private' customers or sellers networks is a capital decision which increases new-comers' advertising and distribution costs (Salop and Scheffman, 1983). In dynamic industries the acquired experience causes a constant reduction in production costs and may be considered, as other intangibles, a capital form. So experience, and in general each specific know-how, represents a competitive advantage and therefore can inhibit new entries.

From a general point of view, the creation of barriers could be related to the distribution of information on the market (Scherer, 1980), and in particular to the creation of cognitive asymmetries. When the spread and the use of information technology differentiate the existing distribution of information and create new asymmetries, new barriers are bound to come in to being. An electronic network, where circulating information keeps the 'wired' firms in an asymmetric condition with respect to the external environment, represents itself as a barrier for the external competitors. More exactly, differential advantages can be obtained in operating as a node of a network if at least a part of the relevant information available inside the network is inaccessible to external firms.

A very important feature of these cooperative organizations is their dynamic efficiency, which is obtained by continuously adapting the number of nodes and the role of the involved subjects. This flexibility is strongly tied to the level of network entry and exit costs. The focal firm has to be able to ensure low enough entry costs for those new units that it is going to cooperate with. From the focal firm's point of view, exit costs of cooperating units would be high enough to constitute a warranty against opportunistic behaviour. Exit costs are due to the fact that all investments for entering and staying in the network become sunk costs and switching costs (Stiglitz, 1988). One may assume in general that the ability of an electronic network to maintain dynamic efficiency depends on an appropriate trade-off between entry and exit costs.

Thus it may be useful to analyze costs that single nodes have to support in entering and leaving a network (see Fig. 2); on the basis of the nature and amount of these costs some considerations will be imposed on entry and exit barriers. We will first try to describe the existing relations between these costs and the characteristics of nodes and organizations; then we will present some considerations on the effects of communication standards on the size of barriers.

4.1. Entry costs

Entry costs pertain to three different phases of the firm's presence in a network: the physical setting and acquisition of telecommunication channels, the management of these channels and their updating. Physical linking up with the network needs the acquisition of hardware devices and software tools. Even in the case of sophisticated electronic devices, these costs can be

ENTRY COSTS	EXIT COSTS
Physical linking up • acquire hardware devices • acquire software tools Information exchange • learn new procedures • adopt communication standards Exploitation of information flows • adapt organization structures • think over again competitive strategies Network dynamic management • learn by learning	Sunk costs • irrecoverable capital investments (physical network, hardware devices, software routines, communication standard, ...) • irrecoverable intangible assets (knowledge, planning procedures, decision routines, ...) Switching costs • unlearn old methods and procedures • recover control over decentralized activities

Fig. 2. Electronic networks: entry and exit costs.

rather accurately evaluated. In some cases these new devices involve a specific and expensive training.

The costs of an effective and profitable *entry and permanence* in the network are more difficult to capture, and are sometimes nearly impossible to quantify. These costs depend on the firm's ability *to learn* how to make the most of communication channels.

First the firm has to be familiar with hardware and software routines which allow communication. Secondly, it must acquire the ability to exchange meaningful information within the network, by adopting a common communication standard. Information processed by firms must also be compatible with that already flowing in the network; this usually requires adjustments in the accounting and reporting methods, and in the measurement systems, in the planning procedures, in the decision routines. For an effective implementation, the design of new relations requires significant changes related to internal operative processes. Sometimes physical resources which yield information must also be changed; i.e.: when a CAM system is acquired in order to exploit numerical data produced in another network unit; EDI, for example, requires the rearrangement of in-house computer applications.

Thirdly, the firm must learn to play an 'intelligent' role in the network. Hence new ways of employing the available knowledge flow must be researched, even changing the organization structure. High costs are the result of changes in the organization and work process involving the implementation of this technology. In particular, firms must learn to effectively manage and coordinate the more complex interdependencies occurring among various companies. Furthermore, firms have to think over again their business scope and competitive strategy: the laying of this telematic network organization on top of existing work processes trivializes its potential (Benjamin et al., 1990).

Finally, as networks are a dynamic entity, every node must learn by learning, that is: it must readily react to the system evolution. Both physical and intangible components of an operating network must always undergo maintenance operations: this is particularly true during rapid technological changes.

4.2. Exit costs

Exit or transit costs include both sunk and switching costs. Sunk costs are due to irrecoverable capital and intangible investments previously sustained. Exit barriers are produced by the need of substituting hardware devices, highly customized software routines, and of replacing communication standards. More specific entry costs generally correspond to higher exit costs. This specificity falls into four categories (Williamson, 1975): *site* concerns assets having high set-up and relocation costs; *physical assets* refer to specialized facilities; *human assets* concern the need of skilled labour; *dedicated assets* refer to incremental investments (not in telematic technologies) necessary to share in the network. Highly dedicated equipment (i.e., network or software) generally has very little salvage value, while personal training and organization procedures may easily be recovered and transferred (Kambil and Venkatraman, 1990).

Switching costs concern the 'unlearning' of old methods and procedures and the learning of new ones. In the transition period a readjustment process is started and in this interval the firm is largely ineffective. Also in this case, switching costs (and readjustment time) increase together with the learning process specificity. A serious problem must be faced when exit or transit means that the firm has to recover control on partially or completely decentralized production phases.

The nature of transactions occurring within the network creates additional exit barriers. When information contents and communication channels are highly specific, the nodes are prevented from communicating outside the system. Furthermore, if corporate decision-making depends only on information produced inside the network, several nodes run the risk of completely losing their autonomy and their response capability to the changing environment. In other words the less independent from the network resources the firm remains, the easier is its exit, and vice versa.

An explanatory framework which tries to schematically outline factors af-

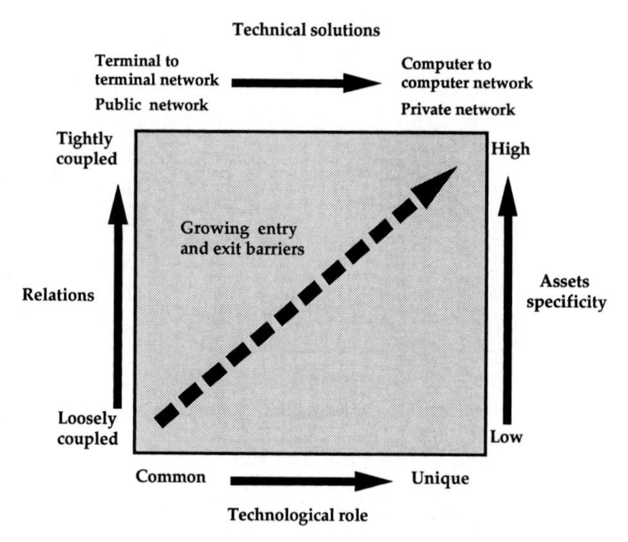

Fig. 3. Factors affecting entry and exit barriers size.

fecting entry and exit barriers and relations linking their nature to the barriers dimension is reported in Fig. 3.

4.3. Effects of communication standards on barriers size

Independently from the technical solution adopted, a telematic network is characterized by the adopted communication standard, that is, the *specific type* of the used devices, protocols, programs and data formats. The necessary acquisition of the network standard represents an entry barrier that is specific for that network, and determines, as already evidenced, to what extent a network is 'open' to the external world.

The two most important factors affecting the size of these barriers are as follows.

The first is the *technical nature* of the communication standards. When a firm decides to enter a telematic network it must acquire the adequate *equipment*. Prices of equipment depend on its complexity and capability of performing sophisticated computations. When a device can be replaced with a different model of the same kind, this means cost savings and reduced network entry costs. The second factor is the *data format* required to communicate correctly inside the network. Data format can vary from an unstructured (e.g., FAX transmission) to a highly structured form (e.g., EDI networks), depending on the complexity and *value* of the exchanged information. The costs of data format standards are closely related to the complexity of the data format itself.

Standards can be either *private* or *public goods* (Fornengo, 1988). Up to today the greatest part of standards are voluntarily developed by private organizations, trade associations or even single companies or consortia of firms. A 'proprietary' standard, even if favouring communications among firms agreeing on it, has also the practical effect of building entry and exit barriers to the network and consequently it creates or increases the market power of the network itself. Hence the adoption of a strictly private standard is used to protect the network boundaries.

The introduction of proprietary standards is more likely to occur in concentrated industries. Here adequately sized firms can sustain the high costs related to standard development. On the contrary 'public' standards are developed to favour the diffusion of telematics among firms, with the aim of preventing both the proliferation of non-compatible standards and the dominant position of a system or a supplier.

The diffusion of private or public standards has different effects on market structures because they establish different network entry/exit costs. Public standards, lowering barriers due to widely interconnectable systems, seem to favour either the market or non-hierarchical networks. On the contrary, proprietary standards favour hierarchical or integrated networks mainly because of sunk costs and learning costs: the former are necessary to switch between different networks and the latter to acquire new standards.

5. Structural consequences and conclusions

The diffusion of information technology, particularly telematics, has various effects on the economic systems. In theory, an increase in the information flows should reduce the market imperfections, namely those due to intrinsic uncertainty, and should lead to a better global efficiency. In the real world, telematics is diffusing asymmetrically and it seems at present to be a source of new heterogeneity. Large and complex organizations have better reasons and opportunities for adopting telematic solutions; moreover, they have greater resources for the developing of vast telematic projects. Hence, large firms and multinational organizations have exploited the opportunities offered by telematics: i.e., reducing coordination and control costs, and consequently strengthening their management structures. We might say that up to now telematics has favoured hierarchic organizations. From this point of view, the use of information technologies seems to have enhanced the structural and cognitive differences and finally the asymmetries in market power distribution.

On-coming changes in economic and institutional environments, e.g., market globalization and higher competition levels, tend to increase the complexity and uncertainty of decision making. When a firm has to face uncertainty and risk, information technology and its applications as telematic networks play an important role. At present, sharing the risk in an organizational network and facing uncertainty with information technology is a more and more used strategy. Thus the analysis of telematics opportunities and effects is now being carried out in the literature, not only when hierarchy is concerned, but also when inter-firm relations and network structures are involved. Thanks to telematics, the reduction of transaction, coordination and control costs tends to favour these inter-firm organizations.

In this issue, telematics effects on network structures have been considered. Even if, under the present economic and technical conditions, telematics seems to favour the diffusion of cooperative organization forms, the introduction of this technology may enhance disequilibria in sharing benefits. This is due to the raising of network entry and exit barriers during the transition towards such an electronic network. Associated costs due to new equipment and to new operating procedures enabling the use of the exchanged information affect the nature of these barriers and influence the firm's profitable activity, as well as the possible transfer to another.

Various empirical evidence shows that at present most networks seem to spread, assuming a star configuration. Actually, the leading or focal firm plays the central role not only in coordinating the activity of the whole network, but also in defining and implementing the informatics project. Telematic linkages produce the raising of entry/exit barriers in the network. These barriers derive from adopting/learning and abandoning/unlearning costs. New strategic opportunities are offered to leading firms. In fact, the latter can appropriately modify the other firms' entry/exit costs trade-off and they can govern, in this way, the adapting and modifying processes of the network. This opportunity is enhanced by the difficulties existing at present in widely imposing public communication standards, and by the prevailing of private ones.

Our analysis suggests that the observed tendency to a particular organizational form, as a *quasi integrated network*, might be encouraged by the diffusion of information technologies and proprietary standards. The development of large network firms and the raising of entry/exit barriers could favour the creation of oligopolistic market structures.

The adoption of information technology tending to exalt the asymmetries in the networks might theoretically cause two opposite consequences. On the one hand power concentration and high exit barriers might increase the *stability* of these organizational structures. On the other hand increasing disequilibria in benefits distribution might in the long run enhance their *instability*. The latter effect might be intensified by technical progress, i.e.: reduced specificity and costs of technical devices; tuning and diffusion of public communication standards. Such conditions, stimulating the reduction of exit barriers, might create new competitive behaviours and promote a new orientation towards the market, as it is stated by the transaction costs approach.

Acknowledgements

This paper arose from a research program 'Trasferimento delle tecnologie dei progetti finalizzati' supported by a contribution from the the National Research Council (CNR).

References

Antonelli, C. (ed.), 1988. *New Information Technology and Industrial Change: The Italian Case* (Kluwer Academic Publishers, Dordrecht).

Antonelli, C. (ed.), 1992. *The Economics of Information Networks* (North-Holland, Amsterdam).

Axelsson, B. and Easton, G. (eds.), 1992. *Industrial Networks. A New View of Reality* (Routledge, London).

Benjamin, R.I., de Long, D.W. and Scott Morton, M.S., 1990. Electronic Data Interchange: How Much Competitive Advantage? *Long Range Planning* 23 (1), 29–40.

Bolisani, E., Gottardi, G. and Scarso, E., 1992. Managing Effective Product Development: Some Operative Guidelines. *Proceedings of International Product Development Management Conference on New Approaches to Development and Engineering*, Brussels, May 18–19.

Brosseau, E., 1990. Information Technologies and Inter-firm relationships: the spread of Inter-organization Telematics Systems and its impact on economic structures. *ITS 8th International Conference, Telecommunication and the Challenge of Innovation and Global Competition*, Venice, March 18–21.

Capello, R. and Williams, H., 1990. Corporate Information Flows, Organisational Change and Computer Network Trajectories. *ITS 8th International Conference, Telecommunication and the Challenge of Innovation and Global Competition*, Venice, March 18–21.

Cash, I., McFarlan, F.W. and McKenney, J.L., 1988. *Corporate Information Systems Management: Issues Facing Senior Executives* (R.D. Irwin Inc., Homewood).

Conctractor, F.J. and Lorange, P., 1988. Why Should Firms Cooperate? The Strategy and Economic Basis for Cooperative Ventures. In: F.J. Contractor and P. Lorange (eds.), *Cooperative Strategies in International Business* (Lexington Books, Lexington).

Easton, G., 1992. Industrial networks: a review. In: B. Axelsson and G. Easton (eds.), *Industrial Networks. A New View of Reality*, pp. 1–27 (Routledge, London).

Etheridge, D. and Simon, E., 1992. *Information Networks. Planning and Design* (Prentice-Hall, New York).

Fornengo, G., 1988. Interorganizational Networks and Market Structures. In: C. Antonelli (ed.), *New Information Technology and Industrial Change: The Italian Case* (Kluwer Academic Publishers, Dordrecht).

Gottardi, G., 1990. Telematics and the industrial district. *ITS 8th International Conference, Telecommunication and the Challenge of Innovation and Global Competition*, Venice, March 18–21.

Gottardi, G., 1991. Diffusion of Information Technology and Firm Spatial Reorganization. *38th North American Meeting of the Regional Science Association*, New Orleans, November 8–10.

Gottardi, G. and Quaglio, E., 1992. *Piccole e medie imprese nel villaggio globale* (CEDAM, Padova).

Jarrillo, J.C., 1988. On Strategic networks. *Strategic Management Journal*, 9, 31–41.

Johnston, H.R. and Vitale, M.R., 1988. Creating Competitive Advantage with Interorganizational Information Systems. *MIS Quarterly*, June.

Kambil, A. and Venkatraman, N., 1990. Information Technology Mediated Exchange Relations. Toward an Integrative Framework for Research. *ITS 8th International Conference, Telecommunication and the Challenge of Innovation and Global Competition*, Venice, March 18–21.

Keen, P.G.W., 1991. *Shaping the Future. Business Design through Information Technology* (Harvard Business School Press, Boston, Mass.).

Pontiggia, A., 1991. Information Technology and Strategic Conduct. Competitive and Cooperative Opportunities: Electronic Networks. *Economia Aziendale*, X (3), 301–325.

Porter, M.E., 1985. *Competitive Advantage* (Free Press, New York).

Rockart, J.F. and Short, J.E., 1989. IT in 1990s: Managing Organizational Interdependences. *Sloan Management Review*, 30, 2.

Rockart, J.F. and Short, J.E., 1991. The Networked Organization and the Management of Interdependence. In: M.S. Scott Morton (ed.), *The Corporation of the 1990s. Information Technologies and Organizational Transformation*, pp. 189–219 (Oxford University Press, New York).

Salop, S. and Scheffman, D., 1983. Raising Rivals' Costs. *American Economic Review, Papers and Proceedings*, 73, 267–271.

Scherer, F.M., 1980. *Industrial Market Structure and Economic Performance*, 2nd ed. (Rand-McNally, Chicago, Ill.).

Scott Morton, M.S. (ed.), 1991. *The Corporation of the 1990s. Information Technologies and Organizational Transformation* (Oxford University Press, New York).

Stiglitz, J.E., 1988. Economic Organization, Information and Development. In: H. Chenery and T.S. Srinivasan (eds.), *Handbook of Development Economics, Vol. 1* (Elsevier, Amsterdam).

Teece, D.J., 1986. Profiting from technological innovation: Implication for integration, collaboration, licensing, and public policy. *Research Policy*, pp. 285–305.

Thorelli, H.B., 1986. Networks: between Markets and Hierarchies. *Strategic Management Journal*, 7, 37–51.

Venkatraman, N., 1991. IT-Induced Business Reconfiguration. In: M.S. Scott Morton (ed.), *The Corporation of the 1990s. Information Technologies and Organizational Transformation*, pp. 159–186 (Oxford University Press, New York).

Williamson, O.E., 1975. *Markets and Hierarchies* (Free Press, New York).

Global Telecommunications Strategies
and Technological Changes
Edited by G. Pogorel

Chapter 3

The impact of EDI-based quick response systems on logistics systems

Jiro Kokuryo

Graduate School of Business Administration, Keio University, 2-1-1 Hiyoshi-Honcho, Kohoku-ku, Yokohama, Kanagawa 223, Japan

1. Background

By 1990, Electronic Data Interchange (EDI)-based Quick Response (QR) had become a commonly used term in the American retailing industry and its vendor community for a set of business and technology improvements that are aimed at developing capabilities to provide fast responses to market changes. It is reported that effectively implemented QR systems are capable of reducing delivery lead time and increasing delivery frequency while lowering the overall costs. These in turn allow the retailers to generate profits by reducing inventory and by increasing sales (via reduced stock-out rate). At the same time, manufacturers benefit from the lower costs in channel coordination.

Electronic data interchange is a core technology in implementing Quick Response. EDI has already been instrumental in reducing the order-placement and order-processing components of the entire product-delivery cycle. While it is possible to implement a QR program without EDI, its scope and depth would be very limited.

While QR has potentials for numerous benefits, there are several important challenges to a successful implementation thereof. Firstly, QR programs usually involve re-organization of business processes not only within a firm, but also across firms within the distribution-channel. Accusations that QR reduces retailers' inventory merely by pushing it up the distribution-channel are symptoms of the inherent difficulties in coordinating across firms.

Secondly, QR programs also require cross-functional re-organization. A QR program requires involvement not only of the IS (information systems) department, but also of many functional areas including accounting, logistics, production, and marketing. The source of the problem is the incompatibility among the goals of the various functional groups. For example, an emphasis on increased delivery frequency (which is usually good for inventory control) often works against the efforts of transportation managers to minimize trucking costs.

Thirdly, and probably the most serious, problem is the lack of frameworks to find the 'right balance' from total systems points of view. As a result, managers often place too much emphasis on a limited scope of performance indicators (notably inventory turnover within their companies) only to sub-optimize the system by paying the price of missed opportunities for total system improvement.

2. *Theoretical foundations*

There were two theoretical foundations for the research. One was the recognition that a business is an integrated group of subsystems. Information systems (IS), logistics systems, financial systems, and human-resource management systems are all parts of a product/service delivery system. The nature of subsystem integration is such that the performances and mechanisms of subsystems are also mutually interdependent.

An important implication of the model is that the success of an emergent IS is dependent upon the capabilities of other subsystems to exploit the opportunities created by the information system. This line of reasoning is based on the recognition that the performance of a system is determined by the most severe constraint within the system. A natural extension of this logic is that when a technology removes one constraint, it will reveal a new constraint. Thus, a good way to assess potentials of an information technology is to consider what binding constraints it creates in related subsystems and exploring the possibilities of overcoming them. (Fig. 1 shows the model used for this research to operationalize the research.)

Such a relation is most apparent between information and logistics systems. EDI, for example, enhances the capacity to communicate not only within organizations but also across firm boundaries. Orders for goods by the buyers can now be transmitted out of the buyers' computers straight into those of the suppliers'. There is a strong expectation that such EDI based systems will enable firms to deliver products to their customers very quickly and very flexibly. Data alone, however, do not bring products to buyers' doors. In such an environment, it is up to the logistics and the production systems to exploit the opportunities to improve the services. In other words, EDI creates pressures for logistics systems to improve.

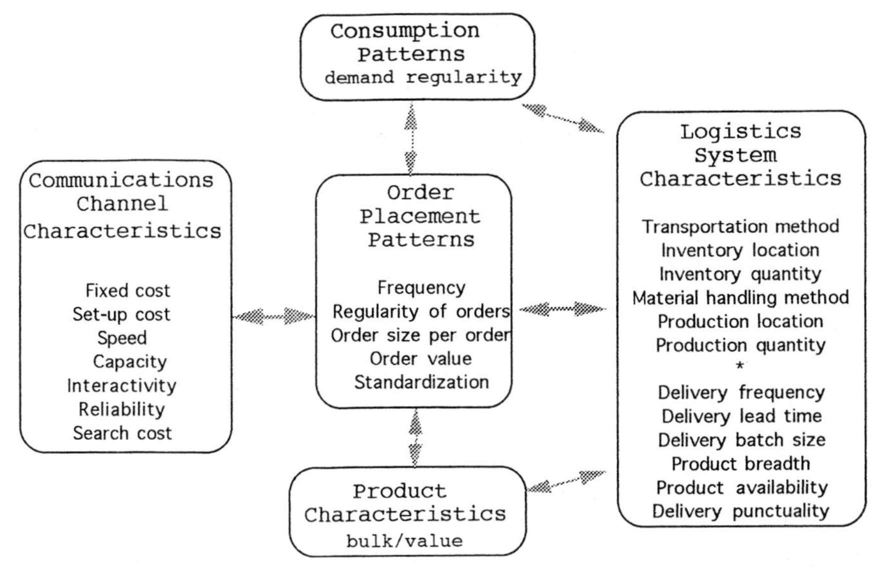

Fig. 1. Analytical framework.

The second theoretical foundation was constructed on an early observation of this research that 1) there are many different ways to design QR logistics systems, and 2) vendors are struggling to meet the diverse requirements of the retailers that strictly impose conformance. To give a flavor of how diverse QR implementation methods might be, the range includes, even within a single (apparel) industry, from a globally centralized supply from Asia using air freight, to a town level, localized production using small 'flexible' production systems.

The theoretical question, then, is whether diverse solutions for QR can exist simultaneously. Is the diversity that we currently observe merely a transitory phenomenon? This question, as we will see later in this paper, is very important because failure to address the issue can lead to an imposition of a 'standard' that sub-optimizes product-delivery systems.

One way to think about the in logistics system design of question diversity is offered by Shapiro et al. (1985). They argue that when product-delivery systems are evaluated by buyers in not one, but in multiple dimensions (e.g., price and speed of delivery), there can be multiple ways of serving the market. Behind this statement is the recognition that there are trade-offs that force companies to compromise on certain goals to achieve others. In fact, standard frameworks of logistical analysis are constructed around the notions of trade-offs.

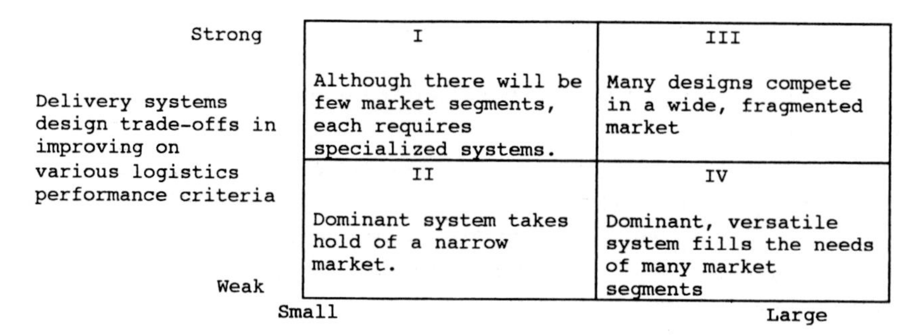

Fig. 2. Trade-off in delivery systems and strategic choices.

The underlying cause for such trade-offs is the diversities in both the demand and supply sides of the equation. On the demand side, customers of a product have different needs for the product. For some customers, speed of delivery is the key. For others, price may be the key. The same person may have different types of demand on different occasions. On the supply side, a product-delivery system is often capable of addressing certain types of demands but not others. Strategic implications of trade-off and demand diversity can be summarized as in Fig. 2.

To facilitate the understanding of the figure, let us review two quadrants. Quadrant I, for example, shows a situation in which there are few service criteria that are appreciated by the customers. Freshness, for example, is an overwhelming factor for the consumption of non-frozen seafood. Here, the logistics system to serve this product becomes quite limited even if many ways to physically transport the goods may be available. Quadrant IV, on the other hand, shows a situation in which there is a large diversity in the type of service appreciated by customers, but then there is one (or a few) general solution that addresses all the needs better than others. Personal computers, because of their versatility in handling information, are being used for a diverse range of tasks that were once done by using many different kinds of specialized tools.

This analysis seems to show that the existence of diversity is justified when there are trade-offs in the design of logistics systems. Verification of the existence of such trade-offs was thus a part of the research agenda. The results indicated the existence of trade-offs, suggesting the legitimacy of the diversities in the QR logistics systems.

3. *Outline of the research*

A multi-method research was conducted on the topic in 1991 . The methods used were as follows:

(1) Case studies of six vendors to a chosen retailer.

(2) Case studies of two retailers.

(3) Simulation analysis of a distribution-channel wide QR cost.

(4) Mail survey of QR vendors (67 responses. 68% response rate) to a chosen retailer.

4. *Increased level of interdependence between retailer and vendor systems*

The research indicated, among other things, that EDI-based QR programs raise the level of interdependence among retailer-vendor systems and among various functional groups within the product-delivery chain. This perception was formed out of a series of interviews with managers of both retailers and vendors.

A frequently heard comment during the vendor interviews was that retailers are becoming more detailed and strict in the way they specify delivery requirements (such as labeling format, labeling location, receipt time window, case size, requirements for ticketing, and so on). At the same time, one of the most frequently heard retailer complaints was vendors' failure to comply with the specifications (such as missing labels, failure to notify partial delivery and so on).

The cause for such increased attention to the tighter delivery specifications is the need to standardize and automate the logistics functions on the retailing side. For example, in order to automate the receipt of the goods from the vendors, the cases containing the goods have to be packaged and marked with barcodes exactly in such a way that retailers' automated equipment can handle them.

Tighter delivery specifications in a short delivery lead time environment are a major challenge for vendors. When long delivery lead times are provided, vendors may have the option of dealing with the tighter specifications manually and off the production lines. This is an unaffordable luxury in a QR environment. Thus vendors have little choice but to integrate and automate the packaging process according to retailers' detailed (and very often proprietary) specifications. Any changes in the retailer specifications now affect the vendor operation to a much greater extent than before.

Another problem for the vendors is the investment necessary to comply with the tight and proprietary specifications. To use the barcoding example again, for those that had been marking cases only with UPC (universal product code which indicates what product is in a certain case), compliance with the retailers' demand to mark cases individually (that is to mark cases with unique identifiers) with the industry standard UCC 128 code requires not only the barcode equipment itself but also total restructuring of the product flow to control products at the individual case level. The investment involved can be substantial. This not only raises the overall cost for the vendors but also changes the cost structure (more fixed cost). Without retailers' willingness to share the cost and the risk, this can make vendors vulnerable.

To summarize the above conceptually, EDI-based Quick Response systems raise the level of interdependence of vendor and retailer systems by eliminating buffers that previously existed between them. As a result, an action that either party adopts affects the other greatly. In this environment, one of the toughest issues is determining how to jointly improve the overall system, as well as how to share the costs and benefits. Retailers and vendors both have to make efforts to understand the operational impacts of their own actions on their business partners.

4.1. The importance of managing uncertainty

In a product-delivery system in which components are highly interdependent, mismanagement of any one part quickly impacts the entire system. For example, the research indicated that some retailers' mismanagement of Quick Response programs resulted in an increased level of uncertainty for the vendors. The vendors in turn are forced to bear the cost of uncertainty through increased inventory and/or excess production/logistics capacity. Indicative of the problems were the following responses by vendors to the mail survey question which asked for challenges in implementing Quick Response. (There were twelve comments in all in this category.):

- (The most challenging aspect of QR is) Lack of information (forecast) from some retailers — their expectation that we will have on hand anything that they may want in any quantity.
- Large fluctuations in weekly order quantities make staffing difficult, especially for accounts whose shipments require pick and pack. Unusually large quantities slow (down the operation) and leave only a little time each week to pack and ship
- Inventory levels have had to be increased in our warehouse to anticipate immediate demands to fill orders with a shorter lead time.
- With the push (by the retailers) to purchase at the last possible moment, it becomes extremely difficult for all concerned (the manufacturer, the truckers, and the retailer) to handle the sizable volume required at the last minute.

To theorize the situation a little, in a non-QR environment in which orders are given sufficiently ahead of delivery, it is easy for the vendors to identify when the peak demand comes and how great it may be. As a result, if the peak demand exceeds the capacity of production and/or the logistics infrastructure, the vendors have the option of leveling the load over a period of time. In a mismanaged QR environment, however, EDI allows the retailers to procrastinate until the last minute and dispatch rush orders. Vendors in turn are hit with a unexpected level of demand but are given a very short time to react. This is a typical sub-optimization situation in which retailers improve performance at the cost of the deterioration of total systems performance.

On the surface, the vendors' desires for reduced uncertainty seem contrary the philosophy of Quick Response which asks the vendors to construct flexible product delivery systems that can quickly meet changing needs. In thinking about uncertainties, however, it is important to recognize that there are at least two kinds thereof. One is uncertainties in the aggregate size of orders and the other is uncertainties in the mix of orders (such as mix of sizes within the same product line). The literature, including Ohno (1978) has indicated, and this research confirms, that in the process of implementing Quick Response, vendors have the means to deal with the uncertainties in the mix, but the tools are not effective in dealing with the uncertainty in the aggregate demand. An important implication is that effective implementation of Quick Response requires a long-term commitment by the buyer to stably purchase goods from a certain vendor so as to ensure efficient use of the vendors' operational capacities. If such mechanisms are well understood, Quick Response systems can be used to replenish the right combination of goods (style/model, color, size and so on) flexibly to meet the changing taste of the end user without hitting the vendors with an unnecessary level of uncertainties in the aggregate demand.

There were indications of a large variance in the retailers' success in managing uncertainty. For example, there was a statistically significant relation between those respondents that said forecasting was important and those that said 'success of QR depends on retailers' programs'. A line that separates winners from losers with QR is clearly drawn around this issue.

4.2. Dangers of functional sub-optimization

Section 4.1 portrayed a possible situation in which retailers sub-optimize their systems at the cost of deterioration in vendor (and possibly total) system performance. We can characterize this as sub-optimization between trade partners. Another form in which sub-optimization might occur is functional optimization. This is not a new message in the systems-analysis world, yet managers nevertheless fall frequently into this trap.

To illustrate, Fig. 3 shows the simulation results from the research. This analysis compares total costs of delivering products of varying value density

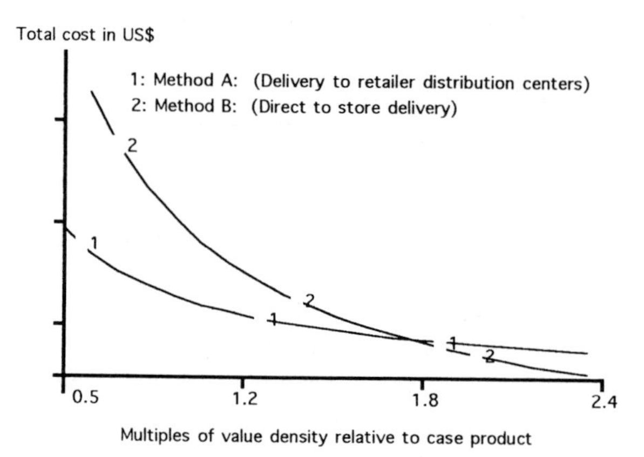

Fig. 3. Total costs of providing a 99% in-stock level service with various distribution-methods.

(value per bulk) using two different delivery methods. Most of the parameters used in this analysis are real life numbers obtained from case studies for a particular product.

Lines in the figure show the costs at which the two methods operate if they were required to perform at the 99% in-stock level at the retail store. The X axis shows varying levels of value density. (The analysis assumes constant product value and varying bulk. The lines are downward sloped because as one goes higher on the X axis, goods become less bulky and therefore less costly to distribute).

Figure 3 indicates that when value density is low, method A has the lower cost while method B excels for high value density products. (That is, it provides the same level of service at lower cost.) Although this looks simple, a very different result can be obtained if the objective is to minimize inventory cost within the distribution-channel.

Figure 4 shows the breakdown of the cost exhibited in Fig. 1. Here we see that method B has a lower inventory cost across the X axis. On the other hand, the transport cost superiority is dependent on the level of value density. Such an underlying mechanism was behind the crossing of the total cost lines in Fig. 3. An important implication of Fig. 4 is that if one focused only on the inventory cost, one would choose method B for all products, while if we looked at the total cost, we would choose method A for a wide range of products.

The bias we see here is extremely important because inventory turnover is one of the most commonly used indicators to assess the performance of Quick Response among the companies researched. Many firms lack the organizational measurement and communication to weigh different portions of the total logistics cost to make informed decisions on the right design of the product-delivery systems.

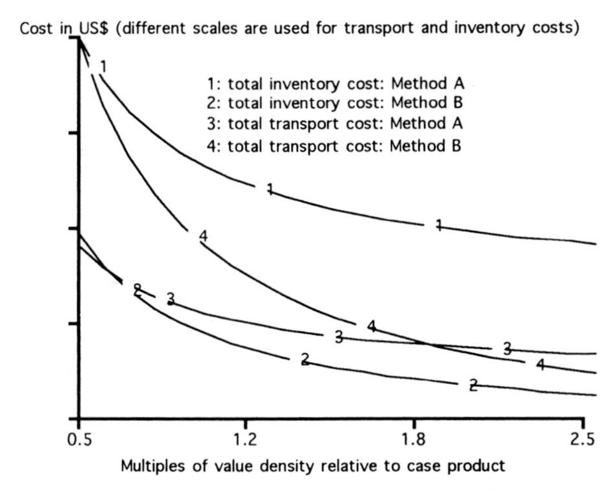

Fig. 4. Inventory and transport cost in the modified simulation results.

This research suggests that managers should be reminded of the dangers of functional sub-optimization. Myopic focus on a single (or a limited number of) performance indicator such as inventory turnover can hurt the overall profitability.

4.3. Dangers of failing to address diverse needs

The last type of danger that this paper points out is the failure to address diversity in the design of information systems. We have already seen that there are diverse ways of configuring QR logistics systems. In a similar way, the most frequently cited (by the vendors in the mail survey) technological challenge in implementing QR was the heterogeneity among retailers in their information systems requirements. The vendors were obviously struggling to prepare information systems that meet the diverse demands of retailers.

A question around this issue of systems heterogeneity was whether it is a transitory state or if there are underlying causes that perpetuate it. If it is a transitory phenomenon, we can expect standardization to progress over time. As far as EDI communications protocols are concerned, comments from the vendors IS managers indicated that there certainly is a convergence to ANSI X.12 standard under way, although some complained of excessive flexibility in the standard which causes heterogeneity within the boundary of the standard.

Interviews conducted during the case studies, however, indicated that the issue goes far beyond sharing a mutual communications protocol. It is not simply a matter of standardizing protocols, but is an issue of coping with very different designs of product-delivery systems.

For example, two commonly seen modes of delivery from the vendors to the retailers are: (1) direct to store delivery of cases that are pick and packed (containing an assortment of goods that the stores ordered), and (2) bulk delivery to retailer distribution-centers in full cases. The information and communication systems involved in serving the two are, however, dramatically different. Pick-and-pack/direct-to-store delivery requires the transmission of store-level orders and vendors have to track the flow and the content of each individual case. Bulk distribution-center delivery, on the other hand, lets vendors manage goods in units of full cases (or even pallets) but instead requires close communications with the retailers on the timing and quantity of delivery to stabilize retailers' distribution-center operations.

What is clear from the above example is the heterogeneity of that information processing and communications systems is a reflection of the heterogeneity in logistics system designs. The danger, then, is in imposing a particular type of information system without considering the nature of the operation of the business partners. Examples were observed during the research in which retailers' imposition of particular transaction formats (both in the logistics and information systems sense) forced vendors to operate at much lower efficiencies than they could potentially perform.

5. Conclusion

A common thread that binds the findings of the research is that EDI-based QR programs are increasing the interdependence among sub-systems in product-delivery systems. As a result, any design and/or operational changes in one sub-system quickly and greatly affect other components of the total system. In such an environment, coordination among sub-systems becomes critically important. In practice, this means greater coordination among trading partners and among functional groups. Unfortunately, such coordination is easier said than done for reasons explained in this paper.

A popular word to use to describe Quick Response is 'partnership'. When one looks at the business literature, both retailers and vendors talk about how cooperation between business partners is critical in making Quick Response successful. The irony is that the list of firms that are eloquently advocating partnership include those that are very unpopular among their partners for being dictatorial and insensitive to the partners' needs.

If Quick Response is to succeed, the 'partnership' between retailers and vendors should go far beyond exchange of goodwill, or the willingness to share costs. Instead, retailers and vendors have to be involved in each other's *internal* operations. This is because in a Quick Response environment, actions taken by one group within a distribution-channel are likely to affect the internal operations of others to a much greater degree than before. In this kind of environment, it requires the vendors and retailers to understand each others' operational details and be willing to cooperate at a very concrete level.

Acknowledgements

This research was funded in part by University of Southern California's Center for Telecommunications Management.

References

Ohno, T., 1978. *Toyota Seisan Hooshiki (Toyota Production System)* (Diamond, Tokyo). Translation available from Productivity Press, Cambridge, Mass., 1988.

Shapiro, R.D., Rosenfield, D. and Bohn, R., 1985. Implications of Cost-Service Tradeoffs on Industry Logistics Structures. *Harvard Business School Working Paper*, 9-785-036, (Harvard Business School Division of Research, Mass.)

Global Telecommunications Strategies and Technological Changes
Edited by G. Pogorel

Chapter 4

The making of a Pan-European network as a path-dependency process: the case of GSM versus IBC (Integrated Broadband Communication) Network

Gabriella Cattaneo

Teknibank, Milan, Italy

1. Introduction

While the projects for a Pan-European Integrated Broadband Communication infrastructure are encountering serious difficulties in taking-off, the implementation of the digital cellular European system GSM is proceeding, showing that it is actually possible to build a new communications network from scratch on standards and cooperation.

The analysis of GSM development helps to highlight the mix of factors needed to ensure the success of such complex projects, especially the role and motivation of the main actors and the importance of right timing in a favourable historical context, typical elements of a path-dependency process.

The common characteristics of path-dependency processes are that *factors of chance*, with an essentially historical character, are combined with the role of systemic forces in the reconstruction of social and economic developments. A particular sequence of choices at the beginning of a process may lead to a self-stimulating mechanism which encloses the development in a growth equilibrium, that *locks out* alternative development possibilities. Path-dependency theory reconstructs the interdependence of individual choices under conditions of positive (or negative) local feedback in a network context.

The important consideration here is that the predominance of a determined technology or system is not necessarily due to its greater efficiency, but simply to critical historical events or coincidences. Also, the persistence of inefficiencies may be explained by lock-in processes created in historical situations.

These historical factors, as will be seen, played a substantial role in the building of a consensus around the GSM standard, and are on the contrary acting against the success of existing IBC projects.

2. The key factors in the development of communication networks

The development of communication networks is a complex, dynamic process of interaction between various partners or actors in a specific historical situation. Such processes are often defined as path-dependency processes, meaning that the outcome is not predetermined, but the path of development and therefore the final outcome, is influenced by certain key events, even casual events.

Another important concept for the development of communication networks is that of *critical mass*, used to define the minimum number of users needed to make it worthwhile for new users to join the network, creating a positive feed-back and a self-sustaining development. Both concepts shed a new light on the key success factors in the development of telecommunication networks, the importance of the historical context and of the approach to demand, emphasizing factors often overlooked in telecommunications development strategies.

Examples of such processes are given by the market-driven institutionalisation of the QWERTY typewriter keyboard (certainly not the best possible solution) or the IBM PC-DOS standard (David, 1985).

The path-dependency scheme of analysis has been also applied to the different developments of Videotex services in France, Germany and the U.K. A combination of choices by the Telephone Operators (TOs) in the early phases of development of the system, influenced by the respective socio-economical context, achieved in France a vicious circle of growth, which was never reached in Germany and the U.K. For example, France Telecom did not expect the *kiosque* tariffing system to become one of the key success factors of the system, by encouraging mass market acceptance thanks to its ease of use.

The most important lesson from the French success in Videotex development is that the demand for such services is not given, but is influenced by and dependent on the technical and social organization of videotex systems, their implementation strategies, and the diffusion process of the system itself.

Some considerations concerning Videotex systems can very well be applied to possible broadband systems. Formation of Demand for new telecommunication services is an interactive process, where user preferences are interdependent and collective learning is involved. The *user value* of the system depends on the size of the user community (the more users and service providers are linked to the network, the more interesting it is to join it). On the other hand, demand formation is a collective learning process, where late adopters learn from early adopters. And the more flexible a system is, the easier it is for new applications to be discovered.

The interactive process of the growth of telecommunication networks can lead to a *vicious circle* similar to the growth mechanism of an avalanche: the bigger it is, the faster it grows. The problem, however, is often located in the start-up phase of the system, when there is a dilemma similar to the famous *chicken and egg* problem: users join only when there is a sufficient number of other users to maximise the network's communication value, while suppliers and operators will invest only if they feel/see a sufficient number of users joining the network.

The same dilemma concerns equipment: terminals/interfaces will sell in large quantities if they are cheap, but the prices will go down only if there are large economies of scale for a mass market. This problem is usual in mass production, and is normally solved by strategic pricing and portfolio management (cross-subsidizing of new products by older, commercially successful products — also called cash-cows). But for new telecommunications services it is sometimes also difficult to judge correctly an acceptable pricing level for the market.

The problem is how to reach the key threshold in the number of users, when the appeal of the system becomes positive for users/suppliers. This is what is called in the socio-economic literature the critical mass of users, that is, the turning point where the growth of the network becomes self-supporting.

When the rate of adoption, on the contrary, is too slow, the feed-back becomes negative and there is a strong risk of a situation of stagnation, where no actor is willing to invest time/resources in the innovation.

The delicate balance of positive/negative feed-back is decided in a rather short time in the early phases of development of a new service: in other words, to achieve success not only cooperation between key actors is important, but also the timing of key decisions.

As several authors have shown, there are at least three possible solutions to the dilemma of reaching the critical mass in a new telecommunication network.

The direct approach: The system operator or the service provider eliminates the cost of access and provides the service free or for a nominal fee, creating the market from scratch. This is essentially what France Telecom did for Minitel. But for Pan-European broadband systems no single actor can shoulder the investment alone, at least not without incentives or some market guarantees.

The sequencing approach: This approach needs less or no subsidies in the initial phase, but presumes a heterogeneous user potential. It consists in segmenting the potential market and singling out user groups for whom the utility of the service exceeds the initial high costs of access. This group may then work as a critical mass for a second group, which finds the size of the terminal park adequate for their applications. If this group also joins, then the critical mass for a third group is reached, and so on. Such a strategy, resembling a *chain reaction*, places a large emphasis on using market research to detect different user groups and on the ability to mobilise them at the right time: in other words, it requires quite sophisticated marketing.

This approach could be fruitful in the case of broadband networks, where the problem is to start materialising a potential demand.

The management of expectations strategy: This is an interesting strategy, which reaches the critical mass by triggering and managing self-fulfilling prophecies. It is actually the way by which the market chose the DOS standard: the general expectation that the IBM PC software would become *the* standard, a classic self-fulfilling prophecy. But it is a strategy that is extremely difficult to apply, by itself.

It is, however, an important element complementary to the strategies described before: to induce growth, some kind of positive expectation must be created in the potential market. This should be a crucial issue for broadband projects, suffering up to now from limited credibility, even in the operators' eyes.

In the case of GSM development, there was no need to reach a critical mass of users, because mobile telephony subscribers get in touch immediately with the huge community of fixed telephony subscribers. But it can be said that a *critical mass* of consensus was reached, when in the interdependent group of decision-makers a number of key actors decided to commit themselves to development of the system.

3. The driving forces of GSM development

3.1. The historical process

The opportunity for a Pan-European cellular system arose when WARC, the World Administrative Radio Conference, set aside (in 1979) the 900 MHz band for mobile telephony systems. In 1982, following the proposal of the Scandinavian and the Netherlands PTT administrations, CEPT decided to create a workgroup to study the outline of a Pan-European cellular system to be implemented in the beginning of the 90's: GSM, Group Special Mobile.

From 1982 to 1986, there was discussion on the choice between an analog and a digital system, with the second eventually prevailing. Cooperation started in earnest in 1985, when the German and French PTT administrations decided to invest in R&D for digital systems, soon followed by other administrations.

In 1987 the crucial events for GSM took place:

- In February, a meeting of the GSM group in Madeira decided on the outline of the basic digital standard, rejecting a French–German technology developed by Alcatel-Sel because it was too *proprietary*.
- In May, a meeting between the TLC ministers of France, Germany, Italy and the U.K. accepted the GSM standard, and the four TOs decided to commit themselves to the development of the system, with a Memorandum of Understanding (MOU).
- In June, an EC Council Directive provided for the reservation of the 900 MHz band to the Pan-European digital cellular system in all member countries.
- In September, the Memorandum of Understanding was signed by 15 countries, with a general consensus that surprised the same promoters.

From 1988 on, work started on the development of national networks, with the contribution of manufacturers, who started to attend GSM meetings. The date of 1991 for the commercial launch of the system has not been respected, but most countries planned it for 1992 and gradual implementation is to follow in the years to 1995, when the European roaming service (the ability to use the same mobile phone in all the European countries while roaming, i.e., going around) must be in place according to the Memorandum of Understanding.

The historical factors working in favour of GSM were the emerging thrust towards a unified European market, and a correct timing in crucial decisions preventing dispersion of investments in conflicting national technologies. The decision by France and Germany to cooperate in digital cellular R&D was also important, because even if one goal — the choice of a Franco-German technology — was not achieved by them, it created the conditions for making an innovative technological choice, considered adequate for the 90's.

Another element was the first results of the Esprit research projects, which seemed to indicate a fast development of chip technologies for mobile communications (faster than it actually happened, because only now has it become possible to put all GSM software on only one chip, which makes first generation GSM terminals bigger and heavier than analog ones).

The key decisions in 1985–1987 happened with a timing such as to generate positive fall-back towards the other actors, creating a widespread expectation of success for GSM.

A key controversy between operators and industries surfaced immediately in 1988, concerning Intellectual Property Rights (IPR).

The group enacting the Memorandum of Understanding tried to establish a common policy, asking that industries having patents relevant for GSM

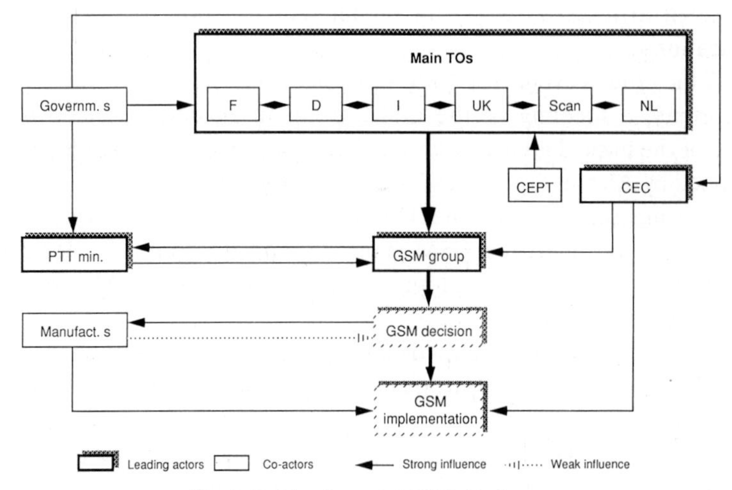

Fig. 1. Driving forces of GSM development.

should agree to license the technology to other suppliers, with a fair and equal sharing of all royalties. Manufacturers, particularly Motorola, refused, and eventually national operators had to short-cut the issue, negotiating the contracts for the development of national networks individually.

Actually, the controversy about IPR has proved one of the weak points of GSM agreements (which did not include any clear policy about patents) and more generally about ETSI standardising activity. At the present moment, manufacturers are not obliged to declare, during the building of a standard, whether they hold a patent relevant to it. The problem surfaced again once the time to build terminals came, with Motorola asking royalties from all the other manufacturers by declaring that a key part of GSM technology falls within its patents, according to U.S. law.

3.2. Motivations and role of key actors in GSM development

The main actors of GSM development can be identified as (see also Fig. 1):

- National Telephone Operators
- GSM group
- National Governments, in this instance mainly the Ministers of Post and Telecommunications and of Industry
- International cooperation bodies, chiefly the European Community, CEPT and ETSI
- Manufacturing industries of infrastructures and equipment (namely Alcatel, Siemens, Motorola, and others)
- Users

3.2.1. National Telephone operators
When the GSM group was created in 1982, the National TOs were still under direct political control in all countries: deregulation started in the U.K. immediately after that, but it was still very far in all other countries. This does not mean that TO managements did not have considerable autonomy in their strategic technical choices and negotiating power in the interaction with the respective ministries.

Secure in their monopolies (liberalization of the services was not yet seriously discussed) the TOs were looking for new services to increase the traffic in their existing telecommunication networks and for new follow-up products for the time when the telephone expansion would reach its saturation point. Mobile telephony, with a growing demand and a strong potential to increase traffic, was therefore considered a very interesting way to enlarge their markets. It must also be remembered that at the time the TOs were also the exclusive sellers of mobile phones, another source of income.

The example of the success of the first Scandinavian system with its international roaming capabilities within Scandinavia and its competitive suppliers was very inspiring for other PTT's. The perspective of a Pan-European system presented several advantages for them: it allowed sharing the risks and costs of development of a new technology, and establishing an open, standardised technology, it allowed freedom from national suppliers, sensibly reducing the prices of equipment thanks to the economies of scale.

The choice of digital technology was also a far-sighted one, but important in guaranteeing the new services (data transmission, encryption, and others) the TOs were looking for. In the larger countries, the differential of investment between guaranteeing national roaming and international roaming was also not very relevant. This is why the larger countries — U.K., France, Germany and Italy — were the key players in GSM development.

Moreover, the development of GSM started early enough, before important investments were made for national technologies. There were then strong motivations of self-interest for TOs to push GSM.

But their behaviour was by no means homogeneous. The Nordic countries and the British operators, handling successful services, were already quite well oriented to the market and to users needs, underlining for example the importance of portable phones. Their priority was the service, and they were ready to choose any technology available on the market with a good price/performance ratio. French and German TOs, on the contrary, were more technology oriented, interested in controlling the key features of the system by autonomous development. They were also under greater pressure from their respective governments to support the development of national industries.

The compromise developed in 1987 for GSM, first in Madeira and then in the Bonn TLC Ministers meeting in May, successfully joined these two *cul-*

tures, creating a system technologically advanced, but oriented to the market. The strong motivation of the TOs, fueled by the involvement of the representatives working in the GSM group, succeeded in overcoming the reluctance of national industries to accept a standardised solution open to competition.

3.2.2. GSM group: the champions of the system

The emergence of a cohesive international équipe in the GSM group was also crucial, because the people working in it, coming from TOs and PTT administrations, were able to lay the basis for compromises overcoming national interests. The GSM group acted as the *champions* of the system, and all innovations need support from determined individuals. Especially in the controversy around the German–French proposal, GSM group members from France and Germany kept working towards the goal of making their countries accept the general choice, which they thought was more fruitful for TOs strategy. They took care to include in the Madeira standard key elements of French and German developed technology, in order to make a compromise acceptable.

3.2.3. National governments

National governments were basically interested in the development of both new telecommunication services, and domestic industries, either in the field of consumer electronics or of telecommunications. The official goals of the Memorandum of Understandong are exactly these. Among the key countries, it was only France and Germany, though, which had an industrial policy strong enough to attempt to impose a certain standard developed by the national champions on the other countries. This attempt, as seen before, did not succeed, because the other countries refused it, and because the German Ministry of Telecommunications in the end preferred to act in the interest of the deployment of the service, rather than in the interest of industries.

3.2.4. The European Community

The European Community's role regarding GSM was a role of supportive, it was determinant especially in securing the legal basis for setting aside the necessary frequencies in all countries (not a banal issue). In parallel with the development of GSM, the Commission was elaborating the key concepts afterwards expressed in the Green Book, thereby fueling the ideal of European cooperation and harmonization of services in telecommunications.

The Community, as often happens, could not however act on resolving the conflicts of interest between the countries before the compromise of Bonn in 1987. Only after the controversy was resolved, it could step in with a Directive in favour of GSM.

3.2.5. Manufacturing industries

The main European suppliers (Alcatel, Ericsson, Motorola, Philips, and Siemens) were not in the beginning favorable to GSM. For some of them it represented the menace of opening the market also to dangerous Japanese competition. Few of them were convinced, especially in the early years, of the potential of the market for mobile services (the first Scandinavian system started without Ericsson, but with Mitsubishi among the suppliers). Some of them also pushed an analog solution as less risky than the digital one. Manufacturing industries did research on digital cellular systems. Nevertheless, once the TOs agreed on the Memorandum of Understanding, creating a market for GSM, they joined the project and started investing in the development of the products.

The controversy about IPR proved also that industries' interests can be in conflict with the implementation of general standards: the problem was actually not solved, but circumvented by single negotiations among operators and suppliers.

3.3. Degree of success of GSM as a Pan-European network

How far has GSM project met its promises? The Pan-European cellular network was scheduled to start operation on July 1st, 1991. The opening of the commercial services has been in most countries in 1992–1993. The schedule of the Mou plans also for coverage of all capital cities by 1993, and transport routes between them (that is, European roaming) by 1995. The areas where the demand for international roaming is stronger (i.e., Northern Europe) will probably implement it earlier than that date.

The main manufacturers showed their GSM phones at the Geneva International Telecom Conference in November 1991. Due to the complexity of the software, GSM portable handsets were heavier and more expensive than analog ones up to 1993.

Most manufacturers believe it will take until 1995 before there are more digital than analog phones in use. But by the end of the decade, phone companies could be exploiting the digital capabilities of the networks to extract extra revenue from mobile facsimile and personal computer services.

The delay in the launch of the service has been due to the tightness of the original schedule for such a complex system, particularly in not building in enough time for systems testing, according to some experts. The delays in setting the specifications have also meant delays in developing the equipment to test the phones.

The agreements for coordinated billing for European roaming are also proving complex, but it is expected they will be in place when it will starts.

In terms of deployment of the service, therefore, it seems that the goals of the Memorandum of Understanding are being achieved. Moreover, the standardized nature of the system has helped to achieve a greater competition than expected in the beginning, with at least two operators providing the service in every country. The ability of customers of one country to use their phones in the other countries will put pressure on operators to keep on schedule with the others.

Still in question, though, is whether the Pan-European collaboration can create a strong European electronics manufacturing base, keeping the Japanese competition at bay, which was the other key goal of the Memorandum of Understanding. For infrastructures, Alcatel, Ericsson, Motorola, Nokia and Siemens are the major players. No Japanese manufacturer is yet thought to have negotiated licensing agreements with manufacturers such as Motorola and Philips, which precludes their right to manufacture them. However, no one doubts that the Japanese will want to participate, once the market starts moving.

European manufacturers, on the other hand, will also be unable to sell their equipment outside Europe unless Motorola (the company holding some of the key patents) licenses them to do so: actual agreements are confined to Europe. Japan and the U.S.A., the biggest exterior markets, are veering towards digital standards different from GSM. Australia, on the contrary, has adopted GSM. Eastern Europe and the Arab countries are also adopting GSM. The potential of export for European manufacturers looks good.

3.4. Key success factors

The key success factors for GSM can be thus summed up:

- A strategic vision of assignation of frequencies to mobile systems, in a European perspective since 1979.
- The existence of a group of *champions* of the GSM concept, the international equipe forming the core of the GSM group.
- The ability of a relatively small group of actors (TOs of the main countries) to make the crucial decisions for the take-off of the system, by committing themselves and creating a critical mass of consensus towards its implementation.
- The way of implementation: no single international network, but national compatible systems.
- Common perception of a strong demand.
- Correct timing: before investments in diverging technologies, with key decisions close enough to build positive fall-back and expectation of success.

4. *GSM-IBC: parallels and differences*

4.1. Present IBC projects: the problems

The analysis of GSM development helps to highlight the mix of factors needed to ensure the success of such complex projects, especially the role and motivation of the main actors and the importance of right timing in a favourable historical context. The analysis of the problems encountered by the present projects for European broadband infrastructures — METRAN, GEN, HERMES — shows also that the pattern of cooperation behind them does not seem adequate to guarantee a rapid implementation.

The rapid change in Europe's telecommunications between the 80's, when GSM was developed, and the 90's, influencing the role and motivation of key actors, makes clear that different patterns of cooperation must be enacted now, to give European broadband infrastructures a better chance of success. The wave of deregulation, particularly, while it forces traditional TOs to change profoundly, on the other hand opens the way to newcomers and emerging actors, such as the utilities or large users transforming themselves into suppliers. These newcomers could be the key to the success of IBC projects.

Moreover, in broadband development, the role of demand and especially of pioneer users requires a much stronger attention than has been the case up to now in new telecommunication network developments. A closer appreciation of the market leads also to redefining the same concept of the kind of European broadband infrastructure now needed and to be developed.

The most important present-day projects for Pan-European broadband infrastructures are METRAN, GEN and HERMES, all of which are facing various obstacles and do not seem to be gaining momentum.

METRAN is a Pan-European multipurpose telecommunication transport capability, which is the goal of a Memorandum of Understanding prepared by CEPT in October 1990, open to signature by all its members, now involving 26 TOs. The purpose of the agreement is to provide a framework for all the necessary measures to be taken by signatories in concert, to ensure a technically and economically optimised plan and a unified strategy for the establishment of METRAN in Europe.

METRAN is defined as a Path Layer transmission network, independent of any specific service, and comprising the totality of digital cross-border transmissions between designated gateways, to provide access from METRAN into national networks and from national networks into METRAN. Based predominantly on optical fiber transmission, the network will utilise Synchronous Digital Hierarchy (SDH) technology. Digital cross-connects will operate at the SDH equivalent of the 140 Mbit/s and 2 Mbit/s levels, with possible intermediate levels where a service demand is identified.

The time schedule of METRAN originally called for implementation of cross-border links with STM (Synchronous Transmission Mode) capability by the end of 1991: this has not been implemented as yet, since the work is proceeding much more slowly than planned. Next deadlines according to the Memorandum of Understanding are 1995 — to implement combined cross — connect under national control — and 1998 — for application of network-wide automated network management.

Since METRAN is a long-term proposition, the five biggest European public carriers (British Telecom, Deutsche Telekom, France Telecom, STET and Telefonica) have joined forces on an interim plan to build a 34 Mbit/s fibre transmission backbone called the General European Network (*GEN*). The original schedule was to put it in place by September, 1991, but again implementation is slow, so that the new work program calls for the backbone to be up and running by 1993–1995.

METRAN and GEN are both initiatives taken in the traditional framework of TOs cooperation, which are facing strong problems of coordination and of reluctance to cooperate, depending on the change of the competitive scene.

Neither programs substantially alter the approach followed by previous failed initiatives, such as MDNS (Managed Digital Network Services), or EBIT (European Broadband Interconnection Trial).

When *HERMES* was announced in 1990 it was saluted as an important step forward in the telecommunications scenery, because it is an initiative to provide pan-European broadband connections promoted by large users (European railways and others) and foreign private operators (such as Nynex and Sprint). It was believed that, besides its main goal of offering a needed service, competition from HERMES would have pushed TOs to move more quickly with METRAN. But after many years HERMES has few results to show.

HERMES is a three-stage plan, calling for the interconnection of existing packet-switched data-networks run by major European rail roads and its upgrading to a fiber-optical international broadband network, over which spare capacity and value-added services would be resold to third parties.

The plan was promoted in 1990 by the Union International des Chemins des Fer. The consortium that won the contract to implement it, called Europa Telecom, included Nynex International, Sprint International, TeleColumbus AG, Tractebel SA, Daimler-Benz AG and Comnpagnie de Suez, while Deutsche Bank AG and Banque Indosuez are the financial advisers. Railway companies from 11 countries are involved, all the main EC countries plus some EFTA countries, represented by Hitrail NV in Amsterdam.

During 1990-91 the consortium led a feasibility study over the technical lay-out of the network and regulatory problems in the resale of international data traffic. It also submitted a proposal to the U.K. Department of Trade and Industry for an international operator's license.

But negotiations among so many different partners proved difficult. In June 1991 one of the main partners, Racal, withdrew, claiming financial pressures due to internal reorganization. Since then also Sprint, Nynex and Daimler-Benz gave up.

It is by now clear that the project will probably not succeed.

The main strong points of HERMES were the financial and political clout of its partners, and the availability of the railways' right of way, which means that the broadband network can be deployed with no prior approval by regulatory authorities. But it faced two main problems:

- Third-party access had to be negotiated with national regulatory authorities, even if the ONP directive should help to open the way and data traffic is supposed to be liberalised.
- Economic justification, in terms of return of investment, was likely to be fraught with uncertainty at a time when market demand is just emerging. The project will be faced by forms of competition which are not limited to the deployment of competing infrastructures, but which will encompass delaying tactics from both established and would-be players.

4.2. Comparison between GSM and IBC

The comparison between GSM and IBC projects can be carried out on the following key points:

- Key actors
- Perceived demand
- Nature of the service
- Pattern of cooperation
- Historical context

The successful implementation of a Pan-European network requires an efficient pattern of cooperation among a small core of motivated actors, answering to a perceived demand in a favourable historical context.

4.2.1. Key actors

The key actors are basically the same for the GSM and IBC projects, but their role and motivation are different.

TOs, which were the driving force behind GSM, had in the case of mobile telephony a strong motivation driven by self interest, and were secure in their role.

IBC projets however matured later, because of a weaker motivation by TOs seeing them as a longer term perspective. By the end of the 80's, TOs had become much more cautious in strategic cooperation and unsure of their role, entering a transition phase from monopolistic, politically motivated entities to public or private companies, oriented to the market and required to operate in a very competitive environment. This is the main problem in the way of METRAN/GEN projects.

Moreover, there is a problem of controlling the migration of users from previous to broadband networks, which concerns TOs and is not yet clearly addressed in their divulged strategies. This problem is certainly not accelerating the offer of broadband infrastructures.

Besides, TOs have less direct control on the development of broadband networks, requiring more cooperation with a larger range of suppliers than mobile telephony, for the development of interfaces, customers equipment and especially of applications.

In IBC projects, also, up to now a group of *champions* of the new system has never emerged, something like the individuals from different countries working together in the GSM group, to lay the basis for the necessary compromises overruling national interests.

The role of Governments as represented by their Ministries of PTT has been of active support in the GSM case, seen as a chance to develop national industries, and of indifference in the case of IBC projects. Even if public carriers are being separated from national PTT Ministries, the remaining connections are such that up to now Ministries do not seem interested in giving support to competing initiatives, such as HERMES, for the sake of providing a useful service to the community.

The European Community, on the contrary, which gave its support to GSM, but was not the driving force behind it, has an even stronger motivation to promote the development of broadband infrastructures and is now in a position to take a more active role, and is considering options to do so.

Manufacturing industries did not have a strong interest in GSM development and basically had to accept the standard once it was decided. In the case of broadband, the range of technologies being developed and investments engaged is even wider, so that the thrust towards a standardised approach cannot come from industries, interested in seeing their own technology win.

Newcomers had no role in the development of GSM, but may be the winning card in the development of IBC infrastructures.

New comers are large users, industrial companies, utilities, and such, turning into suppliers of communication services either by reselling excess capacity on their existing networks, or developing new networks like the HERMES project. Foreign operators also, like AT&T, are now seen as possible partners in European projects, and not only as competition to be kept at bay. Compared to TOs, the newcomers can have a stronger economic motivation to develop broadband networks, less problems with previous investments in older networks, less constraints in cooperating over the European scene not having national markets to protect. Their experience as private companies (or of companies used to competition, such as the American carriers) and their need to make profits should be a guarantee of their ability to develop services oriented to the market in a reasonable time frame efficiently .

Users had no role in the development of GSM, and have up to now been largely passive in broadband developments, but must be taken more in consideration.

Pioneer users must be involved at an early stage of broadband development, because it is the only way to develop applications answering to real needs and to diffuse a culture of the use of broadband, now very scarce.

4.2.2. Perceived demand

The extent of demand has always been a strong point for GSM advocates, built on the rapidly growing subscription of cellular networks, especially in the Northern countries.

The extent of demand for European broadband infrastructures has been in discussion throughout the 80's, and is only now is becoming recognized, especially because of the growth of computer internetworking and the emerging of visual services.

Intra-European traffic, though, still accounts for only 8% of PTT revenues, and it is not therefore a priority for TOs.

Some major operator believe that only about 50,000 broadband connections will be in operation in Europe by the year 2000, and around 100,000 in the U.S.A.

It can be argued that it was easier for traditional telecommunication operators to perceive the demand for a service such as mobile telephony, based on voice transmission and therefore close to their traditional activities, than to estimate correctly the demand for new services such as broadband, intimately connected with the use of information technologies by companies and new information services by residential users.

The weakness of the perceived demand, therefore, has been a major obstacle to IBC development, even if some communities of users (the scientific community, industries strong users of CAD-CAM applications) have arguably been ready to use it since some years at least. The relative lack of field tests made an approach to the potential demand by niches and categories of users, which is now emerging as the easier one to open the market, more difficult.

4.2.3. Nature of the service

Both GSM and IBC are new Pan-European communication networks, built on common standards, but they are very different in all other aspects.

Mobile telephony, as pointed out, is easy to understand, does not need dedicated applications to work, guarantees a high communication value for the subscriber, who connects immediately with the universe of fixed telephony users.

Broadband networks instead are difficult to understand, require substantial marketing, need dedicated applications to be developed, and the communication value for the subscriber rises only after a critical mass of users have joined.

4.2.4. Pattern of cooperation

The approach which worked for GSM was that of an international dedicated group, and then a Memorandum of Understanding between the TOs.

The approach to IBC has also been one of different attempts at Memorandum of Understanding agreements among the TOs, all relying too much on TOs motivation and initiative. No motivated workgroup has yet emerged, as a force to push for cooperation. As METRAN shows, key controversial questions (of technology, tariffs, organization) were not basically solved before the signature of the Memorandum of Understanding (as in the case of GSM). Besides, METRAN calls for an international backbone which is only the first step towards actual provision of services to the users, and requires a much more complex coordination among TOs than is required by national GSM networks.

The HERMES alliance is a new approach which, involving newcomers, seemed very promising. Its present difficulties — not necessarily final — prove that the economic scenery is still very uncertain for a private venture, motivated only by expectations of profits. Added motivation, in the form of incentives or a clearer appreciation of demand, could solve these problems.

5. New options for European broadband infrastructures

New emerging options for European broadband infrastructures, to get around actual problems, differ from past ones in two main aspects.

The first difference is the trend to narrowing the focus of the network itself, by building specialised networks targeted to specific group of users, in order to start developing a real market. This is the pragmatic approach now followed in the U.S.A., trying to stay a step in front of the rising wave of MANs and their interconnections, which is threatening to create a multitude of incompatible technical standards.

The second difference is the decision by several actors — national Telecoms and private operators — to take separate initiatives in the field of high-speed telecommunications. These initiatives vary from France Telecom launching a national broadband network for the scientific community, to Cable and Wireless opening a world fiber-optical network aimed at business users. They create a new scenery where new possibilities of cooperation emerge, not necessarily involving all national TOs or preserving actual market-shares, but laying the groundwork for European-wide networks. Even the possibility of a single long-distance broadband carrier in Europe, with national operators reselling the services, looks now to be a much less far-fetched idea than a few years ago.

The U.S.A. strategy is worth mentioning. The U.S. Federal Government finances, among other projects, five high-speed wide area experimental networks, at speeds in the gigabit-per-second range, aimed at the scientific community. These research networks — named Aurora, Blanca, Case, Nectar and Vistanet — are being used to explore concepts that will form the basis of a proposed National Research and Educational Network (NREN). NREN is designed to provide high-speed, real-time connections between computers at government, commercial and academic research institutions.

The idea of NREN comes from the need of the scientific community for ultra-high speed, long-distance communication links, to ensure the efficient interaction of supercomputers and workstations, and to match the performance offered at local level by LANs. Eventually NREN can be expanded to technical and commercial users as demand develops.

Diverse NREN-related concepts made it appropriate for the National Science Foundation to fund five test beds rather than one. The NSF wanted multiple participants in the projects to share the costs, tap the best minds in industry and academia, and explore diverse research topics. The five experimental networks are taking different approaches and addressing varied research questions.

President Clinton's administration is trying to step up these investments with the 'Super Data Highways' project to build a high speed network across the U.S.A.

The European Community has no formal plans for a program similar to the U.S. one (the performance of the existing Cosine network is about 1/10,000th of what is expected from a Gigabit network and is not considered adequate by current users), even if demand is certainly existing. A major gap is developing with the U.S.A. in terms of communication infrastructures available to the scientific community. But there is strong interest in the U.S. networking community for cooperation with the EC for a joint research gigabit-network. Such cooperation would be beneficial in promoting the deployment of a more effective Pan-European Gigabit infrastructure and improving cooperation between the two scientific communities.

The scientific community is also targeted by France Telecom, who launched a high-capacity network to connect research sites, to be commercially available from 1992. *RENATER*, as it is called, offers a multi-capacity data-transmission service with speed from 64 kbyte/sec to 34 Mbyte/sec. The main network protocols (X25, Internet Protocol TCP/IP and Decnet Phase IV) are available. The initiative follows two years of cooperation with main French research sites, which have already been linked locally. Eventually 300 sites will be linked in, enabling exchanges between 15,000 computers and terminals in laboratories in France. Connection with partners throughout the European Community and in the U.S.A., via NSFnet, are being planned. It is not a Gigabit network yet, but the introduction of ATM technology planned for 1993 should open the way to upgraded performance.

Other private initiatives are also brewing in the same context that gave birth to the HERMES plan. It is rumoured that the electrical companies are thinking of setting up a system with fibre optic cables running alongside their power lines, which could then resell transmission capacity.

The pragmatic approach taken by the U.S.A. in developing focused experimental networks helps to solve two of the main problems of the development of broadband networks: the difficulty to pin down demand and the coordination problems in building an international network answering all needs. By involving a group of highly motivated users, such as scientists, it is possible to start a vicious circle of development of interfaces, equipment and applications, likely to generate fall-out on other groups of users and build a critical mass of users. By experimenting with focused networks, it is possible to solve real problems on a real market, tackling limited operational problems requiring the agreement of a limited group of operators/suppliers and laying the basis for the eventual general purpose network.

These tests help also to overcome the credibility problem broadband has experienced up to now, channelling resources and investments from manufacturing industries and operators (this is happening in the U.S.A.), previously reluctant to get involved without the perspective of servicing real users.

It is important, though, to analyse potential demand for broadband with this new perspective in mind, looking for communities of users with strong needs of high speed intercommunication with each other, needs not yet satisfied or not adequately satisfied by other means. In other words, it is not enough to target large multinational companies as such, or companies of a certain industry sector: what is important are the communication needs of the group of potential users. It must be remembered though that European users are as a whole less active and demanding than U.S. ones, and will probably need more active marketing.

Even if the goal is restricted, a strong nucleus of motivated leading actors cooperating together is needed to drive the development of European broadband infrastructures, as was needed for the development of GSM. Recent events indicate that an alliance likely to succeed could include some of the newcomers to the public telecommunications scene, such as private or foreign operators or large users (such as the European utilities) in partnership with some of the more active national TOs. The new historical context is becoming favourable to such alliances.

National TOs will probably have to accept a different pattern of behaviour than in the past, by cooperating, for example, in the infrastructure and competing instead in the services, without the guarantee of historical market shares. Or their role could be reversed, for example, by becoming the resellers of capacity carried on a fiber-optical network built by international utilities, such as the European railways or Electrical companies.

6. Conclusions

The development of communication networks is a complex, dynamic process of interaction between various partners or actors in a specific historical situation. The concepts of path-dependency and critical mass shed a new light on the key success factors for the development of such networks, stressing the importance of the historical context and of the approach to demand, often overlooked in traditional telecommunications development strategies.

The analysis of GSM shows how a positive feed-back among the leading actors, that is the TOs, led in the crucial year of 1987 to the widespread consensus that GSM was going to be a reality. That was a self-fulfilling prophecy, by convincing all actors involved to cooperate towards the same end in their own interest.

This is a process often described in physics, biology and also economics, when a positive feed-back in non-linear dynamic systems causes certain patterns or structures that emerge to be self-reinforcing (see Arthur et al., 1987). Such systems tend to be sensitive to early dynamic fluctuations, in other words the pattern first emerging is the one able to grow faster and therefore dominate.

The conclusion to draw is that there is a *window of opportunity* in time, when a potential TLC standard is emerging, when it is possible (but difficult) to influence the selection by the system of the winning technology/service. Or, in other words, that once the feedback mechanism starts working, it is extremely difficult to redress or change the balance. GSM (not by any means the only possible digital radiocellular technology) won because a number of positive events in the European telecommunications scenery acted in its favour relatively early in its development.

The other consideration is that radiocellular systems are perhaps the only telecommunication networks not to need a critical mass of subscribers to be attractive for users: paradoxically, there could be only one subscriber, and he would still receive a good communication value from using the mobile network, by getting in touch with the universe of fixed-telephony subscribers. This implies that for mobile telephony networks there is no need of a painful and long initial phase of active persuasion of potential pioneer users. There is possibly some parallel in the degree of geographical/population coverage (a critical value of coverage must be reached before a new mobile network starts attracting the users, according to the Scandinavian experience). But this is a problem that all telecommunication networks have.

The process of development of GSM required the interaction of motivated actors on three aspects, the technological one (development of the standard), the regulatory one (provision of same frequencies for all Europe for the digital cellular system) and the economic/political one (perception of strong demand and political consensus for a Pan-European mobile telephony system).

On the other hand, the analysis of the problems encountered by the present projects for broadband European infrastructures — METRAN, GEN, HERMES — shows that the pattern of cooperation behind them does not seem adequate to guarantee a rapid implementation.

Both METRAN and GEN are initiatives taken in the traditional framework of TOs cooperation, which are facing strong problems of coordination and reluctance to cooperate, depending on the change of the competitive scene.

The rapid change in Europe's telecommunications between the 80's, when GSM was developed, and the 90's, influencing the role and motivation of key actors, makes clear that the pattern of cooperation behind GSM cannot now give a realistic chance of success to IBC projects. The wave of deregulation, particularly, while it forces traditional TOs to change profoundly, on the other hand opens the way to newcomers and emerging actors, such as the utilities or large users transforming themselves into suppliers. These newcomers could be the key to the success of IBC projects.

Moreover, in broadband development, the role of demand and especially of pioneer users requires a much stronger attention than has been the case up to now in new telecommunication network developments. Broadband services need to reach a critical mass of users before a vicious circle of growth can be started, and this has not been taken into the right consideration by the actual IBC projects.

References

Arthur, W.B., 1989. Competing technologies increasing returns and lock-in by historical events. *The Economic Journal* 99.

Arthur, W.B., Ermoliev, Yu.M. and Kaniovski, Yu.M., 1987. Path-dependency processes and the emergence of macro-structure. *European Journal of Operational Research* 30, 294–303.

CEC, 1990. *CEC Directive of 28 June 1990 on the market competition in telecommunication services.*

David, P.A., 1985. Clio and the Economics of QWERTY. *American Economic Review*, 75(2), May, 332–337.

David, P.A., 1989. *A paradigm for historical economics: path dependence and predictability in dynamic systems with local network externalities* (Stanford University, Center for Ecomomic Policy Research).

Denmark, Finland, Iceland, Norway and Sweden PTT Administrations, 1982. Public Mobile Communications System in the 900 MHz Band. *GSM Group Document, CEPT Telecommunication Commission Meeting*, Vienna, June 1982.

Dupuis, Philippe, 1989. *Rôle des facteurs humains dans la cooperation franco-allemande.* Dies Academicus 1988, Hagener Universitätsreden 12, Juny 1989.

EC Council, 1984. EC Council Recommendation of 12 November 1984 concerning the implementation of harmonization in the field of telecommunications.

EC Council, 1987. EC Council Recommendation of 25 June 1987 on the coordinated introduction of public Pan-European cellular digital land-based mobile communications in the Community.

EC Council, 1987. EC Council Directive of 25 June 1987 on the frequency bands to be reserved for the coordinated introduction of public Pan-European cellular digital land-based mobile communications in the Community.

ETCO, 1987. *Updated Detailed Analysis of Second Generation Mobile Communications.* Final Report to the CEC, July 1987.

Failli, Renzo (Sip), 1988. *The International Cooperation for the GSM System.* Speech at International Meeting, 1988.

GSM Group, 1982. Document of Constitution of GSM Group. *CEPT Telecommunication Commission Meeting*, Vienna, June 1982.

GSM Group, 1987. Outcome of the Cept/Cch/GSM Group Meeting in Funchal (Madeira) 16–20 February 1987, concerning the technical standard for a Pan-European digital cellular radio system. *GSM Document.*

Hohn H.-W. and Schneider, V., *Path-dependency and critical mass in the development of research and technology: a focused comparison* (Max-Planck Institut für Gesellschaftsforschung).

Memorandum of Understanding on the Implementation of a Pan-European 900 MHz Digital Cellular Mobile Telecommunications Service by 1991, September 7, 1987.

Netherland PTT Administration, 1982. Public Mobile Communication Systems. *GSM Group Document, CEPT Telecommunications Commission Meeting*, Vienna, June 1982.

Netherlands, Denmark, Finland, Norway, Sweden PTT Administrations, 1982. Proposal for a Study Plan on GSM. *CEPT Telecommunication Commission Meeting*, Vienna, June 1982.

Rogers, E.M., 1990. The 'critical mass' in the diffusion of interactive technologies. *ITC Meeting*, Venice.

Schneider, V. et al., 1991. The dynamics of Videotex development in Britain, France and Germany: a cross-national comparison. *European Journal of Telecommunication*, 6, 187–212.

Science and Public Policy, Vol. 18, No. 2, April 1991 (Beech Tree Publishing, Guildford, Surrey).

Silberhorn, A., 1989. *Deutsch-Französische Zusammenarbeit: der Motor für das europäische digitale Mobilfunksystem.* Dies Academicus 1988, Hagener Universitätsreden 12, Juni 1989.

Technology Investment Partners (Paris), General Technology Systems (London), Telmark (Köln), Teknibank (Milan), Palo Alto Management Group (Palo Alto, California), and Dir. XIII, EC, 1992. *PACE '92, Perspectives on Advanced Communications for Europe.*

Global Telecommunications Strategies and Technological Changes
Edited by G. Pogorel

Chapter 5

Telecommunication strategies: incremental change in Europe

Robin Mansell and Michael Jenkins *

Centre for Information and Communication Technologies, Science Policy Research Unit, University of Sussex, Brighton, UK

1. Introduction

In the 1990s the 'network firm' organised around advanced intra- and inter-firm communication services is a model toward which firms in Japan, the United States and the European Community aspire (see Imai and Baba, 1989; Ciborra, 1992). But the successful network firm requires substantially more than investment in telecommunication and computer networks. The policy, regulatory and industrial environment creates differentiation in the patterns of diffusion and use of telecommunication network systems.

Insofar as learning is a sequential process and institutional conditions differ among the member states of the European Community, investment aimed at stimulating the growth of advanced telecommunication networks is not sufficient to stimulate greater coherence in the European communication infrastructure or the widespread diffusion of advanced services. Case studies of the ways in which larger and smaller telecommunication users in Europe are using telecommunication services suggest that they have found innovative ways of coping with the disparate technical, organisational, and regulatory environment that continues to exist within the Single European Market.

---

* Dr. Robin Mansell is Reader and Head of the Centre for Information and Communication Technologies, Science Policy Research Unit (SPRU), University of Sussex. Michael Jenkins was a Research Fellow at SPRU at the time of writing. The views expressed in this paper are those of the authors and not of any institution. Contributors to the case studies based on Mansell and Morgan (1991) included G. Thomas (SPRU, UK), Rob van Tulder (University of Amsterdam, NL), Volker Schneider (Max Planck Institut, GER), and Godefroy Dang N'guyen (ENST, FR).

Nevertheless coping strategies are resource intensive. When time and financial resources must be committed to establishing and maintaining European telecommunication networks to a greater degree than in other geographical regions such as the United States, European-based firms are likely to be disadvantaged. The European Commission has been a champion of changes which aim to stimulate the provision of advanced telecommunication services throughout the Community market (CEC, 1987, 1992a, b). Changes in network access conditions, pricing strategies, and standardisation procedures are being implemented gradually and they continue to generate controversy in the member states as well as among suppliers and users with a stake in the evolution of the European telecommunication equipment and service market [1].

In spite of these policy initiatives a truly pan-European telecommunication infrastructure had not emerged by the beginning of the 1990s. In January 1990 the Council of the European Communities adopted a resolution which aimed to promote the implementation of seamless, higher capacity telecommunication networks on a trans-European basis. The Council suggested that "... special priority should be given ... to the development and interconnection of trans-European networks, ... in particular the linking of the main Community conurbations by broadband telecommunications networks" (CEC, 1990). In April 1991 the Commission linked the development of these networks to its industrial policy framework.

> "The Community should consider launching a second generation of Research and Technology Development, ranging from projects at the pre-competitive stage to projects geared more closely to the market ... Such projects should cover software, ... high-performance computing and telecommunications" [CEC, 1991a]

---

[1] The Commission has highlighted the following as being of particular importance: Commission Directive of 16 May on competition in the markets in telecommunication terminal equipment (88/301/EEC, OJ L 131/73); Council Directive of 29 April 1991 on the approximation of the laws of the Member States concerning telecommunications terminal equipment, including the mutual recognition of conformity (91/263/EEC, OJ L 128/1, 23.05.91); Council Directive of 17 September 1990 on procurement procedures of entities operating in water, energy, transport and telecommunications sectors (90/531/EEC, OJ L 297/1, 29.10.90); Commission Directive of 28 June 1990 on competition in the markets for telecommunication services (90/388/EEC, OJ L 192/10, 24.7.90); Council Directive of 28 June 1990 on the establishment of the internal market for telecommunications services through the implementation of open network provision (90/387/EEC, OJ L 192/1, 24.07.90); Council Directive of 5 June 1992 on the application of open network provision to leased lines (OJ L 165, 19.06.92); Proposal for a Council Directive on the application of open network provision to voice telephony (COM (92) 247 final — SYN 437); and Proposal for a Council Directive on the mutual recognition of licences and other national authorisations for telecommunications services including the establishment of a Single Community Telecommunications License and the setting up of a Community Telecommunications Committee (COM (92) 254 final — SYN 438, see CEC (1992a, b).

The view of the long term development trajectory from a supply side perspective is one of a 'network of networks' linked together through harmonised technical standards to ensure interoperability and to provide a stimulus to the growth of advanced business communication. In the coming decades the new broadband infrastructure is expected to provide for many forms of 'telepresence' that will be capable of abolishing the constraints of distance and time (CEC, 1992c). Applications that are expected to create demand for ever higher capacity trans-European networks and greater flexibility in their use include: local, wide and metropolitan area network interconnection; connections between high capacity central processing units; supercomputing and multiple access nodes; distributed applications such as CAD/CAM and electronic publishing; large file transfers, videoconferencing, distributed fax systems and Group 4 fax; image processing, cheque recognition and electronic document management; multimedia applications with text, graphics, image and full motion video on high resolution workstations and videophony. Thus, when technical, organisational and regulatory constraints are removed, these and other applications are expected to generate a strong demand for 'broadband' networking in the European business communication market.[2]

This paper considers the experiences of European based firms who have attempted to extend their telecommunication networks across the boundaries of the member states of the European Community. It assesses the nature of their requirements for advanced communication services and the types of barriers that they continue to face. These barriers are of both a technical and organisational nature and they extend considerably beyond the confines of the traditional telecommunication sector to include financial regulations which affect the flow of information within both public and private networks.

2. *Communicating across European borders*

The likelihood that firms will adopt advanced business communication services is affected by more than their awareness of the potential of new technologies. The perception of the cost effectiveness of new information and communication services is influenced by strategic considerations, internal organisation constraints, the mix and timing of investment in computing and telecommunication technologies as well as the characteristics of the telecommunication supply environment. The challenge is to create effective synergies

---

[2] 'Broadband' means the 'high-end' (in terms of bit-rate) portion of services and also designates the total mix of services to be considered, starting from the 'upper end' of ISDN (e.g. including 2 megabit/s accesses), up to what will be required by a realistic introduction of video (interactive and distributive) services (e.g. 140 megabit/s), see CEC (1990).

in all of these areas. The challenge is even greater when networks expand beyond the boundaries of the nation state.

In 1990 the European Commission supported a series of case studies which sought to assess the ways larger companies in Europe were using advanced communication services (CEC, 1991b). The network requirements of large telecommunication using firms in the United Kingdom, France, west Germany, the Netherlands and Sweden were examined. The case studies included companies in the electronics, banking and insurance, automobile, oil and aerospace sectors. The emphasis in these studies was on cross-border telecommunication requirements and on the opportunities and barriers to the expansion of higher capacity networking. Data communication and digital imaging service applications were of special interest.

The analysis pointed to the following trends in the use of advanced communication services (Mansell and Morgan, 1991). In the *Automobile* sector the predominant trend is the emergence of a new production paradigm which has been called 'lean production'. This production style is being driven by the Japanese automobile firms such as Toyota, Nissan and Honda and it appears to be far superior to traditional mass production strategies. Western automobile manufacturer strategies are attempting to catch up with Japanese trends in the design, manufacturing, procurement and distribution spheres. The Japanese have been late as compared to Ford and General Motors in utilising telecommunication applications, but they have shown that other forms of networking and information flows, including face-to-face communication, can deliver enormous advantage.

In the automobile sector in Europe, Ford of Europe has sought to reduce design to conception time from 60 to 48 months and to improve quality. The company's manufacturing plants are linked to Ford's private network which enables sharing of design and manufacturing information. Most of the links in the European network were operating at 2 Mbit/s in 1990. On the supplier and distribution side the company has relied primarily on the Public Switched Telephone Network (PSTN) and in 1990 this network did not support 2 Mbit/s links across national borders within the European Community market. The company is planning the integration of telecommunication-based applications and bandwidth sharing to ensure the optimal use of its networks.

General Motors Europe (GM) has depended on its wholly-owned subsidiary, Electronic Data Systems (EDS), to meet its information technology requirements. Although GM and EDS Europe believe that improvements to design and manufacturing are crucial to competitiveness, EDS has concentrated on networks to support distribution and service support activities. The company has considered the use of two way satellite communication using Very Small Aperture Terminals in order to meet its telecommunication bandwidth requirements. The forerunner of this system was the one-way 'GM Automotive

Satellite Television Network' which has been implemented in the United States to support training sessions using video equipment. The system has created a strong demand for sporadic transmissions over high capacity networks.

The company has made less headway in the development and use of video-conferencing than its rival manufacturer, Ford. In 1990, GM Europe had only recently begun to experiment with a 1.544 Mbit/s link which was also carrying voice and data traffic between the company's operations in the United Kingdom and west Germany. Complementing the company's plans for its distribution and sales network, was a plan to establish a 2 Mbit/s network to support design and production.

In the *Financial Services* sector, one of the most important changes in the market has been the trend toward deregulation. The emergence of a more liberal regulatory regime has encouraged banks to diversify into securities, insurance and a wide array of related financial services. Deregulation and new technology have threatened to make the financial services sector increasingly contestable and competitive. In Europe, the Commission's liberalisation programme has emphasised three main objectives: encouraging the free-flow of capital by removing exchange controls; promoting free-trade in financial services within the Single Market; and encouraging greater standardisation of banking technology across Europe so as to promote the interoperability of services.

Citicorp, for example, with its world-wide headquarters in the United States and its European base in the United Kingdom provides a baseline for judging the extent to which broadband communication is integral to the global and European strategies of financial and related service companies. Citicorp claims to operate as if there were no frontiers between the European Community countries. The company's network supports internal operations with fixed link connections across the borders of the member states. These are supported by connections operating at from 64 Kbit/s to 256 Kbit/s. In 1990, the United Kingdom was the only country within the Community where the network was supported by 2 Mbit/s links.

The availability of higher capacity networks can create an incentive for the introduction of advanced services. This is illustrated by Citicorp's establishment of 'processing factories' which have enabled the company's financial operations to be centralised. Most of these operations have been located in the United Kingdom because of the liberalised telecommunication market and the attractive tariff structure for leased lines. As far as customer services are concerned, Citicorp has built a network to support the Citicorp Customer Access Terminal (CAT). For the volume and speed of traffic generated by the CAT network, 64 Kbit/s lines have been considered adequate.

Reuters is one of the world's leading providers of news and information services. It offers real time information services, transaction products, trading room systems, databases and media products. In 1989, the information carried on its global networks reached 194,000 terminals in 127 countries. There is a close relationships between Reuters' telecommunication strategy and its core products. The company has been developing an 'Integrated Data Network' that will link its previously separate networks. In Europe most of the network in 1990 was comprised of fixed links running at speeds of 14.4 to 64 Kbit/s. Faster connections had been introduced at speeds up to 1024 Kbit/s and there were a few links running at 2 Mbit/s. An overlay satellite network was running at speeds up to 168 Kbit/s.

The experiences of the Amsterdam-Rotterdam (AMRO) bank and the NMB Postbank Group in the Netherlands suggest that there will be variations in the telecommunication strategies of banks that target different segments of financial markets. For example, the AMRO bank serves large corporate clients and has been moving into treasury management services for its international clients. It has been actively building alliances to face the changes that will be introduced by the completion of the Single European Market. Despite its highly internationalised profile and its domestic use of telecommunication links at speeds of up to 144 Kbit/s, internationally the internal Amronet was not operating at speeds as high as 2 Mbit/s in 1990. Although broadband capacity was needed to support connections between data centres, most of this activity was contained within national networks. Within the Netherlands, AMRO's network supported 64 Kbit/s channels in a 2 Mbit/s backbone.

In contrast, in 1990 the NMB Postbank Group, which serves a mix of national private customers and smaller firms, believed that most of its future applications including digital imaging processing centres would be unlikely to require connections in excess of 2 Mbit/s in the near term. Although plans for the future had been informed by an awareness of developments in the United States and the growing use of T1 circuits in that country, there was a 'wait and see' attitude within this European bank.

In west Germany Colonia Insurance had developed relatively advanced data implications within the German market. The company was a co-founder of the network service vendor, MEGANET, which was operating at 64 Kbit/s in 1990. The west German, Dresdner Bank has a strong international orientation in its business strategy. It was pursuing a selective process of forging joint ventures and its telecommunication requirements were applications-led. The Commerzbank had operated an international network since 1983, but its branches in the United Kingdom, France, Spain, Luxembourg, the Netherlands and Belgium were all connected by circuits running at 9.6 Kbit/s in 1990. A link between Frankfurt and New York was operating at 64 Kbit/s.

The *Electronics* sector embraces five main industries: telecommunication, data processing, semiconductors, capital equipment and consumer goods, but there are a number of common pressures facing firms in this sector. All face the problem of how best to deal with the 'globalisation' of markets, a problem that is more pronounced in telecommunication and capital equipment, i.e. defence. In the telecommunication industry, for example, firms are attempting to establish a direct presence in their major markets. The co-location of production and marketing is now deemed to be essential if telecommunication equipment suppliers are to be credible in the eyes of foreign telecommunication operators.

There is also a strong trend towards shorter product life-cycles. This phenomenon is most acute in the data processing industry, where the life-cycle of many products has declined. A priority across the electronics sector is the creation of stronger supply chains. The upstream chain which links component suppliers with electronics manufacturers is being recast to support long-term, high-trust partnerships. The downstream chain between the electronics manufacturers and the final customers is the most critical interface. Logistical systems supporting distribution and after-sales effort are becoming more important. Corporate strategies which embrace the globalisation of markets, product development and supplier relationships are key determinants of success or failure. Each requires increasingly sophisticated communication applications at both intra-corporate and inter-corporate levels.

In the electronics sector Siemens is a pacemaker in high technology development with a comprehensive product range. Its transnational networking began with the Siekom network in the 1970s. This has evolved into a complex mixture of subnetworks. However, the highest transmission speed in 1990 was a 2 Mbit/s line which connected operations in Munich and Vienna. Connections between approximately 600 local area networks and the company's other internal networks were established only within Germany and many of these supported CAD/CAM production 'islands'. Plans in 1990 included the Siemens Corporate Network which would interconnect in-house networks using a packet switched network running at speeds between 64 Kbit/s and 2 Mbit/s. In spite of the fact that Siemens is a producer of advanced telecommunication technology, in 1990 it had yet to extend broadband (>2 Mbit/s) links beyond its national boundaries.

Grundig's use of broadband connections was similar to Siemens insofar as it was using network capacity running at 1.92 Mbit/s. This was limited to the interconnection of separate operations in the domestic headquarters and Vienna. However, the company appeared to have a more forward looking strategy than Siemens. Telecommunication demand was being driven by the continuing automation of its production process, the need to achieve tighter co-ordination of R&D activities, production, logistical systems, and the company's increased co-operation with Japanese companies. The latter was expected to lead to greater use of videoconferencing, high speed fax and a variety of other applications in the longer term.

In the Netherlands, Philips' highest priority has been the development of CAD/CAM and Computer Integrated Manufacturing (CIM) systems using local area networks. There was long-term interest in international CAD applications. Philips, like Siemens, is a producer of, for example, 2.4 Gbit/s cross-connect switches. However, as a user the company had no 2 Mbit/s links in operation between its sites in 1990. Several applications, including on-line administrative tasks, funds transfer applications, PC to PC connections, and the need to achieve higher levels of security and to provide internal redundancy were expected to stimulate demand for higher capacity connections.

These examples of the corporate use of advanced telecommunication applications highlight different strategies and they suggest that although firms clearly see an increasing need for higher capacity and broadband networks their strategies are heavily influenced by experience in their home countries. For example, firms that have experienced the removal of regulatory restrictions on the use of leased capacity and the interconnection of private and public networks in the United Kingdom have tended to implement a greater variety of advanced applications than firms in other European countries.

Similarly, in the Netherlands, where the aim of policy is to become the 'teleport of Europe', there has been earlier use in some sectors of international broadband networks than in Germany or France. Firms in more restrictive environments have less chance to learn to use telecommunication services effectively and they do not take advantage of technological opportunities as early as their counterparts in more liberalised regimes.

Higher capacity cross-border telecommunication initiatives have been slow to develop in Europe and few existing public network providers can even begin to emulate the coherence of supply that characterises the United States market (Thomas, 1992). Table 1 shows selected initiatives which have been under discussion or in partial operation since the late 1980s. The main requirement for cross-border European telecommunication is for a high capacity *intelligent* network which allows users to access bandwidth as needed.

The European Commission has emphasised that the Single European Market will be difficult to forge in the absence of a coherent telecommunication infrastructure. This component of the Community's infrastructure must provide the networking speeds, capacity, flexibility and accessibility that are available to firms in other regional trading blocs. The technical, organisational, and regulatory pre-conditions for improved electronic communication among the member states — and between them and their major trading partners — are important issues and they create constraints for the innovative use of electronic information services. The following section highlights some of the characteristics of electronic trading networks to illustrate the extent to which the characteristics of the telecommunication environment affect the development of innovative communication applications.

Table 1
European cross-border telecommunication initiatives

Initiative	Name	Date initiated
Intercarrier Agreements		
Managed Data Network Service (CEPT)	MDNS	Late 1980s
European Broadband Interconnect Trial, 17 PTOs	EBIT	Agreed 1989
General European Network, 5 PTOs,	GEN	Tenders 1990
Managed European Transmission Network, 14 PTOs	METRAN	Agreed 1991
National Carrier Initiatives		
British Telecom	BT Tymnet	Late 1980s
European PTOs and MCI (25%)	Infonet	Late 1980s
British Telecom	Primex	1990
British Telecom	ABFK Alliance	1990
British Telecom	Global Network Service	1991
British Telecom	Pathfinder/Synchordia	1991
France Telecom	CentreNet LRT	1991
PTT Netherlands and Swedish Telecom	Unicom	1991
Mercury Communication	Switchband Service	1991
Swiss PTT	MegaNet	1991
PTT Netherlands	MegaSwitch	1991
France Telecom and DBP Telekom	Eunetcom	1992
Non-European Carrier Initiatives		
Racal Electronics (withdrew 5/91) NYNEX, US Sprint, Daimler-Benz, Compagnie de Suez, TeleColumbus, Tractebel, European Railways	HERMES	1990
AT&T Istel	Value Added Network	–
US Sprint	Value Added Network	–
GEIS	Value Added Network	–
EDS	Value Added Network	–
IBM	Value Added Network	–

Source: Adapted from Thomas (1992).

3. Electronic trading networks and advanced applications

Electronic trading networks are a new generation of technical and organisational structures for the distribution and exchange of goods and services. By combining the collaborative elements of information networks with the competitive elements of the marketplace, these networks offer users the opportunity to disconnect logistical, commercial and information processes and to execute trading at a distance. In the United Kingdom electronic trading has become synonymous with Electronic Data Interchange (EDI) whereas in continental Europe and elsewhere, electronic trading tends to refer to screen-based markets where the actual negotiation and consummation of an individual contract takes place electronically. Buyers and sellers can negotiate 'many to many' and trade without seeing the goods.

Communication in support of electronic trading has relied mainly on the telephone, fax or telex and a limited number of relatively simple non-interactive data transmission services. However, a growing number of electronic trading networks are incorporating a range of network-based services with requirements for interoperability across national boundaries and for the interworking of public and private networks operating at different speeds. Case studies of a number of these initiatives show that the network operators are devising and championing different technical, legal and financial regulations and standards in support of electronic trading and these have implications for future development patterns as well as for the demand for advanced telecommunication infrastructure (Mansell and Jenkins, 1992).[3]

There is growing interest in the policy issues surrounding network-based trade activity. Independent initiatives are emerging at different levels within the various European policy making regimes, i.e. financial, competition, international trade and telecommunication. At the firm level, the management literature has focused on the ways in which electronic trading can be used to exploit opportunities for competitive advantage and to support changing business practices (Porter, 1985; Keen, 1986). At the sectoral level trade associations have formed mutual organisations to develop electronic trading networks.[4] At the national level, an increasing number of countries are promoting trade strategies based on telecommunication networking.

In the European Community trade facilitation initiatives are predicated upon a more broadly based telecommunication networking strategy. For example, there are relevant projects under the DELTA (Telematic Systems for Flexible and Distance Learning), DRIVE (Dedicated Road Infrastructure for Vehicle Safety in Europe), and RACE (Research and Technology Development in Advanced Communications Technologies in Europe) programmes. The last is examining the implications of a system of tele-marketplaces utilising multimedia, bandwidth hungry trade information applications such as in-

---

[3] The case studies included, PRODEX, UK Oil and Gas Industry Product Reconciliation Scheme; Transvik's London-based NORDEX cross-border electronic marketplace for professional investors; VABA, the Dutch Tele-auction for fruit and vegetable auctions; Westland, an electronic trading centre for auctioning flowers and pot plants, and Marine Cargo Processing Plc, a trade facilitation system which allows direct access to UK Customs data and inventory control system.

[4] SWIFT and SITA are two of the largest examples of a mutual approach to ICT-based trade related initiatives.

teractive electronic catalogues. Developments under the TEDIS programme have led to concerns that information and documentation as well as telecommunication standards are being set in such a way that national boundaries are disappearing only to be replaced by sectoral ones.

The prospects for electronic trading networks as media supporting all forms of communication within trading relationships at the local, national, regional or global levels depend upon the manner in which they create changes in economic relationships over geographical space and time. The pace and ease of movement within an electronic trading network is conditioned, not by the maximum potential of parts of the trading system, but by the minimum characteristics of the individual elements present across the network. Thus 'bottlenecks' created by inadequate underlying telecommunication infrastructure can easily constrain the extent to which electronic trading applications contribute to demand for higher capacity networks. Similarly, the absence of a flexible cross-border telecommunication infrastructure can suppress demand so reducing incentives for much needed investment infrastructure in order to 'pull through' demand.

The conduct and regulation of international trade is a complex and fragmented process. For analytical purposes five distinct phases in the trading cycle can be identified: trade information, trade facilitation, trade execution, clearing and settlement and trade regulation. Each phase involves the production and exchange of information. Each of these phases embodies a logically separate and coherent set of technical, organisational and regulatory design parameters. For example, at the technical level, these include network access conditions, data quality, security, speed, network redundancy, application standards, gateway protocols, cross-border regulations and dispute settlement procedures.

The design parameters for accessing or exchanging trade information, for instance, must take account of the requirement that this information may need to be widely accessible. Records of past trading performance and certain forms of market intelligence are generally not real time critical. These data may be stored off line and require a friendly interface for access by infrequent users. These design parameters can be incorporated into electronic networks that use relatively unsophisticated telecommunication, e.g. slow speed analogue connections, and simple modems.

However, the same technical configuration would not be viable for a trade execution phase. Here the requirement is often for rapid data exchange. Losses of individual bytes of data containing trade execution instructions cannot be tolerated and the telecommunication network must be highly reliable as data are switched across networks. Frequently, these requirements can be met only by dedicated private links among trading partners using advanced digital systems and network management techniques.

The design parameters of an electronic trading network can be considered in terms of three interlocking network elements — technical, information and organisational (Mansell, 1992). The construction of these networks incorporates a series of compromises with respect to each of these elements. For example, network design constraints are set by functional/user requirements. In most markets during the trade execution phase network users need a fluid environment where all parties can negotiate. There are requirements for short data cycles (three seconds in the Dutch Auction; less for others); information is not stored, e.g. when a bid is rejected; and single packets of data (yes/no) are exchanged. Other technical elements must be embedded in the network to match information elements such as the ability to react in real time, not just interactively, the absence of data buffers and protocol converters, the ability to maintain the 'handshake' between remote machines, the ability to refresh only the relevant parts of the screen, and little signal degradation over long distances.

The investment costs associated with telecommunication services that support electronic trading may not make economic sense, even when the services are technically feasible. This increases the importance of the organisational elements of network design. In some sectors and regions the economic resources and electronic infrastructure are not in place to ensure that the technical features are available.

Generally, public telecommunication operators and value added service providers restrict themselves to concerns regarding the technical elements of electronic trading. They do not concern themselves with the content of the information that is exchanged. For example, these organisations do not have dispute and grievance procedures available to support users involved in the trade execution process. Complex debates about the legal implications of electronic trading have yet to be resolved in most countries in the industrialised world and they are only beginning to be addressed in Europe.

Successful electronic trading requires that a minimum number of preconditions be met. These include the opportunity to gain access to the telecommunication network, a favourable assessment of the cost effectiveness of available public and private networks, e.g. with respect to cost, quality, speed and reliability, and the presence of institutions which act as 'gateways' between the technical, information and organisational aspects of these networks. Interface institution are needed to function as electronic clearing houses (Kimberley, 1991). Such institutions function as the equivalent of a 'traffic manager' among trading partners who have no bilateral agreements with each other.

Case study evidence suggests that, at least in Europe, electronic trading network operators have been able to minimise the constraints imposed by the limitations of the telecommunication services that are on offer. For example, few EDI users are aware that messages sent through third party gateways cannot be tracked or audited by the originator. Although a leased line network can meet the requirements of a large well-resourced service provider in a highly profitable market segment, most electronic trading continues to rely on the public switched telecommunication network.

A good fit between the design parameters of electronic trading networks and the public switched telecommunication network in terms of standards, interconnection and interoperability had not been achieved by any of the case study network operators. Several had identified networking strategies that could not be adopted for technical reasons or because of national regulatory constraints.

The sample of electronic trading initiatives showed that there are strong economic and political incentives to tailor the technical, organisational and information elements of these networks in ways that support conflicting local, national, regional and international trade policy objectives. If the strategic and competitiveness goals of firms are separated from initiatives to develop the telecommunication infrastructure and related institutional reforms the challenges presented by cross border electronic trading will continue to grow.

There is little evidence that government organisations in the United Kingdom or the European Community as a whole are developing a coherent perspective on these complex issues.[5] For example, the United Kingdom Office of Telecommunication is not concerned with information content and does not address issues of data verification or authentication. The British Securities and Investment Board has yet to clarify the extent of its jurisdiction even over real-time trading of company shares domiciled in foreign countries. In the Department of Trade and Industry issues with a trade and information technology dimension fall between the Financial Services and Information Technology Branches.

Similarly, in the Commission of the European Communities the Directorates responsible for information technology and telecommunication (DG-XIII) and competition policy (DG-IV) have yet to address the full spectrum of electronic trading issues although considerable attention has been given to EDI, information market and telecommunication policy issues. With the wider diffusion of electronic trading, the European Commission will need to consider what institutions should exist to mediate between electronic trading network operators and the regulatory regimes of the member states.

---

[5] In the United States policy issues related to electronic markets have been dealt with by antitrust legislation, the Federal Communications Commission, United States Department of Agriculture, and the Office of Technology Assessment.

4. Conclusion: prospects for change

The case studies of business telecommunication applications and their use by both larger firms and innovative electronic trading network operators suggest that they are searching for ways to galvanise public telecommunication operators, third party suppliers and new entrants into action to create opportunities for trans-European networking. Despite regulatory differences in the European Community member states, variations in the structure of the telecommunication market, and disparate sectoral and firm-specific requirements, telecommunication user views on the barriers to the trans-national implementation of advanced communication applications across Europe have shown a high degree of commonality.

Most firms see tariffs for higher capacity networks as a major factor inhibiting increased use. The cost of high speed international links has prevented some firms from taking advantage of wider bandwidth where it is available. The firms interviewed for the case studies discussed above cited inflexibility and lack of predictability as major factors restraining demand.

The absence of co-ordinated planning by the public telecommunication operators for circuits operating at higher speeds is a complaint that is echoed by most firms. The absence of one-stop-shopping creates difficulties in extending service applications across national boundaries. The absence of co-ordination of trans-European networks also creates difficulties in extending security and safety regulations beyond national borders. This is particularly important for applications such as EDI where the lack of European Community-wide jurisprudence is widely perceived to suppress demand.

Continuing restrictions on voice and data integration and interworking between public and private networks are generally believed to have a negative effect on the build up of demand for advanced trans-European networks as are restrictions on the use of two way satellite networks and on bandwidth sharing.

Other major inhibiting factors in the use of advanced communication applications include the availability of appropriate skills. This factor affects decisions to develop in-house training schemes or to acquire network management expertise from public telecommunication operators or third party suppliers. Some firms cite technical heterogeneity and the absence of standards and gateways for international connections as barriers to transnational advanced network implementations. Other firms have gained advantage in the European market as a result of their bargaining strength which stems from the sheer volume of their traffic. However, the majority of firms stress that a learning process is underway and this process is rendered more difficult as a result of the growing number of network operators as competition and liberalisation take effect within the European Community market.

The case study firms look to changes in the European telecommunication supply environment as a precondition for changes in their business organisation and strategic plans for advanced telecommunication services. 'Bandwidth-on-demand' is regarded as a major prerequisite for the use of higher capacity networks.

There is clearly a need for European-wide network operators, but most firms are unclear about exactly what functions such operators should perform. Some support a European organisation that would provide a single point of contact to ensure that transnational network interconnection is achieved with a minimum loss of time and at low cost. Others argue for a bandwidth wholesaler to meet demand for greater flexibility and dynamic bandwidth allocation.

Incremental changes are underway in the European telecommunication market. Cross-border networking initiatives are likely to bring partial solutions that cover the more lucrative geographical parts of Europe or specific classes of services that appeal to larger international users. Specific economic sectors or user groups are banding together to form 'club-like' electronic trading networks that replicate the experience of closed telecommunication user groups which have existed within some of the member-states of the European Community. In order to counter trends toward closed networking arrangements which meet highly differentiated advanced communication requirements there is likely to be a need for some form of Community-wide regulatory authority that is capable of integrating telecommunication policy and regulation with appropriate 'rules of the game' for the service applications which use the telecommunication infrastructure.

Acknowledgements

The authors gratefully acknowledge the support of the UK Economic and Social Research Council's Programme on Information and Communication Technologies and the Commission of the European Communities.

References

Ciborra, C.U., 1992. Innovation, Networks and Organisational Learning. In: C. Antonelli (ed.), *The Economics of Information Networks*, pp. 91–102 (North-Holland, Amsterdam).

Commission of the European Communities, 1987. Towards a Dynamic European Economy: Green Paper on the Development of the Common Market for Telecommunication Services and Equipment. *COM* (87) 290 final (Brussels).

Commission of the European Communities, 1990a. Towards Trans-European Networks — Proposal for a Community Action Programme'. *COM* (90) 585 final, January.

Commission of the European Communities, 1990b. *R&D in Advanced Communications Technologies in Europe* — Workplan '91, Brussels, December.

Commission of the European Communities, 1991a. The European Electronics and Information Technology Industry: State of Play, Issues at Stake and Proposals for Action. *SEC* (91) 565 final, 3 April (Brussels).

Commission of the European Communities, 1991b. *Perspectives for Advanced Communications in Europe — PACE 1990*, Vols. I–VIII (Brussels).

Commission of the European Communities, 1992a. *1992 Review of the Situation in the Telecommunications Service Sector*, 21 October (Brussels).

Commission of the European Communities, 1992b. *The European Telecommunications Equipment Industry, The State of Play, Issues at Stake and Proposals for Action*, 25 June (Brussels).

Commission of the European Communities, 1992c. *Perspectives for Advanced Communications in Europe: Impact Assessment and Forecasts*, Vol. II (Brussels).

Imai, K. and Baba, Y., 1989. Systemic Innovation and Cross- Border Networks. *Paper for OECD International Seminar on Science, Technology and Economic Growth*, Paris.

Keen, P., 1986. *Competing in Time: Using Telecommunications for Competitive Advantage* (Ballinger, Cambridge, Mass.).

Kimberley, P., 1991. *Electronic Data Interchange* (McGraw-Hill, New York).

Mansell, R., 1992. Information, Organisation and Competitiveness: Networking Strategies in the 90s. In: C. Antonelli (ed.), *The Economics of Information Networks*, pp. 217–228 (North-Holland, Amsterdam).

Mansell, R. and Jenkins, M., 1992. Networks and Policy: Interfaces, Theories and Research. *Communications & Strategies*, No. 6, deuxième trimestre, pp. 63–85.

Mansell, R. and Morgan, K., 1991. Communicating Across Boundaries: The Winding Road to Broadband Networking, Synthesis Report. *Commission of the European Communities, Perspectives for Advanced Communications in Europe*, Vol. VII (Brussels).

Porter, M., 1985. *The Competitive Advantage of Nations* (Free Press, New York).

Thomas, G., 1992. *Cross Border Telecommunication: The Prospects for Long-Distance Carriers in Europe* (Centre for Information and Communication Technologies, Science Policy Research Unit, University of Sussex, Brighton).

Global Telecommunications Strategies and Technological Changes
Edited by G. Pogorel

Chapter 6

Market support systems: theory and practice *

Claudio U. Ciborra

Institut Theseus, France and Università di Bologna, Italy

1. Introduction

The market has been depicted as an institution enacted by the interactions of people each of whom possesses only partial knowledge for the need to acquire and communicate information that the individual decision maker lacks (Hayek, 1945). The price system can be regarded as a mechanism to communicate knowledge in an economic way, that is the individual participants need to know very little to be able to take the 'right' action. This mechanism is also very quick in registering and communicating changes, and discovering emerging opportunities: it acts as a device to diffuse innovation.

Information technology can be harnessed to create new markets where none existed before (for an example within a hierarchy, see Ciborra (1993)), or support and improve existing ones. In order to identify the main principles in designing computer-based market support systems, one needs to explore a few important aspects of the functioning of markets. Firstly, markets have to be analysed as information processing mechanisms. The economics of information can help us in this respect, especially in pointing out where the perfect market and the price system fail because of information imperfections. Such imperfections represent the grounds for evaluating the potential scope of information technology for supporting markets. Secondly, a number of applications of computers and telecommunications that today support markets, such as EDI, EFT and others that go under the labels of Strategic Information Systems (Wiseman, 1988) or Interorganisational Systems, should

* A larger version of this paper is contained in the book by the same author (Ciborra, 1993).

be framed as 'electronic markets', for which it is possible to identify a precise evolutionary path.

In order to design and build electronic markets, one should first evaluate the scope and nature of an electronic market, in respect to other markets for services, in particular the existing universal services. Secondly, one should deal with the issue of the actors who are supposed to create them; in other words: who should set up and design electronic markets? We conclude that such markets can be roughly divided into two broad classes: those where information services traded can be treated like commodities, and those for information services where externalities are a relevant factor, like in present and future telecommunication services. While the former type can be dealt by the *laissez faire* of market forces, there seems to be some scope for government intervention in markets of the latter type, as the theory, and one successful case, the French Minitel suggest.

2. Markets, information and systems

The market is the social mechanisms that integrates the plans of goal seeking individuals and institutions through the exchange of goods and services. Information is essential to the functioning of markets. Indeed, the price can be looked upon as the information system of a market of a given commodity. Using the price, economic agents make up their mind regarding their participation in the network of exchanges they are embedded in: buy or sell; exit or entry.

The main characteristics of a perfect market are (Hirschleifer, 1980):

- perfect communication, whereby every participant has the same access to information regarding prices and quantities of goods;
- instantaneous equilibrium; thanks to the instantaneous communication of the 'market-clearing' price, all pending purchases and sales are executed;
- zero transaction costs: exchanges take place without friction of any sort.

Extant markets may be more or less distant in their behaviour from such an ideal model, precisely because communication is not fully transparent (there are search costs to find the best price) (Stigler, 1961), equilibrium is not instantaneous (there are haggling processes), and transactions are not costless (Coase, 1937). These information imperfections explain why, in the real world, markets are lubricated by a series of integrative mechanisms, such as middlemen, brokers, dealers, and auctioneers. The imperfections, on the other hand, create a series of opportunities by which information technology can improve the functioning of markets and make them more perfect: by communicating prices faster and to more places; by supporting the work of auctioneers and middlemen in communicating prices and collecting deals, and more in general by decreasing the costs of transacting (Ciborra, 1993).

In order to better appreciate how information technologies can improve the functioning of markets as a knowledge communicating mechanism, it is useful to identify those asymmetries in information that cause the market to fail as a coordination and control mechanism (Arrow, 1974).

Firstly, there can be problems of faulty information channels through which the price is communicated on the market, that is, asymmetric information about *prices*. 'To obtain price information, consumers have to search and therefore incur *search costs*, which may reflect the value of time'. (Phlips, 1988).

Secondly, asymmetric information can be generated by the fact that because of limited rationality and the unique circumstances and experience of each individual participant in a transaction, a particular individual has knowledge that is different from that of the other parties, and it cannot be excluded that he or she will using it strategically or opportunistically. Consider, for example the *quality* of a durable good and the different position the parties (sellers who know the good better than the buyers) occupy. Buyers can appreciate the quality of the good only by using it. Warranty schemes can be set up by sellers to provide protection for the consumer in this respect. But further problems can be generated by the warranties: for example, the consumer feels well protected and feels confident in using the good with less care. This is the problem of *moral hazard*, where one party to a transaction may undertake certain actions that affect the other party in the transaction, but that this party cannot monitor them perfectly (Kreps, 1990); or, suppliers may produce intentionally defective goods in order to sell additional warranty.

Where no warranty is provided at all, as in some markets for used goods, the average quality is lower than that signalled by the price, like, for example, in the used car market, described by Ackerlof as a 'market for lemons' (Ackerlof, 1970). Here, the seller knows things pertaining to the good on sale that are relevant, but unknown to the buyer: thanks to this form of asymmetry, called *adverse selection*, cars of less and less quality invade the market, and there is no incentive to sell used cars of better quality. Whenever there is a gap in the communication of information about prices or quality, intermediaries (retailers) can play a role both in diminishing search costs and in mediating the transmission of information about the quality of goods traded.

Finally, in our brief review, we should mention another type of informational asymmetry: if, say, a monopolist is selling a unique object, for which no market price exists, in order to find out how potential buyers may value the good, the sale takes place through an *auction*. 'Auctions offer the advantage that the participants reveal their valuations or their costs to some extent through the bidding mechanism' (Phlips, 1988).

Market failures can be generated by faulty information channels, by the fact that knowledge is fragmented and information does not flow; and by the information structure of individual economic agents that may cause the phenomenon of 'information impactedness' (Williamson, 1975), or information asymmetry, each time the parties involved in the transaction are willing to use information strategically. The possibility of using the price system is then limited by the structure of the information, to the point where

> 'the value of nonmarket decision-making, the desiderability of creating organizations of a scope more limited than the market as a whole, is partially determined by the characteristics of the network of information flows' (Arrow, 1974).

Information technology can be used to overcome to some extent such limits of real market organisations, and avoid the adoption of nonmarket, hierarchical arrangements. This is the arena of the current designing and building of electronic markets.

3. *Characteristics of market support systems*

According to Malone et al. (1987), there are three main effects of information technology on markets:

(1) Electronic communication effect, whereby technology allows more information to be communicated in the same amount of time, at a much lower cost.

(2) Electronic brokerage effect, whereby computers and communication technologies allow many potential buyers and sellers of a given good to be matched in an intelligent way, that is, by increasing for each party to the transaction, the number of alternatives, filtering them and selecting the best one.

(3) Electronic integration effect, i.e. coupling more tightly (coordinating) value adding stages of production and distribution across different organisations.

These effects contribute in various ways to reduce the information asymmetries mentioned above, and thus improve and expand the functioning of markets. Through the communication effect, search costs for prices are diminished by reducing the cost of accessing relevant price information that may otherwise remain with distant experts or markets. The electronic brokerage effect improves market signalling (Kreps, 1990) and the communication of incentives, so that the risks of moral hazard and adverse selection can be significantly diminished: it is easier to monitor past performance, check reputations, and keep track of customers so as to devise better tailored policies for the customer service (see below). Databases and high-bandwidth electronic communication can handle and communicate complex, multidimensional product descriptions much more readily, and this curbs the opportunism of sellers and buyers in playing games around the quality of goods. Thanks to the higher information processing power, the complex facets of quality can be reduced to a more manageable, and observable set of attributes: in a way, the goods become more standardised, while preserving their complex qualities. Finally, the integration through 'just-in-time' techniques of different production stages through CAD/CAM technology can speed up the overall production time, maintaining at the same time enough flexibility and customer orientation.

To be sure, many of these advantages can be captured by hierarchical firms, as well as markets. However, some of these are specific to markets: for example, a number of Strategic Information System applications fall in this category (Wiseman, 1988), especially those labelled Interorganisational Information Systems, of which Electronic Data Interchange (EDI) is one of the latest examples. These concern linking suppliers and customers or manufacturers and retailers, or in general independent organisations that regularly exchange goods and services, allowing for the following:

- providing benefits of vertical integration (more control, coordination at a lower cost) without requiring actual ownership of other organisations;
- increasing sales of company products due to ease and efficiency of ordering;
- removing levels of the distribution chain by going direct to customer or user rather than through an intermediary (though middlemen are not eliminated, but shifted, see below);
- evaluating quickly the effects of advertising, rebates and other marketing programs;
- extending market reach to customers who could not be economically served by conventional field sales;
- shipping in more economical lots and receiving incoming goods more efficiently by communicating with transportation companies;
- delivering products or services electronically, monitoring compliance with policies related to customers;
- making products easier to select, order handle, use and account for;
- providing immediate feedback on product availability and price;
- improving customers's ability to shop for a third-party products that generate fee income or additional sales of primary products;
- getting the best available prices on purchased commodity materials (Johnston and Vitale, 1988).

Malone et al. (1987) call markets where systems like those just mentioned play a relevant role, 'electronic markets'. An electronic market may emerge because of the single or joint action of a variety of actors, that are always present in a market in general: buyers, sellers, distributors (middlemen), financial service providers and information technology vendors.

Sellers, or producers, for example, are interested in tying customers: information technology, in particular network services related to the sale transaction, has been employed to provide a sort of electronic tie-in. For example, in the early reservation systems, travellers were induced to buy tickets from the airline that provided the electronic reservation system.

On the other hand, buyers would like to see their range of alternatives expand in order to be less and less dependent upon one or few suppliers. Buyers can access systems like the electronic yellow pages of many other services available on successful videotex systems, like the French Minitel (see below), to get information about alternative products or services.

Or, a major firm can have its suppliers linked through a network, so as to be able to choose more quickly and reliably where to allocate a certain job. Sometimes it may be a distributor, such as McKesson, to establish these electronic links with the customers. Financial services can take advantage of their role of middlemen in almost any transaction (where they provide money, credit, means of payment to the parties): they can use the information in their hands to set up databases and networks to become electronic brokers of many more transactions, at a lower cost (Wiseman, 1988).

In the various cases where an electronic market has had sufficient time to fully develop, specifically in the most known strategic information systems applications like the airlines reservation systems, Malone et al. (1987) identify a sequence of stages that generalise the evolution of electronic markets.

The first stage is characterised by the application of information technology to an existing non electronic market or to the externalisation of internal departments of a hierarchical business organisation, typically as a result of a buy versus make decision.

The investment in setting up an electronic link is justified by the possibility of tying in buyers, raising switching costs, building entry barriers and in general modifying market structure to one's own advantage (Porter and Millar, 1985): for example, thanks to the electronic communication effect, buyers will find it advantageous to use only that supplier, while switching costs will increase correspondingly. The earlier reservation systems did contain some features that created such biased markets. In a subsequent stage, however, competition, antitrust intervention and buyers pressure tended to eliminate much of the bias in the system: either other suppliers feel compelled to adopt similar systems, not be cut out completely from the market, or legal forces will mount because of vocal pressures to eliminate the bias.

On the other hand, unbiased markets reduce margins significantly, do not justify any further investment in the technology, and indeed may backfire, since they overwhelm the consumer with too many standard alternatives. A further stage, made possible by the new technology, is to move towards a stage of personalised markets (Jackson, 1985). Thanks to the technology, it is possible to profile the customer, track closely his or her buying behaviour and provide a host of electronic services, from special offers to programs, tailored to the exigencies of the specific client (the systems that lie behind the frequent fliers programs are a familiar example). Personalised markets can bring some technology to the buyers, too: intelligent aids that can help buyers select products from a range of alternatives; or, agents than can act as representatives for the buyers on the electronic market, providing information about their customer's preferences directly to the suppliers, so that better tailored sale policies can be devised.

This evolutionary model suggests the following implications for setting up strategic information systems that create electronic markets:

- firms should review their make or buy decisions, and thus evaluate whether to externalise some activities, i.e., get them through the market;
- all market participants may envisage the possibility of setting up an electronic market and value the benefits they can reap;
- at the same time participants should be aware of the evolution from biased to unbiased and possibly to personalised markets; they thus should carefully plan their investments, and their timing, so as to be able to reap first mover advantage, then, oligopoly membership, and be ready to reinvest for a personalisation of customer relationships.

In summary, given the evolution of electronic markets, the successful applications of information technology rest with those that build strong links and high switching costs (biased markets). These systems can cause pain to other vendors, and may be used simply as a preemptive move (Jackson, 1985). But competitive applications of information technology are not all equally desirable; some are considerably more defensible against competitors' reaction than others (Beath and Ives, 1986): thus in order to maintain a biased electronic market to a firm's advantage, a strategic application must be valuable, inimitable and rare (Ciborra, 1991)

4. *The service idea*

On a perfect market, commodities are exchanged by faceless buyers and faceless sellers. The more the market is imperfect or intentionally biased, the more transactions include a flow of customised benefits over a period of time derived from physical goods or human activities, i.e., they include the production and consumption of services. Designing a biased electronic market, means, then, conceiving and implementing a service idea. Some marketing relationships, however, will stay based on large volume transactions of standardised commodities, and as we have seen above, biased markets evolve eventually because of competitive or legal pressures into such unbiased markets. Finally, we should not forget the existence of so-called 'universal services', where services are often provided outside a real price system and in an uncustomized way (think of the universal phone service; national medical care and other services delivered by the public administration).

Building an electronic market means, then, not only positioning oneself along the evolutionary path described by Malone et al. (1987), but also defining how to carry out the electronic transactions according to one, or a combination of, the three models: relationship-based service; transaction-based contact; and universal service. Vepsælæinen and Mækelin (1987) provide an interesting framework to tackle this design exercise, by identifying the main dimensions and a typology of the service relationship. In order to build such a typology they list the following generic components of a service:

- type of service, or service package, i.e., the content of the exchange relationship;
- delivery channel, i.e., the more or less permanent arrangement that makes the service possible;
- the governance of the service transaction, whether long term or short term, personalised or anonymous.

The object of the design endeavour is to find the best fit between the three dimensions, exploiting the opportunities offered by the technology available and the competitive situation at hand. This requires analysing each dimension and the links between them in more detail.

Specifically, it is possible to build typologies of markets and services based on the first two dimensions of a service, while using the third as an intervening variable when exploring the other two.

Service packages and delivery channels are defined in terms of scope and governance. The service package can be defined on the basis of the scope of the service contract, whether broad or narrow (the service to be provided is left largely unspecified versus the service that is strictly predefined), and on the governance structure, or customer contact strategy, whether relationship based, when major investments are made to strengthen the link between the firm and its customers, or transaction based, when costs vary according to the volume of transactions (Jackson, 1985). The emerging typology identifies three types of efficient service packages: customised services that have a broad scope and are supported by an adaptive, close relationship; standard services, that balance transaction costs and production costs, and mass transactions, simple services in which the identity of the customer does not matter.

Delivery channels are defined by the channel system and its governance structure, or channel organisation. Channel system can vary from narrow band decentralised channels (that usually rely on some 'back office') to generalised network infrastructures. Channel organisation varies from independent agents to internal and field employees, or self service by direct customer ac-

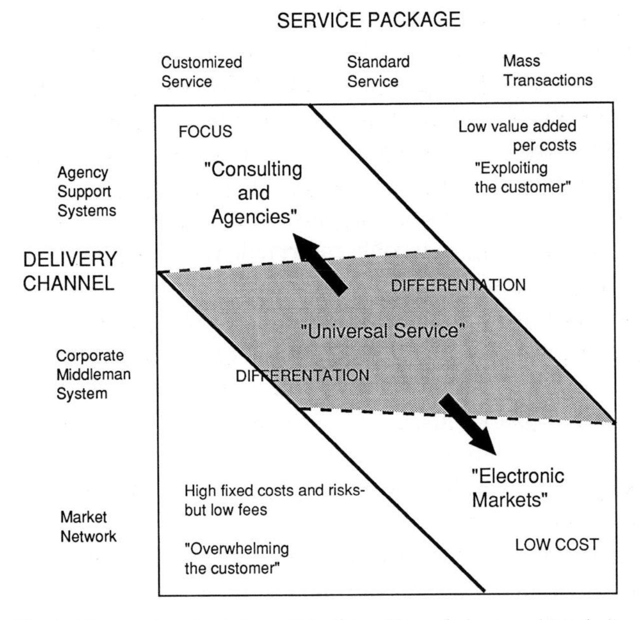

Fig. 1. The service–strategy matrix (from Vepsælæinen and Mækelin, 1987).

cess. The typology of efficient delivery channels consists of agency support, corporate middlemen and market networks, respectively. A typology of service markets emerges if we combine the two tables into a third one, called Service-Strategy Matrix. Along the diagonal of Fig. 1 a variety of service markets, ranging from electronic mass markets (bottom right), where human intermediaries are by-passed; to consultants (upper left), where centralised service systems support an agent's face-to-face contact. In the middle of the diagonal there are services like medical centers, banks or public services such as the employment office, that could be shifted in either direction along the diagonal itself, i.e. towards more standardisation (unbiased markets) in the bottom right or more customisation (personalised markets) in the upper left. The inefficient solutions can be found off the diagonal: 'overwhelming the customer' (providing complex advice without adequate support) with high fixed costs, or 'exploiting the customer' with low value-added services (a heavy organisation that delivers simple transactions).

The new typology leads us to identify the potential role of information technology in transforming the structure of the market.

As a mediating technology (Ciborra, 1993), computers and telecommunications can first impact the delivery channels. They provide more effective distributed networks, thus in general they push towards the right all the markets along the diagonal of Fig. 1: intermediaries can be by-passed, so more electronic markets can be created. On the other hand, new intermediaries may emerge: those who provide the value-added networks to buyers and sellers. Vepsælæinen and Mækeling (1987) suggest that the channels will evolve first from channels developed around specific products and services to integrated channels that will deliver a variety of services surrounding a specific good or service; and then towards a new stage characterized by a division of labour along the channels, between organisations that will perform massive back office activities, and more flexible intermediaries targeted to customers and specific media. Regarding the service package, the trend is to standardise the information content of some services, so as to transform them in commodities for self service (Gershuny, 1978), or to enrich them with knowledge. Systems for sale support, bidding systems and service hot lines are examples of such applications in marketing.

As mentioned above, however, new service markets are not created in a vacuum, they are the result of strategic change implemented in existing markets. And almost any policy for change in this area must cope with, or even challenge the existence of universal services. Designing electronic markets has much to do with transforming universal services. A universal service is based on providing many generic services, with trivial customer relations (no segmentation of customers by needs, costs or profitability) and passive marketing: customers call in or stop by. Often a universal service provider employs a heavy bureaucratic apparatus, with a large back office, high fixed costs, and a constant need to expand offerings in response to local pressures. The ideal information systems support for a universal service provider is the integrated system, whereby one staff member can handle all customer matters through the same terminal which accesses large data bases and centralised procedures. Many PTT's, utilities, banks, insurance companies and most of public services, but also some large manufacturing companies, can be classified under the label of universal service providers.

These providers are often stuck in the middle of the service–strategy matrix shown in Fig. 1. Information technology can support the differentiation of the universal service towards the electronic market end, by offering direct access to standardized services, for example through videotex systems, ATMs in banking etc., or to more customised service providers like consultants. The mediating technology can make services more efficient, by establishing direct inter-organisational customer services; focusing when needed on professional support for customised services, or establishing electronic markets for fast self service. Probably, the universal service provider has to implement some internal changes in organisation (e.g. decentralisation); establishing alliances with private intermediaries and network service providers; deregulation in the use of information technology (moving from unilateral integration to shared systems that match complex, differentiation strategies); targeting different customers on the basis of their needs, opportunities of access and profitability; and finally, abandoning the policy of total ownership of channels to participate in multi-partner value adding channels.

5. Enacting electronic markets

The use of the term 'market' in the case of electronic markets hides some of the complexities of the functioning, and especially of the creation of a market *for information or knowledge-based goods*

In order to find what the correct policies are to set up such markets for the various markets participants (see above), it is necessary to better appreciate their special nature.

A market for information differs from an ordinary market for commodities at least on two counts: the type of commodity, that is information, and the rules governing the market. On the surface, information can be considered a factor of production. Actually, the markets for commodities, or skills depend on the transmission of information about scarcities and needs in the economy. Both consumers and suppliers will find it worthwhile to invest in the acquisition of market information in order to choose according to quality. This explains the emergence of a thriving private sector information market (job agencies, advertising, market research, corporate communication, etc.). However, 'despite the willingness to pay for information, private competitive markets are not likely to provide as much of the commodity as consumers require' (Hartley and Tisdell, 1981). Given the scope of such private sector information markets, one could indulge in plotting quantity of information demanded against price (Emery, 1969); unfortunately, information possesses some idiosyncratic features that make such markets only a part of a more complex picture. For one thing, information can easily be reproduced and sold without destroying it; it is difficult to establish property rights for it; it also shares the qualities of a public good (for example, government statistics), i.e. a good the consumption of which by any one economic agent does not reduce the amount available for others (Demski, 1980; Hirschleifer, 1980).

Because of such characteristics, it may well be that less information is collected and disseminated by individuals than is collectively of value. There is scope, then, for government to interfere with the failing information markets, by subsidising the collection and provision of information (for example, through publicly funded R&D), or by making the information the legal property of the discoverer through the patent law, so that at least some type of information become salable like a commodity (Hartley and Tisdell, 1981).

When we turn to the actual functioning of a market for information goods, we may observe that the rules governing such markets are quite different, also given the special nature of the 'commodity' traded. In the model of the perfect market, each agent makes his or her choice independently from other agents. In particular, the utility of each consumer is unaffected by the consumption decisions of other consumers, if prices are held fixed (Kreps, 1990). However, in the case of information goods, for example telecommunication services, one can observe what economists call externalities, i.e., the consumption pattern of a good by one consumer can affect positively or negatively the buying behaviour and the utility of another.

On the information service market, the willingness of a consumer to buy more of an information good, say a videotex service, may depend on how many other consumers are using the network (installed base or critical mass effect (Gordon, 1989)). Also the market, specifically the electronic one, requires an infrastructure, a market place, which may be another public good to be provided. The crucial design question is: by whom? By the invisible hand? Or by government? Finally, the 'medium' and the 'message' may be interconnected, in a container-content relationship, so that delivering one may require delivering, to some extent, the other (there is little reason for selling CDs, if CD players are not available, and vice versa).

Thus, the electronic markets we have in mind are an institution that allows the satisfaction of a collective need (i.e., where the satisfaction of the need takes place among agents that have to be interdependent), but the functioning of which does not require that the agents act collectively. The agents can act separately, since they expect in their individual acts the institution itself to provide 'the service of interdependence' with the other potential partners.

Creating an electronic market is equivalent to creating a dance hall, or a veritable marketplace or market square, where people can meet, and make a profit out of it. This is precisely the job of Value Added Network service providers. Gordon (1989) distinguishes three types of possible VAN designs, according to a consumer group served, application provided and the role of the VAN operator:

- since the electronic market needs an infrastructure a first type of VAN consist in supporting an infrastructure to a user community already defined and willing to pay for it; the operator does not intervene in enlarging the community or in the applications;
- since the electronic market needs a critical mass, the operator may, beside providing the infrastructure, contribute to the enlargement of the customer base, by 'creating interdependence';
- finally, to address the container-content issue, the operator may intervene also in the content of the applications running on the infrastructure, as a condition to promote the use of the service.

In the case of the French videotex Télétel, the French PTT (now France Telecom) was able to intervene at the three design levels; and the intervention at each level, it can be seen at least after the fact, was essential for the take off of the system:

(1) France Telecom built an infrastructure, a public network, based on a X 25 network, called Transpac, to handle the traffic, and provided the terminals, or Minitels free of charge.

(2) In order to promote the use of the system, thanks to the 'kiosk' idea, it put third parties in a position to provide services through the network; with a flexible billing mechanism the PTT just held the accounting of the transactions between consumers and suppliers of the various thousands of videotex services.

(3) In order to make the system useful from the very beginning, before a critical mass of service suppliers was there at all, the PTT put on line the national directory books of all telephone owners; in fact this was the only content that the telecom operator was proprietor of, and could put on the network at relatively little cost.

Compared to the French Télétel, videotex services launched in other countries, often according to the logic that free market forces would have taken care of the growth of the system (see for example the USA and UK cases; Schneider et al., 1990), failed completely. This seems to be in agreement with the idea that for value added infrastructures that support electronic markets (think of the present and future generations of videophone, ISDN and B-ISDN) there will be a renewed scope for public, government intervention (Egan, 1991).

References

Ackerlof, G., 1970. The Market for Lemons: Quality Uncertainty and the Market Mechanism. *Quarterly Journal of Economics*, 89: 488–500.

Arrow, K.J., 1974. *The Limits of Organization* (W.W. Norton & Company Inc., New York).

Beath, C. and Ives, B., 1986. Competitive Information Systems in Support of Pricing. *MIS Quarterly*, 110 (1) (March), 85–96.

Ciborra, C.U., 1991. The limits of strategic information systems. *Information Resource Management*, 2 (3), 11–16.

Ciborra, C.U., 1993. *Teams, Markets and Systems* (Cambridge University Press, Cambridge).

Coase, R., 1937. The Nature of the Firm. *Economica*, (November), pp. 387–405

Demski, J.S., 1980. *Information Analysis*, 2nd ed. (Addison-Wesley, Reading, Mass.).

Egan, B.L., 1991. *Information Superhighways; The Economics of an Advanced Public Communication Networks* (Artech House, Boston, Mass.).

Emery, J.C., 1969. *Organizational Planning and Control Systems Theory and Technology* (Macmillan, New York, N.Y.).

Gershuny, J., 1978. *After Industrial Society? The Emerging Self Service Economy* (Macmillan, London).

Gordon, P., 1989. *La Place du Marché* (La Documentation Française, Paris).

Hartley, K. and Tisdell, C., 1981. *Micro-Economic Policy* (John Wiley, New York, N.Y.).

Hayek, F.A., 1945. The Use of Knowledge in Society. *American Economic Review*. September, pp. 519–530.

Hirschleifer, J., 1980. *Price Theory and Applications*, 2nd ed. (Prentice-Hall, Englewood Cliffs, N.J.).

Jackson, B.B., 1985. *Winning & Keeping Industrial Customers* (Lexington Books, Lexington, Mass.).

Johnston, H.R. and Vitale, M.R., 1988. Creating Competitive Advantage with Interorganizational Information Systems, *MIS Quarterly*, June, pp. 153–165.

Kreps, D.M., 1990. *A Course in Microeconomic Theory* (Harvester Wheatsheaf, New York, N.Y.).

Malone, T.W., Benjamin, R.I. and Yates, J., 1987. Electronic Markets and Electronic Hierarchies. *Communications of the ACM* 30: 484–497.

Phlips, L., 1988. *The Economics of Imperfect Information* (Cambridge University Press, Cambridge).

Porter, M.E. and Millar, V.E., 1985. How Information Gives You Competitive Advantage. *Harvard Business Review* 63 (4) (July–August), 149–160.

Schneider, V., Charon, J.M., Miles, I., Thomas, G. and Vedel, T., 1990. The Dynamics of Videotex Development in Britain, France and Germany: a Cross-National Comparison. *Proceedings 8th International Conference of The International Telecommunications Society*, Venice, March.

Stigler, G.J., 1961. The Economics of Information. *Journal of Political Economy* 69, 213–285.

Vepsælæinen, A. and Mækelin, M., 1987. Service-oriented Systems and the Economics of Organizational Transactions. *Proceedings 10th IRIS Conference*, Vaskivesi, August.

Williamson, O.E., 1975. *Markets and Hierarchies: Analysis and Antritrust Implications* (Free Press, New York, N.Y.).

Wiseman, C., 1988. *Strategic Information Systems* (Irwin, Homewood, Ill.).

Part 2: Technological Changes and Standards

Coordinated by Cristiano Antonelli, *Università di Torino*

Global Telecommunications Strategies
and Technological Changes
Edited by G. Pogorel

Introduction

Increasing returns: Networks versus natural monopoly. The case of telecommunications

Cristiano Antonelli

Laboratorio di Economia dell'Innovazione, Dipartimento di Economia, Università di Torino, Via S. Ottavio 20, 10124 Torino, Italy

1. Natural monopoly: economies of scale and subadditive cost functions

The standard definition of natural monopoly is based on average cost curves with a negative slope along the section of effective demand. In such conditions it is clear that it is advantageous to limit production to the smallest number (unit) of firms, so as to take advantage of all the benefits which can be derived from a constant negative sloping cost curve. Baumol et al. (1982) have further extended this definition by introducing the concept of 'subadditive' cost functions.

Every cost curve has areas of 'subaddativeness' simply when in a Marshallian U-cost curve there are long stretches of falling costs. The cost curve in this case is 'subadditive' up to the lowest point on the classic U-curve and 'superadditive' in the rising section. When demand is lower than the minimum point on the cost curve, the relevant portion of the latter is 'subadditive'; but this does not necessarily imply that the average cost curve has a constant negative slope. A formal definition of 'subadditiveness' might at this point be useful. There is 'subadditiveness' when given $C(Q)$, the cost curve can be expressed as follows:

$$C(Q_1 + Q_2) < (C(Q_1) + C(Q_2) \qquad (1)$$

where Q_1 and Q_2 are two different levels of output.

On the basis of the concept of 'subadditiveness', Baumol et al. (1982) elaborated a much broader concept of natural monopoly: "an industry is in conditions of natural monopoly if a firm's cost curve is 'subadditive' for the whole arc of effective outputs" (Baumol et al., 1982, p. 17).

The concept of a 'subadditive' cost curve is significantly different from the traditional theory of a cost curve where there are economies of scale and therefore of increasing returns in the production function. When there are economies of scale, average costs are constantly falling because there are increasing returns in the production function. In such a case in fact, the increase in output is greater than the increase in input and consequently the increases in costs are less than the increases in output. Increasing returns are determined exclusively by the character of technology. This is because there are increasing returns when there is indivisibility and the factors of production are used disproportionately. Disproportionate use of factors and factor indivisibility are due to the technology adopted; that is to say increasing returns are determined by the specific design of the equipment and the conditions in which it is used. [1]

When subadditiveness is not the outcome of technological indivisibilities the notion of networks becomes pertinent.

2. *The economics of networks*

A network is an organized set of partially separable productive units, with increasing overall returns which can be attributed not so much to economies of scale as to an overall 'subadditive' cost function which reflects the contribution of relevant technical and pecuniary externalities, superfixed costs and density economies as well as the effects of important demand externalitities. So far the notion of network is intended to incorporate the concepts of interrelatedness and imperfect divisibility, complementarity and hence externality.

The notion of network seems especially promising in analysing a number of complex industries which are characterized by a high degree of integration and interdependence. Networks would appear to be a particularly appropriate way of analyzing the organization of industries providing public services such as electricity, postal services and telecommunications, rail and air transport, water and gas supplies. The large network infrastructures which provide public utilities can, in fact, be analyzed using the economic concept of network

---

[1] Formally, as has been noted, this happens only when the production function has the following characteristics:

$$Q = (K^{\alpha}, L^{\beta}) \tag{2}$$

for $(\alpha + \beta) > 1$.

as they are naturally complex systems with high sunk costs and a long life. They are also composed of different elements that are both synchronically and diachronically interdependent, having economic and technical links which are difficult to separate and therefore they should be considered as quasi-divisible.

Such network-industries are characterized by numerous special factors:

(1) they offer heterogeneous products;

(2) they are formed by infrastructures with a high degree of irreversibility;

(3) they are made up of separate elements which are linked by a high degree of technical compatibility and complementarity;

(4) their design is based on a high degree of integration due to technical, pecuniary and demand externalities.

The high level of investment in fixed capital which is normally required to create a network has numerous implications above all in terms of irreversibility. It involves highly specific investment goods which can seldom be used in other ways, so that leaving the market is very costly. As well as this basic factor of irreversibility there is the fact that technical choice is severely limited when there are changes on the demand and/or on the technological side. Returns from new technology can seldom be maximized without considering the technical characteristics and the cost conditions of the existing technology.

Large networks basically serve three closely related functions:

(1) distribution functions;

(2) switching and supply functions;

(3) transmission functions.

The distribution function consists in both the collection and delivery of information, passengers, energy flows (hereafter known as input), in a local area. However, a distinction should be made between active and passive distribution. The transmission function consists in linking both the passive and the active distribution activities which take place in a distant area. The switching function consists in selecting the destination of the input and in organizing the flows of local and distant traffic. In this case the division between local switching functions and distant switching functions must be introduced. The distant switching function regulates the act of connecting distant traffic and therefore interacts on the distribution function.

There are very high and differentiated degrees of technical and economic linkage among the three large functions which have been identified. Economic analysis usually considers these effects as externalities. On the supply side, this refers to the concept of technical externalities when interdependence assumes a technical form, and pecuniary externalities when it consists of effects on the relative prices of the factors. Demand externalities will arise when: (a) the demand for services supplied by each activity is affected by changes in the demand of other complementary activities; (b) the demand for services of each consumer is affected by the aggregate levels of demand for the same service; (c) the demand for services of each consumer is affected by the number of other consumers.

Technical, pecuniary and demand externalities together represent a series of phenomena of interdependence and integration generated by interrelatedness and imperfect divisibility which are a result of technological factors (production functions) and of consumers' needs (utility functions).

According to Baumol et al. (1982) natural monopoly should be considered when the supply curve has a negative slope at least in that section which corresponds to possible demand and the supply curve lies below the demand curve so as to exclude an equilibrium position. In fact, the economic characteristics of the network make it possible to distinguish whether the negative slope of the curve depends on:

(1) increasing returns in the production function at the plant level;
(2) increasing returns in the production of separable services;
(3) technical externalities;
(4) effects of joint production;
(5) economies of density;
(6) planning economies; and
(7) purchasing economies.

The relevance of each of these factors will help to determine the extent to which to notion of 'natural monopoly' applies and this in turn leads to significantly different regulations being introduced. In fact, natural monopoly actually asserts itself only when the production curve indicates increasing returns that can be attribute to technical economies of scale. On the other hand, if more weight is given to interrelatedness and consequently to externalities, then supply can be organized on such institutions as clubs and regulated competition.

3. *Increasing returns within networks*

3.1. *Economies of scale and subadditiveness. The case of telecommunications*

If we define the network as the aggregate of the different elements, the form of the aggregate supply curve is clearly influenced by the form of the supply curve of each plant as well as by the effects of aggregation. It is therefore obvious that the supply curve of the network exhibits an aggregate negative slope when it is the result of the aggregation of local supply curves with a negative slope because of economies of scale at the plant level, but not at the industry level.

Generally, it is reasonable to assume that there are large technical economies of scale at the plant level for each element operating in a network and in particular in telecommunications. It is possible to express the production function of each element as follows:

$$Q_i = A(K^{\alpha}, L^{\beta}) \tag{3}$$

where $\alpha + \beta > 1$.

Consequently the plant's total cost function will be:

$$C_{\mathrm{T}_i} = h(Q) \tag{4}$$

where $h' > 0$, but $h'' < 0$.

A network's supply curve defined as a number of productive units with cost functions as (4), will normally be expressed with an equation of the type:

$$S = \sum c' \tag{5}$$

where $c' < 0$.

In fact, if the equation which represents total costs (4) has a negative derivative, the separate marginal cost curves which make up the supply curve must have a negative slope and consequently the supply curve will be found to have a negative slope, too. This elementary result confirms that overall increasing returns can be caused by factors other than the indissolvability of production. It follows therefore that a natural monopoly initially thought of as the only and necessary solution to the market problems posed by increasing returns, is called for only at the local level and not necessarily at the overall, aggregate level. In fact, this is exactly the situation which is found in the local supply of water and gas and, to a certain extent, of electric energy.

In fact a monopolistic supply might be the dynamic outcome of an unregulated market competition where there are economies of scale at the plant level. In this case, however, monopoly would be the ex-post outcome of market selection rather than the effect of technical indivisibilities. An ex-post monopolist operating on a multiplant structure would have little cost advantage on a plurality of local monopolists. Economies of scale at network level and economies of scale at plant level must be differentiated if the relevance of the concept of increasing returns in network industries is to be evaluated.

Moreover, economies of scale at the functional level and economies of scale at the network level must also be differentiated. Once more an aggregate supply curve with a negative slope at the network level may be simply determined by the horizontal aggregation of the supply curves of plants operating in well defined specific subindustries or functional components of the network. Specifically in the telecommunications network it is necessary to differentiate between: (a) economies of scale in intra-exchange services at the level of the single exchange plant; and (b) economies of scale for interexchange services at exchange and transmission level.

3.2. External economies of production and economies of scope

The production of goods provided by networks is strongly characterized by technical and pecuniary externalities. A large technical literature shows that the working of a single unit operating in a network, such as the infrastructure for distribution, switching and transmission, is characterized by high levels of interdependence. The activity of distribution cannot exclude the activities of switching unless a geodesic network is imagined where all the terminals of the network are physically connected to channels of continuous communication which are at the complete service of each point. At the same time, it is difficult to imagine that the transmission activity can exclude either the distribution activity or the switching activity at least if it is not a system which links all the points and assumes unlimited capacity to transmit and communicate where the points are only points of access to the local geodesic network.

When the interdependence of the functional components results in technical complementarities, technical externalities arise. The activity carried out by each of the other functions has in fact a direct bearing on the production function of the others: in telecommunications networks the single productive units has a direct bearing on the production function of the other unit (LATA A distributes telephone calls transmitted by the adjoining LATA B).

Such a situation is clearly the classic case in which plant A's output is based on inputs produced by B

$$Q_a = F_a(K, L, Q_b) \tag{6}$$

$$Q_b = F_b(K, L, Q_a) \tag{7}$$

In a simplified industry with only two firms that are perfectly equal, each firm has the following cost curve:

$$C_1 = \alpha q_1^2 + (\alpha + \beta)^2 q_1 + \beta q_1 Q \tag{8}$$

$$C_2 = \alpha q_2^2 + (\alpha + \beta)^2 q_2 + \beta q_2 Q \tag{9}$$

where $Q = q_1 + q_2 = K$ (for K = a constant).

When $\beta < 0$ there are external economies. The shape of the supply curve will therefore depend on first order conditions:

$$p - 2aq_1 - (\alpha + \beta)^2 - \beta Q = 0 \tag{10}$$

$$p - 2\alpha q_1 - (\alpha + \beta)^2 - \beta Q = 0 \tag{11}$$

The maximization of profit for the two firms for $q_1 = S_1$ and $q_2 = S_2$ requires that:

$$S_1 = S_2 = \frac{P}{2(\alpha + \beta)} - \frac{\alpha + \beta}{2} \tag{12}$$

from which it is seen that the aggregate supply curve is

$$S = S_1 + S_2 = \frac{P}{\alpha + \beta} - (\alpha + \beta) \tag{13}$$

It is therefore sufficient that $|\alpha| < |\beta|$ for $\beta < 0$ to have a supply curve with a negative slope. Then the supply curve of an industry made up of a number of firms characterized by strong structural interdependence which causes technical externalities, will be found to have a negative slope.

The relevance of external economies for the analysis of networks significantly increases when the typical characteristics of a network which intrinsically can supply a considerable variety of products with equal ease is considered. As has already been noted, this is particularly true for telecommunications networks which can provide many different forms of communication. However, it is also true for the railways, for the gas supply and more generally for any non-differentiated product distributed by a network as long as it is assumed that the characteristics of final demand are sufficiently distinct from those of intermediate demand.

It is possible therefore to distinguish between specialized networks and universal networks on the basis of the specific product on offer. Specialized networks will be able to provide a limited range of products while universal networks are those which can provide a wide range of products. The characteristics of the demand for products provided by a network in terms of quantity and quality, as well as in terms of elasticities of price, and of income are differentiated so much that it is necessary to consider them as separate products.

It seems important here to contrast the notion of technical externalities with that of economies of scope. Economies of scope occur when the joint cost of producing two goods is lower than the sum of the total cost of producing the goods separately. Formally there are economies of scope at the level of the firm or plant when:

$$C(Q_1, Q_2) < C(0, Q_2) + C(Q_1, 0) \tag{14}$$

where Q_1 and Q_2 are two different outputs.

It seems clear that economies of scope and the effects of technical externalities have important overlappings. The difference in fact consists of: (a) the characters of the organization that coordinates the production of the two products; and (b) the extent of technical divisibility in the production of the two goods.

It seems possible to argue that economies of scope may apply to technical externalities that take place within a single corporation and with low levels of divisibility in the production processes.

3.3. Economies of density and the use of productive capacity

The theory of economies of density can be considered only when there is an appropriate definition of the short and the long run. This is particularly true for networks when they provide universal services. The universal service of a network calls for a simple distinction to be made between the moment the whole network is built, at least in general terms, and the moment when the service is provided. The separation of these two moments is relevant both for the maximum volume of the traffic and for the maximum distance which can be linked together. Supplying a product on a high quality network basically assumes two things:

(1) Productive capacity sufficient to satisfy the peaks of demand and therefore designed to be underutilized for long periods.

(2) Extensive coverage so as to be able to satisfy users' demand even when it is distant and in areas with scarce density of use.

Consequently, the costs of a network can be classified mostly as fixed anticipated costs and in such conditions the analytically relevant cost curve must be the short term one. It should be noted that the short run cost curve of a productive process characterized mainly by fixed 'sunk' costs is certainly negatively sloped for long stretches of demand.

Actually a negatively sloped total cost curve can be the result of short term economies of density, in the presence of 'superfixed' factors of production which are modified only in the very long run:

$$Q = A(SK, K, L) \tag{15}$$

where SK denotes the 'superfixed' factors of production which have a life longer than the normal fixed factors.

Superfixed production factors in the production function affects of the average fixed cost curve as it is clearly shown by the following equation:

$$AVC = \frac{CC + SK}{Q} = \frac{CC}{Q} + \frac{SK}{Q\,T} \tag{16}$$

where AVC are the total average costs, Q is the output, CC are the current costs, SK are the initial outlays on superfixed infrastructure, and T is the length of the economic life of the superfixed capital.

Equation (16) shows clearly that the average fixed costs for a network characterized by superfixed capital will be lower the larger the output and the length of the economic life of the superfixed capital. Hence in the medium term, following particularly significant investments in 'superfixed' productive factors which are characterized by very slow technical and economic obsolescence, firms which manage the network aim at increasing the quantity produced and benefit from the flow of returns generated by 'superfixed' capital; that is to say, they generate apparent increasing returns of scale. This means that the recapitalization of 'superfixed' capital at current values and write-off procedures which are different from those used for fiscal purposes could give us the actual shape of the network's cost curve.

From what has already been said it follows that, when the increase in the size of demand for relevant periods of time is not accompanied by an equal rate of growth in productive capacity, the increase in the overall volume of demand increases the efficiency of the system and, effectively lowers average costs. This confirms that the useful analytical approach to analyze the workings of networks is the funds and flows approach. A fund, characterized by indivisibility and in this case anticipated, is paid for over time only through an increase in flows. A network, especially one which has to provide a universal service which in turn increases the amount of funds which has to be advanced, will certainly enjoy significant economies of density. Such economies will be rather special in that they are in no way related to changes in productive capacity but are determined by the progressive increase in the intensity of use of the productive capacity.

Economies of density are particularly relevant when the increase in demand is mostly intensive and not extensive. This is because an extensive increase would mean that the basic infrastructure would have to be increased. Similarly, economies of density are affected by changes in technology which force changes to be made to the infrastructure and this by definition requires a long run analysis.

Economies of scale are closely linked to economies of density. A network which has ample surplus capacity can become an economically advantageous resource if it enables additional supplies of the service to be provided without first having to expand the productive capacity. Re-routing, in fact, permits erratic demand to be satisfied by compensating the excesses. Re-routing is particularly relevant in the management of telephone networks, rail transport for goods, gas distribution and the electricity grid. (Re-routing is seldom welcomed by passengers of air companies.) Increasing the overall size of a network, as is well known from the basic law of numbers, tends to reduce density and so it becomes increasingly easier to adopt practices which will offset peaks in demand by ceding productive capacity in one part of the network to another. If only moderate increases in variable costs are incurred when re-routing is called for and only modest increases in the productive capacity of transmission are necessary, then the size of the network can be significantly reduced. The network can be designed for levels of production which are lower than the peaks of demand.

Economies of density are also closely related to economies of scope. In the case of extensive infrastructures, the creation of polivent systems which can satisfy the needs of radically differentiated demand leads to a much larger initial investment than would be necessary to create a specialized network. At the beginning of the curve there are significant diseconomies of scope and it would appear to be much more advantageous to build a specialized network. However, as demand increases, and above all, as it intensifies, a universal network would appear to progressively become more advantageous.

In fact, economies of scope become more important as existing capacity is used more. Finally, as the saturation level of the existing capacity is reached, there is a third phase in which diseconomies of scope begin to be felt and the advantages of semi-specialized networks, that is to say, weakly connected networks become apparent. Actually, economies of scope depend (mostly) on particular combinations of different levels of supply and demand. In the long run neither of them can be considered as exogenous.

Ex-post economies of scope will exist every time supplies can suitably balance non-differentiated productive capacity by combining (even stimulating or anticipating) differentiated sections of demand.

Economies of use of the productive capacity are particularly high when the network is working at levels close to the maximum levels of expansion planned for; that is to say, when the peaks of demand reach the levels of maximum productive capacity. By definition, during the initial phases of a network's life, re-routing will be minimal. A sort of trade-off takes place between economies of density and economies of use over time. Such a trade-off can be taken as the basis of an analysis of the actual life cycle for the network. The presence of economies of use of the productive capacity, however, is again independent of the production function and yet it makes a significant contribution to the 'subadditiveness' of the short run cost curve and therefore to the negative slope of the supply curve. More precisely, the economies of use of productive capacity make it possible to extend the 'subadditiveness' of the cost curve to the sections in which the economies of density reach saturation point.

3.4. Long run economies of size

The traditional distinction between the long and the short run in economic analysis is notoriously based on the capacity and the possibility of changing the amount of fixed capital available. Further, as it has already been noted elsewhere, only increasing returns and economies of scale are long run phenomena. An economic analysis of the network must instead take into account the fact that there is significant irreversibility due to the long and sometimes very long economic life of the capital and the stage by stage growth of a network. Such factors mean that decision taking has a sequential nature, thus a network architecture is strongly influenced by its evolution. So far a network is clearly the outcome of a path-dependent, non-ergodic process of growth. Networks of equal size, but which evolved in different ways, can have a significantly different architecture because they followed different patterns and different strategies. Networks of equal size can therefore have significantly different costs because of their different evolution.

It is possible to argue that by giving a single firm the responsibility of building a network in the long run, a more rational network architecture will be created than would be the case if the network were left to the unplanned spontaneous growth of local networks run by independent operators. Of course in this case the monopolist must have very long term goals; and this is more likely to be the case because the growth of demand for the firm coincides with the evolution of the network.

Planning economies essentially consist in the careful use of the transmission facilities; in other words, in the connection of local networks. Planning economies are therefore especially relevant where the transmission facilities are particularly costly.

Empirical evidence (Kahn, 1971) emphasizes the economies of transmission in networks characterized by transmission facilities with high capital intensity (gas distribution, electricity grids and the railways). The fall in telecommunications transmission costs engendered by the introduction of new transmission technologies is consequently likely to reduce the role of planning economies in telecommunications networks

Planning economies can generate particularly important economies of scope when a single transmission infrastructure can be used to connect many local and even heterogeneous networks. This is the case of a widely defined telecommunications industry which covers the transmission of television signals via cable and air, the transmission of voice signals and the transmission of data.

In the case of an electricity network, the economies of planning are not limited to transmission; but they can also include the production of energy. The distribution of the production plant can therefore be optimized ex-ante in the long run by allowing for the characteristics of the regional distribution of demand and by also allowing for possible offsetting needs by providing transmission facilities for distant sources of production.

Economies of planning can be the cause of important rigidities in the management of a network in the medium term because they can induce the manager of the network to subordinate the satisfaction of local needs to the wider goal of ensuring the rational evolution of the whole network. Economies of planning can be particularly important when costs, prices, and the efficiency of the capital goods which make up the networks are not constant over time or are not constant in relation to the quantity purchased, but vary because of important pecuniary economies of scale. The shape of the long term cost curve reflects, in a significant way, the costs at which the factors of production are purchased and in particular the cost of the plant and machinery which form fixed capital. If the plant and machinery is produced in conditions of increasing returns, then as the quantity produced increases, costs will fall and with this the average purchasing costs of the firms using this machinery and plant may fall, too.

In the same way, if, on the production side, there is significant initial long run investment (for example, the investment which is necessary to improve models of generating stations or switches) then the increased quantity demanded will lead to relevant reductions in the average costs of production of that plant and of the production of the goods and services which are based of their use. Thus, there is a further important reason to assume that the average cost curve of the firms which operate in network sectors has a negative slope: purchasing economies of scale in imperfect markets for intermediary and capital goods.

The special nature of demand and supply in industries which produce goods and services destined for the construction of a network infrastructure deserves further comment:

(1) The number of producers worldwide is very small because of barriers to entry created by the very large minimum optimum size (in the field of switches it is reduced to seven or eight large firms).

(2) Technological change is characterized by short periods of intense activity and by long periods characterized by the introduction of incremental innovations.

(3) Demand is limited to a fairly low number of large buyers or in other words, a small number of managers of national monopolies.

(4) Firms which produce plant and machinery (which, as has been seen, make up a large part of the cost of distributing the products of the network) have negatively sloping supply curves.

Generally these upstream industries, where there are few firms with a high degree of specialization, have all the characteristics of strongly imperfect markets. Sectorial analyses confirm that upstream industries are dominated by strong international cartels of multinational firms, based on evident collusion. Faced by such an arrangement of suppliers, creating a monopsony for a network appears to be a suitable way of countering such a situation with the logic of bilateral monopoly; or more precisely, if we assume the international market to be the natural point of reference, bilateral oligopoly. The importance of the bargaining power that monopsonist conditions confer on a buyer, operating in at least national monopolistic conditions, of plant and machinery necessary to form the fixed capital of a network, can be crucial in a period of radical technological change. There is the choice between the dominant design and the successive one. In such a period, in fact, the producer who is forced to make massive investment in research and development, could adopt a form of price discrimination related to the bargaining power of the buyer (the price elasticity of the derived demand curve) in order to transfer a large part of the long run fixed costs onto the buyer whose price elasticity is lower. Fragmented buyers (managers of pluralistic networks) would in this way bear the suppliers' research and development costs leaving them high future profits.

3.5. Demand for network products and tariff fixing

The consumer's utility curve that include goods distributed by a network is characterized by a high degree of interdependence. The consumers of telephone and transport services clearly benefit from the growth of the total quantity 'consumed' through linking because this means that they can reach an increasing number of destinations.

In this case the utility curve of two consumers can be written as:

$$U_1 = U_1(q_{11}, q_{12}, q_{21}, q_{22}) \tag{17}$$

$$U_2 = U_2(q_{11}, q_{12}, q_{21}, q_{22}) \tag{18}$$

where $q_{11} + q_{12} = Q_1$ and $q_{21} + q_{22} = Q_2$.

When the utility curve can be expressed in this way, a fall in the quantity consumed by each agent reduces total utility of all the other agents taken together. In such conditions only when the service is objectively universal can the effects of interdependence on the consumption side be eliminated when forming the aggregate demand curve. Alternatively, it is clear that when the network is growing, the producers can extract increasing shares of the consumers' surplus which is generated by the increase in utility due to equilibrium demand increasing.

This somewhat unwelcome result is possible when it is assumed that consumers' utility can be expressed as follows:

$$U = X^{f(A)} Y^{\beta} \qquad \text{where } \beta > 0 \tag{19}$$

here $f(A)$ is a function of A (the aggregate demand) which can be assumed to be quadratic with a maximum, A^* defining the level of congestion. Obviously, the relevant part for the example being examined is not limited to the portion where $A < A^*$.

The demand equation for a network product can be derived following the accepted procedure:

$$\max \; U = X^{f(A)} Y^{\beta} \tag{20}$$

$$\text{s.c.} \quad P_A X + P_n Y = R \tag{21}$$

$$X = \frac{R}{P_A\left(1 + \dfrac{\beta}{f(A)}\right)} \tag{22}$$

with

$$\frac{\partial X}{\partial P_A} = -\frac{R}{P_A^2\left(1 + \dfrac{\beta}{f(A)}\right)} < 0$$

$$\frac{\partial X}{\partial A} = \frac{R\, f'(A)\beta}{P_A\,(f(A) + \beta)^2} > 0$$

when $f(A)$ is still in the increasing section.

With this kind of demand curve, the quantity demanded can increase with the growth of P_A as long as A grows suitably. Such a result obviously will depend on the levels of A and P_A, on the shape of $f(A)$ and on the link joining changes in P_A to changes in A; that is to say, the slope of the aggregate supply curve.

The close relationship that characterizes a network also has a significant influence on the formation of the demand curve as well as on the conditions of use. Linking up local networks, in fact, significantly affects the formation of demand. Firstly, the local network on being linked up to one or more other networks, receives demand coming from the other networks. Secondly, the demand for entry and exit from each local network forms the transmitting capacity of the network. This is all the more true if the network is hierarchic in the sense that the local networks are not directly linked with all the other local networks, but only with certain transmission points. The demand for the products between each point is, in fact, strongly affected by the demand expressed by various subordinate local networks that must pass through the transmission point so as to gain access to the general network.

If the transmission point is also a higher level local network, the coordinating and controlling function carried out at that point can create significant economies of scale and economies of scope at the same time. Frequently, a large transmission center can benefit from economies of scale only above a certain level of traffic and this level can only be guaranteed by organizing the network hierarchically and therefore through a process of aggregating demand. It is, in fact, clear that a part, sometimes a substantial part, of the demand of traffic between existing transmission points can be achieved only because the network links up the single local networks to the transmission center.

Fixing the prices and the quality at which the products of the local networks are sold has a significant influence on the formation of total demand which involves the traffic and communication between the different points. Likewise, sections of the network that carry out the function of transmission are economically viable only because they link up transmission points which receive demand from various local networks. The physical separation of the local networks from the general network can, however, have a significant effect on the position of the demand curve and therefore on the management conditions of fairly important sections of the network, which provide the transmission and control functions in the whole network. In fact, for significant parts of traffic, demand is a sequential derived demand. Values for flows of demand and consequently the design of the whole architecture of the network must take into account the interdependence of the quantity demanded; particularly in hierarchic networks

4. Conclusions

Increasing returns and indivisibilities in the production function provide the classic case for natural monopoly. When subadditiveness of the cost function, rather than economies of scale matters, the notion of networks becomes relevant. A network is defined as a set of partially separable productive units, with an aggregate subadditive cost function that reflects the important contribution of economies of scale at the plant level, technical and pecuniary externalities, superfixed production factors and economies of density, which delivers products for which demand is affected by relevant consumption externalities. In such conditions supply can be organized with a mix of competition, clubs and regulatory agencies.

As we tried to elaborate, a network industry made up of complementary elements, enjoys increasing returns for a variety of different reasons which strengthen each other:

(1) because the supply curve is the result of the aggregation of the plant's marginal cost curves which have negative derivatives;

(2) because the cost curves of two complementary productive activities are interdependent, the supply curve is found to have a negative slope even when there are no economies of scale at plant level;

(3) because the network is characterized by superfixed infrastructure that exhibit significant economies of density;

(4) because of purchasing economies of scale;

(5) because of planning economies.

At this point therefore it is possible to introduce demand externalities into the analysis. The external effects of a network integration will bring the users in each local network benefits which derive from the fact that they have access to the users in all the other local networks. Once more we see that some institutional arrangements have to be established in order to provide the necessary coordination to the different agents on the market: prices are not sufficient to convey all the relevant information. In an unregulated market significant asymmetries among operators are likely to emerge also on the demand side (David and Foray, 1994).

None of these conditions imply the notion of technical economies of scale: in a network, in fact, divisibility between the different components of the network is always possible albeit imperfectly.

If the presentation of a network as a structurally interdependent system is accepted, it is easy to demonstrate that the supply curve of a network can have a negative slope because of factors which analytically correspond to the concept of 'subadditiveness' but not to the concepts of increasing returns for the production function. In fact, the origins must be traced in the systemic analysis of the rules that shape the aggregation of the marginal cost curves of the single unit and therefore go beyond the factors that shape the supply curve of each unit. Only when the interdependence of the functional components of the network results in technological indivisibility there are technical economies. The most important novelty is that 'subadditiveness' can depend on economies of scale at the industry and firm level but not necessarily on economies of scale at the plant level.

The network analysis of telecommunications systems so far developed makes it possible to distinguish between the technical features of a network and the organizational ones.

It seems, in fact, possible to raise the question as to which is the most suitable form of organization of production in a network.

It is clear how, because of high levels of interrelatedness among agents and falling sections of the supply curve for single operators, the organization of production cannot be left to unregulated, spontaneous markets. So it is necessary to establish rules of behavior to govern relations between the productive units which make up a network and preserve the advantages generated by externalities of an integrated network so to prevent the break-up of the system. These rules can be either informal or subject of specific regulations and institutionalized membership in clubs. The alternative options are:

(1) a bureaucratic structure (a firm) which through integration is given the exclusive management of the whole network, and

(2) a set of specialized clubs for tariff setting, purchasing activity, long term planning implemented by a regulating board which limits itself to defining the rules of behavior for the individual members of the network.

It is clear that a firm will be suitable only when the costs of coordinating the bureaucratic structure are lower than the cost of the regulating agencies. On the other hand, a single firm could be favored when the single management of a network provides economies of size at a firm level which would not be possible with a regulatory agency.

Traditionally, emphasis has been laid on economies of management so that a solution of integration has been favored. However, it has been shown above that the same result can also be achieved through regulation and clubs and at first sight it would appear to be unlikely that the cost of a regulatory agency and clubs could rise to the level of coordination costs of the traditional single management. A variety of institutional forms is emerging to-day to organize the productive complementarity between the productive units which make up the telecommunications industry:

(1) The pluralistic network, i.e. a group of interdependent specialized firms coordinated by a central agency, as in the telecommunications industry in Italy until the end of 1993.

(2) The organized network such as the system of a regulated competition with a variety of firms as in the United Kingdom and in the United States.

(3) The monolithic network based upon vertical and horizontal integration as in France and Germany.

The choice between alternative organizations for networks is significantly influenced by the dynamics of technological change.

Technological change within a network raises an array of important problems. Technological change that concerns each factor separately creates significant problems for the network as a whole and therefore for the technical systems which oversee its operation. Only generic, as opposed to localized, and homogeneous technological changes that involve all the components of the system at the same time could be introduced, in purely theoretical terms, without creating coordination problems for the whole system. In general, it can be argued that technological change raises two significant problems for the network: (a) complementarity between innovations; and (b) compatibility between the new technology and the existing technology.

Adopting a single non-complementary innovation can generate localized increases in productivity and quality of service which are fairly big; but the overall efficiency of the network is compromised. Adopting a single innovation which is not compatible with the existing stock of capital imposes early replacement of a large part of that stock. Such a replacement can result in economic inefficiency when the technical obsolescence is still small. An uneven 'vintage' distribution of capital stock in an area causes some rather complex problems of differentiated advantages.

Problems of differentiated advantages arise when economic obsolescence, that is to say the cost-opportunity of postponing the replacement, is not homogeneous for every class of user. Such heterogeneity is particularly important when economic obsolescence is much greater than technical obsolescence so as to make it advantageous for some to replace equipment but not for others and then only after a long time (Antonelli, 1991).

Thus, the analytical aspects of the basic trade-off between static efficiency and dynamic efficiency in network industries come to the fore. The monolithic network will be particularly aware of the need to maintain a state of static efficiency and because of centralization will diffuse innovations faster. A pluralistic network the operation of which involves a number of agents will, on the other hand, be better able to capitalize on the opportunities offered by dynamic efficiency in generating technological innovations (Sah and Stiglitz, 1986).

When technological change is localized in that it impinges upon the specific and idiosyncratic characters of the production process in place and it results from a choice between many alternative technologies, a pluralistic network offers better possibilities for experimenting, learning and selection. Economic systems characterized by pluralistic networks will be more efficient on a dynamic level also (and above all) because they enable technological superior innovations to be improved (Metcalfe, 1989).

A monolithic network can turn out to be efficient from the dynamic point of view, yet still guaranteeing high levels of interconnection and universality when it takes the technological lead and it is able to compensate the delays in the generation of new technologies with the advantages in the diffusion of technological advances. The generation of new technologies in pluralistic networks is likely to break total demand up into smaller segments. If, as is usually the case, the cost of the innovative product is significantly affected by the gains derived from economies of scale, the new capital goods chosen by a monolithic network could be produced much more efficiently than a variety of goods adopted by a pluralistic network. Further its introduction into the network will minimize the problems of interrelatedness with the existing capital stock both because the technical obsolescence is sufficiently advanced and because the problems of technical interfacing have been overcome or have been reduced by new technical versions of the basic model.

Instead, it is evident that in a pluralistic network, run by many operators, technical changes characterized by the introduction of non-complementary and non-compatible innovation can set off reactions which can lead to the break-up of the network. The single operators of the various sections of the network will separately find it more advantageous to adopt new technologies which favor the sectors of demand with which they have a priority and privileged interaction. Yet for the same token a network with many operators may from a dynamic point of view may turn out to be more efficient overall, at least at a local level.

Such arguments lead to some general considerations:

(1) The generation of technological innovations is likely to be higher in plurastic networks because of the variety of learning and trial opportunities.

(2) A pluralistic network will encourage innovators to improve technology much more radically than is probable in a monolithic network.

(3) The process of the spread of new technology will begin much earlier in a fragmented network with many operators, while saturation would be reached first in the monolithic network.

When the technical features of the new technologies being diffused are not exogenous but are the outcome of the institutional and organizational characters of the systems by which they are generated, complementarity and compatibility both among new technologies and between old and new technologies, also can be organized (Grindley and Toker, 1994).

The definition of technical standards is, in fact, much more relevant than the number of managers in the network. Defining a technical standard means that the advantages of the static efficiency of a monolithic network can be merged with the advantages of the dynamic efficiency of the pluralistic network (Swann, 1994).

Of course defining technical standards has a direct effect on the degree of competition between firms. Hence mutual agreement on definition is rather difficult so that the intervention of a specific authority to regulate the market can be of strategic importance and, at the same time, can resolve the contradictions between pluralistic and monolithic networks (Curtis and Oniki, 1994).

Acknowledgements

The comments of Paul David, Dominique Foray and Stan Metcalfe on the preliminary versions of this paper are acknowledged as well as the support of MURST national and local funds.

Bibliography

Antonelli, C., 1991. *The Diffusion of Advanced Telecommunications in Developing Countries* (OECD, Paris).

Antonelli, C., 1992a. *Reti e regolamentazione*. Rapporto all'Autorità Garante della Concorrenza e del Mercato (Roma).

Antonelli, C. (ed.), 1992b. *The Economics of Information Networks* (North-Holland, Amsterdam).

Antonelli, C., 1993. Externalities Complementarities and Industrial Dynamics in Telecommunications. In: D. Foray and C. Freeman (eds.), *Technology and the Wealth of Nations* (Francis Pinter, London).

Bailey, E.E. (ed.), 1987. *Public Regulation: New Perspectives on Institutions and Policies* (MIT Press, Cambridge, Mass.).

Baron, D.P., 1989. Design of Regulatory Mechanism and Institutions, In: R.R. Schmalensee and R.D. Willig (eds.), *Hand-book of Industrial Organization* (North-Holland, Amsterdam).

Baumol, W.J. Panzar, J.C. and Willig, R.D., 1982. *Contestable Markets and the Theory of Industrial Structure* (Harcourt-Brace-Jovanovich, New York, N.Y.)

Bradley, S.P. and Haussman, J.A. (eds.), 1989. *Future Competition in Telecommunications* (Harvard Business School Press, Boston, Mass.).

Braeutigam, R.R., 1989. Optimal Policies for Natural Monopolies. In: R.R. Schmalensee and R.D. Willig (eds.), *Hand-book of Industrial Organization* (North-Holland, Amsterdam).

Brown, S. and Sibley, D., 1983. *The Theory of Public Utility Pricing* (Cambridge University Press, Cambridge).

Buchanan, J.M., 1965. An Economic Theory of Clubs. *Economica* 34, 1–14.

Coase, R.H., 1988. *The Firm, the Market and the Law* (University of Chicago Press, Chicago, Ill.).

Courville, L. (ed), 1983. *Economic Analysis of Telecommunications* (North-Holland, New York, N.Y.).

Crandall, R.W. and Flamm, K. (eds.), 1989. *Changing the Rules: Technological Change International Competition and Regulation in Communications* (The Brooking Institution, Washington, D.C.).

Crew, M. (ed.), 1982. *Regulatory Reform and Public Utilities* (Lexington Books, Lexington, Ky.).

Curien, N. and Gensollen, M., 1987. De la théorie des structures industrielles à l'économie des réseaux de télécommunication. *Revue Economique* 38, 521–578.

Curtis, T. and Oniki, H., 1994. Economic and Political Factors in Telecommunication Standards Setting in the U.S. and Japan: The Case of BISDN. In: G. Pogorel (ed.), *Global Telecommunications Strategies and Technological Changes*, pp. 000-000 (Elsevier Science, Amsterdam).

David, P.A., 1987. Some New Standards for the Economics of Standardization in the Information Age. In: P. Dasgupta and P. Stoneman (eds.), *Economic Policy and Technological Performance* (Cambridge University Press, Cambridge).

David, P.A. and Bunn, J.A., 1988. The Economics of Gateway Technologies and Network Evolution: Lesson from Electricity Supply History. *Information Economics and Policy* 3, 165–202.

David, P.A. and Foray, D., 1994. Percolation Structures, Markov Random Fields and the Economics of EDI Standards Diffusion. In: G. Pogorel (ed.), *Global Telecommunications Strategies and Technological Changes*, pp. 000-000 (Elsevier Science, Amsterdam).

Demsetz, H., 1968. Why Regulate Utilities. *Journal of Law and Economics* 11, 55–65.

De Palma, A. and Arnott, R., 1989/90. The Temporal Use of a Telephone Line. *Information Economics and Policy* 4, 155–174.

Evans, D.S. (ed.), 1983. *Breaking Up Bell: Essays of Industrial Organization and Regulation* (North-Holland, New York, N.Y.)

Evans, D.S. and Heckman, J.J., 1984. A Test for Subadditivity of the Cost Function with an Application to the Bell System, *American Economic Review* 74, 615–623.

Farrel, J.J. and Saloner, G., 1985. Standardization Compatibility and Innovation. *Rand Journal of Economics* 16, 70–83.

Faulhaber, G.R., 1987. *Telecommunications in Turmoil: Technology and Public Policy* (Bollinger Publishing Company, Cambridge).

Finsinger, J. and Vogelsang, I., 1981. Alternative Institutional Frameworks for Price Incentive Mechanisms, *Kyklos* 34, 388–404.

Fuss, M.A. and Waverman, L., 1981. Regulation and the Multiproduct Firm: the Case of Telecommunications in Canada. In: K. Flamm (ed.), *Studies in Public Regulation* (MIT Press, Cambridge, Mass.).

Gabel, H.L. (ed.), 1987. *Product Standardization and Competitive Strategy* (North-Holland, Amsterdam).

Grindley, P. and Toker S., Establishing Standards for Telepoint: Problems of Fragmentation and Commitment. In: G. Pogorel (ed.), *Global Telecommunications Strategies and Technological Changes*, pp. 000-000 (Elsevier Science, Amsterdam).

Henriet, D. and Volle, V., 1987. Services de télécommunication: integration technique et différenciation économique. *Revue Economique* 38, 459–474.

Joskow, P.J. and Rose, N.L., 1989. The Effects of Economic Regulation. In: R.R. Schmalensee and R.D.Willig (eds.), *Hand-book of Industrial Organization* (North-Holland, Amsterdam).

Kahn, A.E., 1988. *The Economics of Regulation Principles and Institutions* (MIT Press, Cambridge, Mass.).

Kay, J. et al. (eds.), 1986. *Privatization and Regulation. The U.K. Experience* (Clarendon Press, Oxford).

Kiss, F. and Lefebre, B., 1987. Economic Models of Telecommunications Firms, a Survey. *Revue Economique* 38, 307–374.

Loeb, M. and Magat, W.A., 1979. A Decentralised Method for Utility Regulation. *Journal of Law and Economics* 22, 58–73.

Mansell, R., 1988. Telecommunications Network-Based Services: Regulation and Market Structure in Transition. *Telecommunications Policy* 12, 243–255.

Mathios, A.D. and Rogers, R.P., 1989. The Impact of Alternative Forms of State Regulation of AT&T on Direct Dial Long Distance Telephone Rates. *Rand Journal of Economics* 20, 437–453.

McGuire, M., 1972. Private Good Clubs and Public Good Clubs: Economic Models of Group Formation. *Swedish Journal of Economics* 74, 84–99.

Metcalfe, S., 1989. Evolution and Economic Change. In: A. Silberston (ed.), *Technology and Economic Progress* (Macmillan, London).

Nadiri, M.I. and Shankerman, M.A., 1981. The Structure of Production Technological Change and the Rate of Growth of Total Factor Productivity in the U.S. Bell System. In: T.G. Cowing and R.E. Stevenson (eds.), *Productivity Measurement in Regulated Industries* (Academic Press, Cambridge).

Noam, E.N., 1983. *Telecommunications Regulations Today and Tomorrow* (Harcourt-Brace-Jovanovich, New York, N.Y.).

Noll, R., 1989. Economic Perspectives on the Politics of Regulation. In R.R. Schmalensee and R.D. Willig (eds.), *Hand-book of Industrial Organization* (North-Holland, Amsterdam).

Peltzman, S., 1976. Toward a More General Theory of Regulation. *Journal of Law and Economics* 19, 211–240.

Pfeiffer, G. and Wieland, B., 1990. *Telecommunications in Germany. An Economic Perspective* (Springer-Verlag, Berlin).

Phillips, A. (ed.), 1975. *Promoting Competition in Regulated Markets* (The Brookings Institution, Washington, D.C.).

Phillips, A., 1980. Ramsey Pricing and Sustainability with Interdependent Demand. In: B.M. Mitchell and P.R. Kleindorfer (eds.), *Regulated Industries and Public Enterprise* (Lexington Books, Lexington, Ky.).

Phillips A., 1982. The Impossibility of Competition in Telecommunications: Public Policy Gone Awry. In: M. Crew (ed.), *Regulatory Reform and Public Utilities* (Lexington Books, Lexington, Ky.).

Pool, I. (De Sola), 1984. Competition and Universal Service. In: H.M. Shoosham (ed.), *Disconnecting Bell. The Impact of the AT&T Divestiture* (Pergamon Press, New York, N.Y.).

Posner, R.A., 1975. The Social Cost of Monopoly and Regulation. *Journal of Political Economy* 83, 807–827.

Rohlfs J., 1974. A Theory of Interdipendent Demand for a Communications Service. *Bell Journal of Economics and Management Science* 5, 16–24.

Sah, R. and Stiglitz, J.E., 1986. The Architecture of Economic System: Hierarchies and Polyarchies. *American Economic Review* 76, 716–727.

Schmalensee, R.R., 1979. *The Control of Natural Monopolies* (D.C. Heath, Lexington, Ky.).

Schmalensee, R.R., 1989. Good Regulatory Regimes. *Rand Journal of Economics* 20, 417–436.

Schmalensee, R.R. and Willig, R.D. (eds.), *Hand-book of Industrial Organization* (North-Holland, Amsterdam).

Sharkey, W.W., 1982. *The Theory of Natural Monopoly* (Cambridge University Press, Cambridge).

Shepherd, W.G., 1971. The Competitive Margin in Communications. In: W.M. Capron (ed.), *Technological Change in Regulated Industries* (The Brookings Institution, Washington, D.C.).

Sherman, R., 1989. *The Regulation of Monopoly* (Cambridge University Press, Cambridge).

Singleton, L.A., 1983. *Telecommunications in the Information Age* (Ballinger, Cambridge, Mass.).

Snow, M.S., 1987. Telecommunications Literature. A Critical Review of the Economic Technological and Public Policy Issues, Telecommunications Policy Issues. *Telecommunications Policy* 12, 153–183.

Spulber, D.F., 1989. *Regulation and Markets* (MIT Press, Cambridge, Mass.).

Stigler, G.J., 1971. The Economic Theory of Regulation. *Bell Journal of Economics* 2, 3–21.

Swann, P., 1994. Reaching Compromise in Standards Setting Institutions. In: G. Pogorel (ed.), *Global Telecommunications Strategies and Technological Changes*, pp. 000-000 (Elsevier Science, Amsterdam).

Taylor, L.D., 1980. *Telecommunication Demand. A Survey and Critique* (Ballinger, Cambridge, Mass.).

Tullock, G., 1956. Public Decisions as Public Goods. *Journal of Political Economy* 64, 414–424.

Waverman, L., 1989. U.S. Interexchange Competition. In: R.W. Crandall and K. Flamm (eds.), *Changing the Rules: Technological Change International Competition and Regulation in Communications* (The Brooking Institution, Washington, D.C.).

Wenders, J.T., 1987. *The Economics of Telecommunications: Theory and Policy* (Ballinger, Cambridge, Mass.).

Williamson, O.E., 1966. Peak-Load Pricing and Optimal Capacity under Indivisibility Constraint, *American Economic Review* 56, 810–827.

Zajac, E.E., 1990. Technological Winds of Creation and Destruction. In: A. Heertje and M. Perlman (eds.) *Evolving Technology and Market Structure. Studies in Schumpeterian Economics* (Michigan University Press, Ann Arbor, Mich.).

Global Telecommunications Strategies and Technological Changes
Edited by G. Pogorel

Chapter 7

Percolation structures, Markov random fields and the economics of EDI standards diffusion

Paul A. David [1] and Dominique Foray [2]

[1] *Visiting Fellow, All Souls College, Oxford, and Stanford University, Stanford, CA, USA*
[2] *Ecole Centrale, CNRS, Paris, France*

1. *Introduction and overview*

Convergence between telecommunications and computing technologies requires a viable means of automating the transfer of information between organizations. The immediate aim of the collection of hardware and software technologies and the implementation techniques associated with Electronic Data Interchange (EDI) is to enable the electronic transmission of intelligible business documents — such as electronic funds transfers (EFT) — among trading partners in the 'seamless', user-initiated way that telephoning is done today. More profoundly, EDI is viewed by many information industry observers and business management experts as the most promising method for catalysing the fundamental changes in business practices which will be necessary to realize the full benefits of inter-organizational information transfer (see, e.g., Kimberley, 1991).

The economic consequences for firms that become effectively linked in this way by high-speed data networks, and for the countries whose workers they employ and whose customers they serve, are likely to be very far-reaching. The intra-organizational effects of changes in the use of information made possible by advances in communication and control technologies are only now starting to be appreciated fully and studied systematically. [1] An eventual thor-

[1] See, e.g. Beniger (1986) and, for techniques introduced early in the twentieth century more specifically, Yates (1989).

ough transformation of firms' internal organizational structures and modes of business is a likely concomitant of the formation of extensive linkages among corporate data networks. It is to be expected, at least on the basis of analogous historical experience with network technologies, that only through such long-run intra-organizational restructuring will the full productivity-enhancing potential of the 'information revolution' eventually be reaped (see David, 1991).[2] While productivity-enhancing technological and organizational changes have occurred within the firm, improvements of information-linkages among organizations have come much more slowly. As a result, the inter-organizational link has become the bottleneck slowing innovation in business practices and productivity advance within the organization by perpetuating the dependence upon inefficient translator and bridge functions, whose size has had to expand as internal information-generating and processing capabilities have grown.

In this respect, the establishment of *inter*-organizational networks for electronic data interchange seems to promise great opportunities while posing serious challenges.[3] Inter-organizational networks, like other 'network technologies' require standards that assure the compatibility and inter-operability of the network's components.[4] Special problems arise in assuring inter-operability when the autonomy and scope for extra-network operations on the part of components or 'nodes' has to be preserved. Still another group of complications present themselves in the case of EDI document standards, which are supposed to govern the terms, forms, and procedures used in both inter-firm and intra-firm transactions, thereby permitting the data processing apparatus of distinct business entities to interact with one another directly. Standards for EDI documents are simply a means of reducing ambiguities and diminishing translation costs entailed in such transactions.

In the section of the paper following this one, we address the nature of the special obstacles to EDI document standardization by examining the current situation with regard to the adoption of various standards, focusing especially upon the recent European experience of EDIFACT (the United Nations document standard). Because EDI deals with the transmission of intent through expression, we propose to draw an analogy between the economics of EDI and the economics of language standardization. Further, in the subsequent sections, we seek to identify analytically the general conditions that could result in

---

[2] A concise presentation of the conclusions of this study is given in David (1990).

[3] Furthermore, from the little that is known about their workings once effectuated, they appear capable of occasioning complex and unexpected developments — which may be so problematic for the autonomous functioning of the firms involved as to cause a withdrawal from full inter-organizational integration of information flows. For a striking instance, see Hart and Estrin (1990).

[4] On the general topic of compatibility standardization, see David and Greenstein (1990).

the *de facto* emergence of a single, universal EDI document standard through a decentralized, diffusion-like or, more accurately, a 'percolation' process.

Decisions to implement EDI with specific document standards are likely to spread in a population of firms through positive feedback effects from an initial core of users; direct transactions cost savings and indirect benefits are realized by marginal (potential) adopters as positive externalities generated by the intra-marginal group of users. Analysis of the class of so-called 'Polya urn models' that have been introduced to economists as a vehicle for studying competitive technology diffusion under conditions of positive feedbacks (see Arthur, 1989, 1990; David, 1985, 1988), leads to the conclusion — in section 3 — that the mere presence of potential positive network externalities is not sufficient to induce the emergence of a unique, global standard. Reference to those probabilistic models, however, suggests little more than that the source of the problem lies in the positive feedbacks not being 'strong enough'; it is also pointed out that they do not adequately correspond to certain key structural features of the EDI document standards situation.

For those reasons we turn, in section 4, to another class of stochastic models which appear more suitable for the analysis of the behavior of economic agents embedded in local networks of transactions. Positive network externalities can be limited, in such models, to subsets of the entire potential user community, a situation that is readily described using the terminology of Markov random field theory. Drawing specifically upon insights from the related and comparatively new field in stochastic process analysis referred to as 'percolation theory', we are led to account for the observed co-existence of multiple EDI document standards as a manifestation of the 'sub-criticality' of the inter-organizational transactions system. A particular standard fails to 'percolate' throughout when the connection probabilities between typical nodes in the system remain below a critical level.

It is suggested that the pre-existence of various finite 'clusters', or sub-groups of firms among whom there is a history of close commercial relations, will greatly influence the design and diffusion patterns of documents standards by multiplying the occurrences of idiosyncratic forms (like natural languages and local dialects thereof), and leading, at least at the outset, to even further fragmentation or 'balkanization' of the transactions system. The sub-criticality of the system as whole, thus, is likely to be the result of the historical process that created antecedent routine of coordination; EDI networks, as a practical matter, must be implemented so as to conform with local coordination structures that have arisen historically from firm and sectoral specificities, geographical neighborhood, and industry conventions in describing the highest frequency forms of transactions.

In the concluding section (5) of the paper some basic analytical results in percolation theory are drawn upon for insights as to the nature of policies that would be most efficacious in removing the condition of sub-criticality and fostering a decentralized process leading to more universal standardization. But given that such policies would be intrusive in the internal organization and business operations of firms, resistance to that approach points in the direction of a strategy of promoting the spread of EDI by developing EDIFACT so that it can 'supplant' local document standards on efficiency grounds. In other words, like a universal language, EDIFACT is an 'a posteriori language' which has to face the 'competition' of pre-existing, locally entrenched linguistic standards. Here the ambiguous goals of a so-called 'universal' standard become evident. With respect to the existence of multiple idiosyncratic standards, EDIFACT may be considered as a 'meta-standard' (a standard for setting standards) which does not constitute a pure substitute designed to supplant the local standards in all substructures, but rather as a broad framework for integrating these local disparate forms. According to our assessment of the European situation, the main mission of EDIFACT should not be that of displacing the local standards, but rather one of assisting their absorption within a unified framework. Increasing the flexibility of the language and tolerating a sacrifice of aspirations to rigorous universality may therefore prove to be the most effective long-run strategy. However, due to the persisting ambiguity that surrounds the goals of those charged with developing the standard — are they creating a substitute for local standards, or a meta-standard? — EDIFACT policy continues to oscillate between conflicting design criteria. This situation, in and of itself, should be seen to be working against the formation of greater momentum for EDIFACT's adoption and the diffusion of EDI.

2. *EDI standards — the current situation in the world market*

According to Payne and Anderson (1991, p. 1), "EDI is a technique by means of which formatted, transactional information is moved electronically from one organization's computer to another's". By transactions the authors mean the great variety of information exchanges related to conducting business between two autonomous organizations. On this view, EDI cannot be seen as a simple extension of long-established internal business system to a firm's suppliers or customers. Indeed, the challenge of EDI lies precisely in the autonomy of the two parties exchanging transactions electronically: allowance must be made for the fact that new, and in some instances unprecedented transactions will be involved (e.g. a firm might not actually buy and sell internally); formats for communications between novel transactions must be agreed upon, contractual agreements must be changed and new rules made.

Why should one go to the trouble of providing electronic implementation for all such transactions? What logistical problems are likely to be solved by EDI? According to different surveys, users can expect to enjoy benefits from transaction costs savings, joint synchronization of production, quality control information and bypass of information transformations within the organization (Commission for the European Community, 1989).[5] It is clear, however, that the actual impact of EDI as a technology of data transmission depends upon the ability of firms and other trading partners to 'forget' paper methods, i.e. to use EDI in specific ways which do not simply reproduce paper and clerical procedures for the exchange of information. In particular, EDI should not involve the mere electronic transcription of existing paper based procedures, but should lead to a substantial rationalization of the information flow. For example, EDI can support the 'progressive transfer' of information.[6]

2.1. The need for standards

Use of standard formats is especially vital for firms that seek to exchange transactions with most if not all of their trading partners in a consistent manner. The standard considered here involves the terms, forms and procedures for business — i.e., business documents and conventions. Of course, the need for standard business forms is not new in business and industrial practices: early in the 1920s great efforts were undertaken to provide a standardized invoice form,[7] as well as standard forms for governmental purchase procedures.[8] The present-day effort involves the selection of a unique standard

---

[5] See also the survey of Pfeiffer (1990). Payne and Anderson (1991) describe the five key target area for EDI to enhance logistics: (1) shorten procurement administrative lead time; (2) broaden and hasten access to the industrial base; (3) allow for tighter and more dynamic control over vendor performance and actions of all sorts; (4) provide short-term, accurate 'heads up' to logistics pipeline actors; and (5) enable better responses to unpredictable surges in demands for goods or services. Illustrative anecdotes reported in the American trade press during 1989–1991 suggest substantial savings in costs and time: at a Digital Equipment Corporation manufacturing plant the cost of processing purchase orders was cut from $125 to $32, and delivery time was shortened from 5 weeks to 3 days; Navistar International is said to have reduced inventory stock levels from a 33 day supply to a 6 day supply (a cut of $167 million in their value of inventories) within the eighteen months following the implementation of EDI. See Trauth and Thomas (1992).

[6] An example of 'progressive transfer' of information might concern booking informations sent electronically in several phases.

[7] According to Chandler (1928), "Formerly, because of the great need for standard invoices, many large corporations designed their own individual private forms and required all bills rendered against them to be on these forms which they provided". On the early development of business forms as part of the apparatus of managerial control in the U.S., see Yates (1989).

[8] See Erck (1928): "Up to February, 1927, such Standard Government forms as Invitation for Bids; Form of Bid, Instructions to Bidders; Construction Contract; Bid Bond, had been adopted and made effective".

(or a very small number of variants) in order to reduce conversion costs. Providing for conversion between different EDI formats as well as between external and internal data interchange systems is likely to become very costly where the existence of users with many differing standards must be allowed for. [9] According to the European Community's 1989 TEDIS report on EDI use,

> "This is strong evidence for what is now becoming a generally accepted situation — that until the common international standard is sufficiently well defined users must adopt more than one standard to have effective EDI communication with a range of trading partners."

Document standards constitute a piece of 'network technology' and, accordingly, their users enjoy network integration benefits: the performance of the standard as well as its utility increases with the growth of the community of users. [10] Therefore, the individual choice of a given standard will generate positive feedbacks as the choice of a standard by some users raises the probability that other firms will adopt the same standard in the future. Such (positive feedback) effects in reinforcing the adoption of a standard supporting inter-organizational transactions can be expected to stem from the growth of the volume of external transactions carried on by a business enterprise that has introduced techniques which enhance its efficiency as a vendor, or purchaser, or in both activities.

In the presence of sufficiently strong network externalities there will be a tendency for market pressures on the decentralized technology choices made by firms to result in the emergence of a single, *de facto* standard (see Arthur, 1989; David, 1987). [11] Under certain conditions this would represent a socially better outcome than the persistence of different and incompatible standards. [12] In other words, social benefits derived from a single standard which dominates the market tend to exceed the sum of the benefits derived

---

[9] In fact, the most important EDI barrier is the need to change intra-organizational computer applications so as to fully harness the advantages offered by EDI in linking applications directly, rather than sending and receiving messages in door-to-door mode through the use of simple electronic mailboxes. See Pfeiffer (1991). On the economics of 'converters' and 'gateways' between otherwise incompatible systems, see David and Bunn (1988), and further references in David and Greenstein (1990).

[10] On network technologies and network integration benefits, specifically, see David (1987).

[11] For references to the rapid expanding literature on the implications of positive network externalities, see David and Greenstein (1990), Greenstein (1993).

[12] These are conditions in which chosers do not have ex ante preferences regarding the identity of the standard that would emerge as the collective choice, other than the preferences that they maximise at the time of their selection, and, further, conditions concerning the discount rate and the possibility of experimentation with the various alternatives. Losses in network externalities otherwise obtainable by eliminating diversity, may be traded off against the benefits

by the users of several incompatible standards that co-exist in the market — unless the dominant standard is technically so inefficient compared to one of the alternatives that the population would be better off living with diversity. In the latter case they would either forgo some network externalities, or bear the ex post conversion costs and possible performance degradation involved in resorting to 'translators', 'converters', 'gateways', and the like. There is, thus, a general presumption among economic analysts that "the benefits of EDI would increase if the many standards could converge" (Payne and Anderson, 1991, p. 12). But, if the 'benefits' are to be understood as referring to the net benefits (producer and consumer surpluses) involved, this conclusion must presuppose that the fixed costs of designing, testing and maintaining a universal document standard are either not substantial or are not directly and indirectly borne by the users; it assumes, further, that the 'switching costs' entailed in migrating to the common standard would not be significant for existing EDI users. Both assumptions must be evaluated against the realities of the present situation.

2.2. A brief description of the EDI standards scene

Many document standards — with varying pretensions to universality — currently are competing in the EDI arena. Although there is a tendency on the part of EDI enthusiasts and proponents of a universal document standard to speak as if the curtain were just about to rise on the tableau of ubiquitous high-speed data interchanges, it is important to bear in mind that this is a drama already well along into its second act.

Electronic transmission of digital information was initiated as early as the 1960s in Europe and the United States.[13] A number of systems were created, linking branches of the same company, and counterparts within the same industry (e.g. banks, airlines) or small groups of operational partners (e.g. customs offices with freight forwarders; customs with airlines; container operators with sea carriers, etc.). In numerous industries where there were substantial numbers of firms, competitors recognized that there were advantages in 'clubbing together', i.e., cooperating in the development and operation of

---

from concurrent experimentation with a collection of alternatives, in order eventually to select the most efficient design (assuming the conditions prevailing during the experimental phase are expected to persist) as the one on which to standardize. On this relationship between diversity and standardization, see David and Rothwell (1993).

[13] The earliest use of a primitive EDI system is said to date from 1948, when it was employed during the Berlin airlift to track cargo on the many flights into the blockaded city (Trauth and Thomas, 1992, p. 7). Material for the historical sketch given here is drawn largely from Kimberly (1991), Payne and Anderson (1991), Pfeiffer (1990).

a specialized data network, because there was a high degree of similarity in the nature of the transactions in which they engaged, and the fixed costs of creating a network and its messaging standards could be distributed among them without affecting their competitive positions. Within the computerized 'islands' corresponding to those communities of interest, interchange partners who already shared a history of frequent routinized transactions found themselves readily able to agree on a common format for the codes to be used, the message structure, the technique for error correction, and so forth.

Protypical of this situation, the road transportation industry pioneered the use of EDI among trading partners in the U.S. by forming the Transportation Data Coordinating Committee (TDCC) in 1968 to develop a common set of formats for electronically disseminating transportation documents. The first set of TDCC standards was published in 1975 and established a far-reaching precedent. When the first movement to bring conformity to EDI standards got under way, in 1979, with the establishment of the American National Standards Institute (ANSI) X.12 Committee on Electronic Data Interchange, the TDCC standards provided the basis for the cross-industry document standards for order placement and processing, shipping and receiving, invoicing, payment, and cash applications that ultimately were adopted by the X.12 Committee. In turn, these 'open standards' of ANSI X.12 subsequently became the guidelines adopted by many industry and sector-specific EDI networks in the U.S., and imposed upon vendors by federal government procurement agencies.

Another early transportation-related field of EDI application, this one involving transactions with a public agency that were common to the international transport business, was the transmission of customs documents. This saw the development during the 1960s of the London Airport Cargo Export Scheme (LACES) by the British Customs Service, the Système d'ordinateurs pour le fret international (SOFI) of the French Customs, and the US Customs Automatic Commercial Systems (ACS). The most extensive international network in the transport sector, however, is SITA the airlines' worldwide telecommunications and information services, which serves 320 airlines in 173 countries.

In the banking sector, to take a further example from the international sphere, the Society for Worldwide Interbank Financial Telecommunication (SWIFT) was created in 1973 to develop an international network for settlements management that could overcome the proliferation of incompatible national systems.[14] SWIFT has since become a highly reliable and standardized

[14] For a case study of the economics of SWIFT, viewed as a value added service (VAS) network, see Dang-Nguyen (1990, pp. 6–9).

mechanism for electronic interbank exchanges of transactional information over a dedicated network (since 1977). In addition to providing an international inter-organizational network — which entails providing (proprietary) protocol between its center and the member bank, whereby SWIFT controls access, maintains the conformity of messages to standards, guarantees delivery, and provides for audit trails, etc. — SWIFT is an interbank messaging standard. In fact, its messages standards now are in use throughout the banking community even for data interchanges that are not implemented through the SWIFT network.

Although an enormous array of diverse and incompatible proprietary and industry-specific EDI document standards currently are in use, a much smaller number of major alternatives appear to be evolving within the main markets and organizational contexts. In North America, as has been noted, most companies have directly or indirectly adopted standards conforming to the ANSI X.12 format. By contrast, the majority of British users are employing TRADACOMS (a standard developed for the retail trade), or some kind of proprietary standard.

As a result of an early governmental commitment to promoting the sectoral adoption of EDI in British industry, beginning in the late 1960s, the United Kingdom now has the legacy of a plethora of industry-oriented national document standards organizations. Unlike the many industry-specific standards based on ANSI X.12, however, they do not conform with any set of generic, uniform cross-industry specifications (see, United Nations Economic Commissions for Europe, Trade Facilitation, 1988).[15] In the same pluralistic

---

[15] The following are the main nationally based EDI projects in the U.K. today:

ANA TRADACOM — A U.K. Article Number Association initiative, which pioneered the TRADACOMS standards for EDI documents, this standard was initially designed for the retail sector, and became the standard for use over the TRADANET service of I.N.S.;

BACS — Banker's Automated Clearing Services Limited provides a national batch electronic funds transfer (EFT) service for automated payment of EFT transactions. It is owned by the British Clearing Banks and two building societies;

BEDIS — The Booktrade Electronic Data Interchange Standards Committee was set up by the Machine Readable Cataloguing User's Group to further promote common message standards in the book trade where several standards already existed;

DISH — Data Interchange for Shipping is a pilot project set up with four exporters, five shipping lines, a freight forwarder and I.N.S., to enable the exchange of data electronically between importers/exporters, shipping lines, transport operators and freight forwarders;

EDIA — The EDI Association covers most activities and participates in the shipment of goods including sea, air and surface transport, banking and financial services including insurance, customs and government;

EDICON — EDI in Construction Ltd is an initiative for developing the use of EDI in the construction industry;

SHIPNET — A pilot project set up by companies in the freight industry in collaboration with

vein, the British motor car companies played a leading role in creating the ODETTE network, the outcome of a project launched in 1984, which has essentially set EDI document standards for the automotive industry in Europe. All the major European vehicle manufacturers and their main suppliers are now represented in ODETTE. The firms in the British automotive industry came to this cooperative approach only after having developed several incompatible EDI systems (which have not altogether disappeared), including FORDNET (an in-house system from Ford), the ISTEL system (originally designed for the Rover Group), and the SMMT system — a joint venture of I.C.L. (U.K.), and General Electric (U.S.A.) — run on the I.N.S. Because international or European standard messages were not available in the initial phase of the ODETTE project, high priority was placed and continues to be given to the development of the industry's own commercial message set (invoice, delivery, instruction, etc.). [16]

On the Continent the situation is that EDI users tend either to be implementing one of the emerging European-wide industry-specific standards, such as ODETTE, and RINET, a computer-based transactions system for the insurance and reinsurance industry (Oliver, 1990; Dang-Nguyen, 1990, pp. 9–11), or they are considering adopting the United Nations document standard known as EDIFACT.

2.3. EDIFACT versus ANSI X.12

For more than 10 years, the UN/Working Party in Facilitation of International Trade Procedures (WP4) has been developing essential standards covering data elements, codes and syntax rules for EDI. WP4 is a sub-group of the United Nations Economic Commission for Europe, whose membership includes countries and organizations from North America, Western Europe and Eastern Europe. [17] As early as 1975, aware of the interest in the matter and the urgency of a solution, the WP4 set up two ad hoc task teams. One was tasked with 'establishing the standard data elements required in international trade and to identify those which need coded representation', the other with establishing 'a set of standards for data exchange between

IBM. It was started at around the same time as DISH but differed in that it was mainly for smaller/medium freight forwarders.

[16] The ODETTE standard's syntax and general directories are said to be kept 'in line' with international developments, in order to maintain the opportunity of being able to cross industry boundaries.

[17] This development is based on information presented in the document *Introduction to the United Nations Rules for EDI* (1988). Personal communication from Mr. Bellego (U.N. E.C.E., Geneva, May 1991) was also useful.

international trade partners over data communication links, and for computer exchange using various media'. During a certain period, however, owing to the limited amount of experience gained in electronic data interchange on an international basis and the different requirements and capabilities of the various national entities taking an interest in the subject, the co-existence of a number of trade data interchange protocols (each corresponding to particular user's needs) had to be accepted while efforts towards a single standard continued within WP4.

A first set of interchange rules was developed and published in 1981 in the form of 'Guidelines for Trade Data Interchange' (GTDI) which offered potential users a basis for developing their systems. Later on, in the light of experience gained by the many users of GTDI and the various projects under way, the Guidelines were enhanced with a view to permitting their wider acceptance. The UN/ECE generic data directory was transmitted to ISO, to be processed as an international standard and published as ISO 7372.[18]

The final stage in the work towards a common universal set of interchange rules for trade data was a reconciliation, carried out by a joint European/North American ad hoc Group commissioned by the WP4, to bring together the enhanced GTDI and a set of standards for Electronic Business Data Interchange developed in the United States (mainly based on standards developed by ANSI ASC X.12).[19] The recommendations of the ad hoc Group were agreed by the WP4 in September 1986. They led to the development of the UN/EDIFACT syntax rules, which were subsequently transmitted to the ISO to be processed as an international standard. In July 1988, the UN/EDIFACT syntax rules were published as ISO (9735).

According to Payne and Anderson (1991, p. 16), the current situation now seems to favor the emergence of EDIFACT as the de facto international standard. Two years ago, those involved in EDIFACT made up a clear minority in the worldwide community, but, today the picture is very different, as many countries and industries have boarded the EDIFACT bandwagon and it is gaining speed. In spite of this long-run tendency, the regional and sectoral situations remain very diverse. It is true that continental Europe is firmly behind EDIFACT, which is expected to continue to grow in importance and gradually replace other standards. The situation in the UK is different — as has been noted already — in that national standards are deeply entrenched there, giving rise to the expectation that a universal migration to EDIFACT will

---

[18] The name of the standard — UNTDED — stands for United Nations Trade Data Element Directory.

[19] For our present purposes, the reader at this point needs only understand that ANSI X.12 and EDIFACT differ in syntax as well as data elements — i.e., in their structure and how units of information are defined.

take much longer, if indeed it is to happen at all. But the future of the competition depends critically upon the North American choice of a document standard, which still clearly favors ANSI X.12. Indeed, today the majority of the world's EDI users are located in the U.S., and ANSI X.12 is probably the most widely-used public, non-industry standard in the world. Roughly 5,000 to 7,000 U.S. firms now use EDI, but, according to Payne and Anderson (1991: p. 21), very few of them use EDIFACT standards, know much about them, or are interested in them. There are three reasons for this:

(1) Roughly a third of U.S. firms using EDI report using proprietary standards, neither X.12, EDIFACT, nor industry standards.

(2) Most U.S. firms using EDI begin with domestic trading partners, and, understandably, there has been little call such from trading partners to use EDIFACT.

(3) There is a clear differentiation in the supply of standardized documents, caused by the earlier start of ANSI X.12 on the market: twenty-six X.12 transaction sets are available against two EDIFACT messages.

Institutionally, however, the U.S. already is an important player within the EDIFACT standards process: for example, the U.S. Customs Agency has strongly endorsed EDIFACT.[20] In 1988, the North American EDIFACT Board was created and its members, although still few in number, do include major U.S. firms.

The situation is thus today very fluid and the eventual outcome highly uncertain.

> "EDI transaction formats will take years to mature and become adopted as a cross-industry business standard. Even though ANSI X.12 appears to be the most widely used standard, within the U.S., EDIFACT following a different standard process and syntax, is attracting users and attention" [Payne and Anderson, 1991, p. 18].

Will uniformity increase? If so, what competing alternatives will eventually dominate the market? Given that in 1989 ODETTE and TRADACOMS committed themselves to migrating to EDIFACT in Europe, the following analysis will focus on the competition between two 'global' standardization movements: X.12 and EDIFACT. But, instead of conceptualizing this as a competition for adherents *de novo*, it is essential to recognize the pre-existent conditions which have resulted in the formation of 'local' networks (industry- and sector- and region-based) among EDI users.

---

[20] Encouraged by the U.S. Customs Service, a new Federal Information Processing Standard mandates federal agencies use either X12 or EDIFACT.

2.4. The analogy with language standardization

Inasmuch as EDI involves the transmission of intent through expression, it is tempting to draw an analogy between EDI standardization and language standardization. The latter phenomenon, it happens, was the subject of much interest and analytical inquiry among linguists some three decades ago.[21] As with language standardization, the intensity of interactions between agents is a major influence on the design of a document standard.

Each subset of interactions — or, transactions, as we might say, in the present context — forms a 'reference group' whose boundaries might be defined according to whether or not the frequency of interactions exceed some critical value. Within such reference groups, it is presumed, agents are likely to be very interested in the design of a common language (document), in order that they may gain benefits from easy access to local communications networks (local positive network externalities). It is useful here to quote Leonard Bloomfield, the author of one of the most important modern treatises on language (Bloomfield, 1984, p. 10):

> "The most important differences of speech within a community are due to differences in density of communication... We believe that the differences of communication with a speech-community are not only personal and individual, but that the community is divided into various systems of sub-groups such that persons within a sub-group speak much more to each other than to persons outside their sub-group. Viewing the system of arrows (between any speaker and each one of his hearers) as a network, we may say that these sub-groups are separated by lines of weakness in this net of oral communication. The lines of weakness and, accordingly, the differences of speech within a speech-community are local — due to mere geographic separation — and non-local, or as we usually say, social".

Obviously, the intensity of interactions among the economic actors (business firms) with whom we are concerned is the result of the historical process that created routines of coordination. Thus, long before the emergence of EDI, certain forms of transaction standardization had occurred that contributed to forming and reinforcing 'reference groups': local standards for paper interchange, local conventions and agreements regarding tests, measures and quality control, sectoral just-in-time practices, etc., are excellent illustrations of such historical processes of standardization in the field of business practices coordination.

---

[21] For a general treatment, see Marschak (1964). Other noteworthy contributions deal with the questions of language standardization and language planning: Haugen (1966), Ray (1963), Rubin and Jernudd (1971). The question of preservation of minority languages is explored in economic terms by some 'économistes québecquois': see, in particular, the most recent paper by Grin (1992).

Physical implementation of EDI networks could be expected to occur first among the members of high transaction-density reference groups, and the line of least resistance would be for document formats to conform closely with pre-existing local coordination structures — whence they would reflect historical firm and sectoral specificities, geographical neighborhood and industry traditions.

The system of electronically supported inter-organizational network use thus displays the property of 'locally bounded positive feedbacks': as with language standardization, micro-decisions about the adoption of protocols, message formats, and so forth, are influenced by positive feedbacks from local structures. When these local structures are not coupled, but are separated by 'lines of weakness' in the communications network, and further isolated by the consequent reinforcement of distinctive, idiosyncratic business practices, it is not surprising that multiple document standards can become entrenched. In the case of EDI format choices, firms can be expected to formulate decisions by taking into account the sole reference group whose members' transactional requirements and practices are perceived to be those most relevant for maximizing the economic returns on the entailed investment in the implementation of a document standard. The uncertain future is likely to be discounted heavily in such decisions, especially when they must be justified on the grounds that some cost-savings will be quickly forthcoming; perceptions as to the nature of the transactions that it will pay most to expedite via EDI are likely to be colored most strongly by the firm's own recent transactions history, and not by consideration of the universe of conceivable future transactions, involving agents of types and identities with whom there have been few if any previous dealings. Thus, the design and diffusion of a language standard of the sort with which we are here concerned, no less than that of natural language conventions, are processes whose spontaneous occurrence tends to be confined within structured (rather than randomly selected) sub-groups of the potential population of adopters.

3. Properties of the diffusion of EDI document standards: analytical preliminaries

The question that now arises is the choice of a suitable analytical framework, or modelling approach for studying the competitive diffusion of the two main EDI document standards, ANSI X.12 and UNEDIFACT. It should be quite evident from the foregoing discussion that the proximate subject of the proposed analysis is not the functioning of any particular EDI network through which messages and data can pass among the participating business entities.

Rather, we seek to characterize the dynamic process whereby such a network can be implemented; whereby firms independently will come to adopt a common set of EDI document standards, as a consequence of decisions shaped by the particular configuration of their respective prior transactions patterns. Those transactional relationships may be conceptualized as connecting groups of firms in 'network-like' sub-systems (see Antonelli, 1991, pp. 5ff. and Mansell, 1991, pp. 217ff.). But, at the same time, it is important to keep clear the distinction between the metaphoric 'network' of business relationships among firms, and the physical, electronically implemented networks supporting EDI. Adoption of document standards (and the physical telecommunications infrastructure and protocols) permitting electronic implementation can lead to enhancements of the degree of *intelligibility of the relational network*, or transactions system — by enabling the firms involved to send and receive messages and data among themselves in a consistent, more efficient form.

Since the primary interest here lies with the way in which an economic transactions system attains the minimum level of intelligibility represented by the establishment of an EDI network, our attention has been directed to the configuration of the transactional connections established among firms, viewing such 'relational networks' as structures within which acceptance of a standard will spread. The initial configuration, and the particular dynamics of these structures will determine whether spontaneous diffusion of a single, universal standard is possible and likely; or, in the case of breaks in the structure of connectivities, a multiplicity of local standards may co-exist in the absence of more concerted, hierarchically directed policies of standardization.

It will in all likelihood already have occurred to readers — especially to those acquainted with recent development in the microeconomic analysis of competition among alternative technologies under conditions of increasing returns to adoption — that the problem in hand lends itself to the application of a particular class of dynamic resource allocation theories, that is, models of stochastic systems that are characterized by positive feedbacks and possess a multiplicity of locally stable equilibria, or 'absorbing states'. Among these, one well known model is the generalized Polya urn scheme, which has been introduced to economists recently by Arthur (1989, 1990). Formal analysis of this model derives theorems about the existence and properties of the limiting distribution of the contents of an urn initially containing balls of different colors, when the urn is randomly sampled repeatedly with 'over-replacement' (of balls of the same color as the one drawn) in each round. The principal property of this model is that when the degree of over-replacement is sufficiently strong, the contents of the urn will converge in the limit to a population of balls that is uniform in color, with one of the colors initially present having been 'selected' by the accidents of the sequence of random draws to emerge as the limiting uniform type (i.e., the universal standard) for the urn. This implication, and collateral propositions regarding the possibility of sub-optimal outcomes, has been quite illuminating in suggesting how de facto standardization occurs in some contexts. Some cases in point are the emergence of QWERTY as the standard for typewriter keyboards, the victory of the VHS format over the Sony Betamax format in the market for VCRs, and the dominance of the light water reactor technology in the population of nuclear power plants of the world's electric utilities (see David, 1986, chapter 4; Cusamano et al., 1990; Cowan, 1990). But, as will be seen shortly, the generalized Polya urn scheme is not so suitable for the case of EDI document standardization, or other de facto standardization processes that share its principal structural features.[22]

3.1. EDI features and the Polya urn scheme

The essential dynamic described in the Polya urn scheme is driven by the continuous growth of a population, the members of which have to make their choices sequentially in an order that is random with respect to their *inherent* preferences among the set of available alternatives. By contrast, our purpose is to study the properties of a sequence of choices within a *finite* population, i.e. some initial configuration of (actual and potential) adopters, which allows one to know the distribution of the agent's initial choice is observable. Indeed, the technico-organizational decision of a firm in this area can be viewed as a sequential process consisting of (1) the decision to implement an EDI system, and (2) the selection of a standard. Thus, our study will focus on the population of firms already committed in terms of EDI infrastructures but having to make (or reassess) a choice between (the two) alternative standards. We can use the concept of 'collective territories' from Thomas Schelling (1971) in order to include within the finite population the *potential* users who can be unambiguously assigned to a particular 'standard territory', i.e. an inter-organizational network where one and only one EDI document standard is actually being used. Thus, the concept of collective territory allows the finite population to grow in some predictable proportions among the different variants of the standard (see below).

As we wish to characterize the evolution of the share held by each variant standard within a finite adopter population, we must focus on the revision and reassessment policies of the adopters. For this purpose, a second property of the Polya urn scheme makes it poorly suited as an analytical vehicle: each agent decides only once, and irreversibly, at the time she enters the popu-

[22] An extensive discussion of this topic, and comparison of the properties of the Polya urn process with the random Markov field model described below, is provided in David (1992).

lation. Hence, the share of each variant in the total installed base can only change if there are additions made to the stock. Although this assumption (of infinite durability) might be suitable in case studies dealing with long-lived physical infrastructures, it is out of place where the firms plainly are able to alter their choice of a document standard with the passage of time (at some cost, of course). The present phase of EDI standardization belongs exactly in this class of situations: despite the existence of switching costs (Greenstein, 1991), revisions of choices are very popular in the EDI world. However, let us be clear that while the choice of a standard may be subject to revision, the firm's commitment to EDI as a technology of inter-organizational network is treated here as completely irreversible.

From both EDI features described above, we can infer that the essential dynamic of the EDI standard competition is driven by *the process of revision of choices within a finite population*. Clearly, the properties of the Polya urn scheme do not accommodate these features. A third feature further impairs the Polya urn scheme's ability to capture the essential dynamic of the EDI document standards competition: in applying the Polya process to represent the outcome of a sequence of decentralized decisions involved under conditions of positive externalities, it is assumed that micro-decisions are directly influenced by positive feedbacks from the *macrostate* of the system. Thus, the probability of adding a ball of one color rather than another to an urn at each moment is assumed to improve as a function of the *global* proportion of the balls in the urn that currently are of that color. There most certainly are real world situations that fit reasonably easily into this framework. For example, Cowan (1990) has made a persuasive case for the view that, among the agencies responsible for the selection of the designs for nuclear reactors to be installed in electric power plants in the industrial nations of the West, from the late 1960s onwards, considerations of the distribution of global experience with the construction and operation of various reactor types proved to be paramount over local conditions; what reactor technology might have been adopted in the neighboring province or country carried comparatively little weight.[23]

3.2. The locally bounded nature of network integration benefits

There are many other technological circumstances involving networks, and the case of EDI standards is one such, where specifying the dominance of feedbacks from the macrostate in micro-decisions-making is not appropriate.

---

[23] But such is not the case universally, by any means. It is recognized in the industry that any particular utility has a strong motive to achieve internal standardization across the designs of the nuclear power plants it builds, especially when these are constructed on a common site.

A railway engineer selecting the gauge of the track gauge would give prime consideration to the opportunities of making easy connections with railway lines in the immediate surrounding territory, and would be likely to care little if at all about the gauge choices made for railways on different continents. Likewise, the communications manager for a firm choosing an EDI document standard will be most concerned about the choices made by business entities located in the present and immediately foreseeable 'transactional territory'. The firm is likely to care only about the ease of direct electronic data exchanges with its major suppliers, subcontractors, important customers, collaborators in R&D projects — and not with parties drawn at random who are representative of the entire population of EDI-users. In other words, individual decision-agents each are limited to considering only the actions of the members a particular reference group, or subset of 'significant others' within the collectivity (David, 1992). This property, namely, *the locally bounded nature of the positive feedbacks that stem from network integration benefits* should be assumed to characterize the individual decision-agent in the case of EDI standards.

Thus, the process of competition between X.12 and EDIFACT displays the properties of (1) recurrent reassessments of individual choices within a finite population, and (2) the locally bounded nature of positive feedbacks. Both features oblige us to depart from the generalized Polya process as a framework for analysis. Because people do not care about the global proportions of the competing alternatives, one might suggest simulating the process of competition by reproducing as many mini-urn processes as there are local structures. But, apart from the fact that such an analytical procedure does not accommodate the critical feature of the reversibility of choices within a finite population, it disregards another central point of the dynamics: the existence of possible interconnections between the different local structures. Reference groups can be non-disjoint, such that intersecting networks of local relationships allow for the transmission of a decentralized choice from one group to some others. Thus, the argument developed above — that positive feedbacks are limited to subsets of the entire user community so that a particular standard fails to percolate through the entire structure — has some additional nuances. Disconnection between reference groups does not necessarily imply the absence of a continuous chain of links among them. For analysis of EDI, it should be possible to represent graphically a continuum of relations, say from the logistics function of nuclear plants to the logistics function of the automobile industry (not to mention that of a grocer's shop!), through raw materials, packaging, chemical, engineering, etc. While the positive feedbacks induced by a decentralized choice of an EDI format have only a limited area of influence within the connective chain, there is a probability of a more-or-less high pervasiveness of this choice. Whether it is pervasive or not depends on the structure of social, technical and economic networks pre-dating the dominance of a particular technology or reference.[24] It also depends on the capacity of the technology under consideration to modify, through its very diffusion, the existing networks of inter-organizational relationships. This latter consideration, however, would seem to call for application of a set of analytical tools different from the ones we shall introduce in the following section. A more adequate treatment of the interdependence between the spread of EDI technology and standards, and the evolution of new patterns of inter-firm transacting must therefore be left for a future occasion.[25]

4. Percolation and predictability in a ring inter-industries structure: an introduction to Markov random fields

Simply put, the structure of the 'competitive diffusion' process we would like to study is the following: members of a finite population of agents each make (recurrent) standard selections at random intervals in time (corresponding to a stochastic replacement process), subject to the positive influence of the currently prevailing standard usage among a subset of neighboring agents; the neighborhoods are not completely segregated. This structure is that of a heuristic model, initially known under the form of the 'voter model' elaborated by Harris and presented in Kindermann and Snell (see Harris, 1978; Kindermann and Snell, 1980a). This heuristic model is based on a particular branch of probability, called Markov random field theory (Kindermann and Snell, 1980b), the extension of which to the economics of competition between technological alternatives is introduced in David (1988, and 1992). It displays several important features:

- Firstly, this model is concerned with the features of an adoption process made up essentially of inter-linked local structures. The crucial limitation of this model, which must be borne in mind when interpreting the results is that interconnectivity within a structure and between structures are assumed always to be equivalent.
- Secondly, in this model, the share of each standard will vary on account of choice revisions made by agents and not because of the arrival of new agents. Despite the recurrent reassessment of choices, the global process of

[24] The same argument is developed in the field of language standardization: "Success in language planning depends on the already existing network of social communication, that is on the established channels of commerce in material and intellectual goods" (Ray, 1963).

[25] The mathematics of random graph theory appear to be more suited than is Markov random field theory for analytical work on the co-evolution of information technology and interfirm trading.

sequential decisions will be seen to display the property of non-reversibility.

- And, thirdly, unlike the Polya urn scheme, it is possible ex ante to assign probabilities of eventual 'victory' to each of the competing standards, simply on the basis of information concerning the initial distribution of the agents' choices.

4.1. The model

Leaving aside technicalities, it is possible briefly to describe this framework in order to consider the competition as a stochastic path-dependent process in which knowledge of the initial conditions enables one to make predictions on the limiting macrostates of the system.[26] Following the notation by Kinderman and Snell (1980b), we may begin with the basic definitions relating to Markov random fields.

Let $\mathbf{G} = (\mathbf{O},\mathbf{T})$ be a graph with $\mathbf{O} = (o_1, o_2, \ldots, o_n)$ the vertices (i.e. the set of organizations) and $\mathbf{T} = (t_1, t_2, \ldots, t_m)$ the edges (i.e. the set of transactional lines). Figure 1a shows a simple connected graph with five vertices and five edges. A *configuration* $\mathbf{x}$ is an assignment of an element of the finite set $\mathbf{S}$ to each point of $\mathbf{O}$. We denote this configuration by $\mathbf{x} = (x_o)$ where x_o is the element of $\mathbf{S}$ assigned to vertex o. If we let $\mathbf{S} = [u,a]$ represent assignments of the two possible standards (u standing for U.N. EDIFACT and a for ANSI X.12), then a configuration would be an assignment of either o_u or o_a to each of the points in $\mathbf{O}$. A *random field* p is a probability measure $p(\mathbf{x})$ assigned to the set $\mathbf{X}$ of all configurations, such that $p(\mathbf{x}) > 0$ for all $\mathbf{x}$. By the neighbors $\mathbf{N}(o)$ of the point o we shall mean the set of all points o' in $\mathbf{O}$ such that $(o'o)$ is an edge. A random field p is called a *Markov random field* if:

$$p\left[\mathbf{x}_o = \mathbf{s} \,|\, \mathbf{x}_{O-o}\right] = p\left[\mathbf{x}_o = \mathbf{s} \,|\, \mathbf{x}_{N(o)}\right].$$

That is, in trying to predict the value at o (either u or a in the example), given the values at all other points of $\mathbf{O}$, we need know only the values assigned to the neighbors of o.

Assume now, following the 'voter model', that associated with each point of a graph we have a firm and with each firm we have a reference set of other firms (possibly including the firm itself). Then the model may be described informally as follows: each firm selects a standard u or a. At random points in time it will reassess its choice. At these times it will commit to the choice u with a probability equal to the proportion of u-assigned firms in its reference set. Consider the graph consisting of the points $[-N, -N{+}1, \ldots, 0, 1, \ldots, N]$.

---

[26] The reader with some interest in the variety of applications of this model in economic science could refer to David (1992).

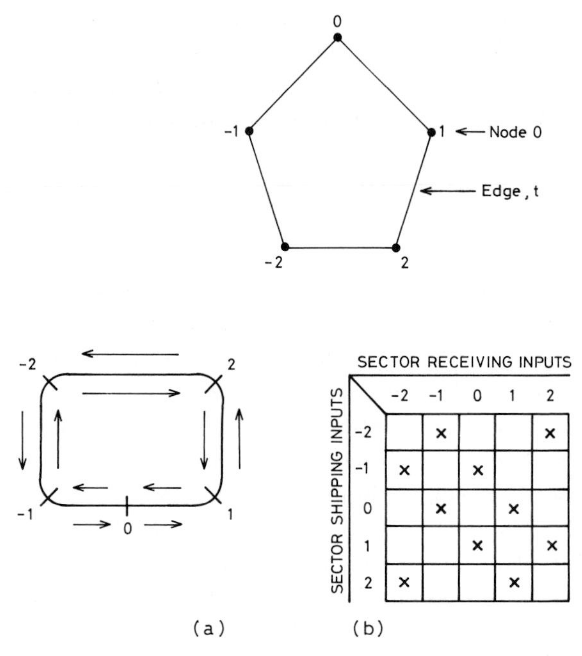

Fig. 1. Simple connected graph with five vertices and five edges.

The reference set of i consists of the points $i-1$ and $i+1$ for $-N < i < N$. The reference set for N consists of the points $N-1$ and $-N$, and for $-N$ it consists of $-N+1$ and N. According to the view presented by Fig. 1b, which gives the graph in 1a an interindustry commodity flow interpretation, this particular neighborhood arrangement could represent a transaction matrix that has a 'ring' structure: a firm gets impulses from a preceding one and sends impulses to a succeeding one, the last firm being connected with the first in the array. The closed nature of this production structure proves particularly convenient for heuristic purposes, in that it avoids the problems of having otherwise to deal with the special influence exerted by sectors that lie at the boundaries of the system — primary suppliers, and/or final demanders, so to speak. One important feature of the local structures described in the interindustry 'ring' model is that the reference sets or relational 'neighborhoods' are not disjoint. Because one of the neighbors of an agent is also a neighbor of another agent, individual unit's decision become linked indirectly by the interlocking networks of local relationships, which serve to transmit their effects.

The dynamics of this model are quite simple, as it is assumed that the times between successive reassessments made by any given organization (in this case of its choice of an EDI document standard) follow an exponential distribution with mean duration 1. Thus, the occurrence of micro-level re-evaluations at each node is independent of the timing of reorientations that might take place among the other firms that share the same neighborhood. Where a significant number of symmetrically influential organizations constitute the subsets of neighbors, this simplifying assumption does not seem unreasonable. The role of 'dominant users' of EDI in influencing the standardization process, however, cannot be well represented within the framework of this model — except through the device of assigning satellite organizations of the dominant firm to a 'collective territory' that is completely under the dominant firm's control as far as the choice of standards is concerned.

The global process of migration between competing standards now can be described by a finite state continuous time Markov chain, with states being configurations of the form $\mathbf{x} = (u, a, u, u, a, \ldots, u, a, u)$ where $x(i)$ is the choice of firm i. Several properties follow from this:[27]

(1) It is evident on even the briefest consideration that the extremal states $\mathbf{x}^u = (u, u, u, \ldots, u, u, u)$ and $\mathbf{x}^a = (a, a, a, \ldots, a, a)$, in which there is a perfect correlation of standards choices throughout the population, constitute *absorbing states* for this system. Once such a state is entered, there can be no further change.

(2) A somewhat less obvious proposition, also true, is that for any starting state $\mathbf{x}$ the chain eventually will end up in either $\mathbf{x}^u$ or $\mathbf{x}^a$. Thus, *in the limit*, the process must become 'locked-in' into one of its extremal solutions.

(3) There exists a limiting probability distribution over the macrostates of the system which is non-continuous, such that, starting in $\mathbf{x}$, the probability that the chain will end in $\mathbf{x}^u$ is equal to the proportion of u in the initial configuration $\mathbf{x}$ (without regard to their position in the array); and the probability that it will end up in $\mathbf{x}^a$ is equal to the proportion of a in the initial configuration $\mathbf{x}$. Therefore, although subject to random influences, *the asymptotic macrostate of the system can be readily predicted from information on the initial conditions*.

What do these results (lock-in into an extremal solution, predictability of the eventual outcome on the basis of information concerning the distribution of the agents' initial choices) mean in the context of our case study? This model foresees the emergence of a single standard (in certain conditions which will be examined more closely below) in the space defined by the entire body of EDI users. The certainty that the outcome will be one of those extreme solutions means that the network integration benefits will be fully exploited in the future — which should increase the expected value for an

[27] A simple graphical device (the dual graph) has been applied by Kinderman and Snell (1980a) and David (1992) in a way that makes propositions (1), (2) and (3) quite easy to confirm for the cases of finite Markov fields in one and two dimensions.

individual firm of belonging to it. Further, an outside observer who knew the actual distribution of the agents' initial choices among the available standards would be able to calculate the odds of each standard emerging as the eventual victor in the competition, so long as such information remained irrelevant to the individual agent's actions.

There are, however, two important qualifications that render application of the 'voter model' to the EDI standards competition less than completely straightforward. These qualifications arise from the assumptions of the model with respect to the 'finite population' condition, on the one hand, and the 'mixed percolation probabilities' of the global system on the other.

4.2. Finite population, 'entry' process, and collective territories

The finite nature of the population enables one to know precisely the initial distribution of the agents and hence the proportion of u and a at the outset of the process — knowledge of which has been seen to be an essential prerequisite for predictability.[28] In other words, the model's properties would cease to hold during a growth phase in which the configuration was being perturbed by the entry of new members of the network, each coming with their initial orientation towards standard a or u. During such a phase, the macro-state would tend to undergo marked random fluctuations between the extremes. But, once entry slowed to occur at a rate lower that the mean time-rate at which agent's 'polled' their neighbors to decide what standard to use, reassessment would become predominant in shaping the dynamics of the system. Does this preclude being able to say anything conclusive about situations where entry and network growth is taking place? The qualifications stemming from the finite population condition actually are less severe than they would at first seem. For, what matters most model's assumptions is the stability of respective proportions of (u) and (a) at the start of the process, rather than the invariance of the number of agents making up the population. It is therefore possible to include from the outset in the finite population those potentially adopting agents who had no effective option regarding standards other than to conform exactly with the one that was preponderant in their neighborhood. For example, consider the case of small firms which belong to an inter-organizational network where a EDI standard is actually used. Those firms belong to *collective territories*, defining industrial (or regional) spaces that are unambiguously committed to using a particular standard.

---

[28] The qualitative property of asymptotic lock-in to one of the extremal states does not obtain when the population in the voter model is infinite, and very large finite systems may behave the same way. See Kinderman and Snell (1980a), and David (1992) for further discussion of the implications.

This notion of 'collective territories', introduced by Thomas Schelling (1971) in his dynamic models of residential segregation, allows for expansion of a finite population to occur without the initial proportions being disturbed. Schelling reported some simulation results with an initial random distribution of agents, composed of two different populations (whites and blacks) located on a checkerboard (but without any alternating colors). It was posited that 25% of squares initially were vacant. The parameters to be defined were the size of the neighborhood (8, 24), the proportions of the populations (equal, unequal), the level of demand for the like-color proportion within a neighborhood (no fewer than one-half or one-third of one's neighbors are of the same color). According to certain rules, the agents discontented with their location move to the nearest satisfactory vacant square. These first movements, however, can produce new discontents, who will in turn move, until an equilibrium is achieved. When the segregation is sufficiently marked, Schelling claimed, blank spaces can be imputed unambiguously to 'white territories' or to 'black territories'. This observation introduces a notion of collective territory, as a necessary supplement to the individual neighborhood.

In the present context collective territories are territories 'belonging' to a certain network standard, wherein potential entrants will adopt the local standard with virtual certainty. Thus, the finite population is made up of the actual users of EDI and the collective territories where potential users can be imputed unambiguously to a given standard. Through the device of defining 'collective territories' we can allow for the entry of new users under conditions which would prevent the process from being perturbed in a way that would cause it become unpredictable and undergo continuous random fluctuations between the extremes.

4.3. Percolation probability and the 'voter model'

The stochastic feature of the 'voter model' described in the preceding subsection is confined to the re-orientation of the standard selected by the node-organizations, and is made conditional on the orientations of the set of their neighbors. The population of firms portrayed in this set-up is homogeneous in a two-fold sense:

(1) All of the firms are taken to be equally ready to implement a particular standard in response to the external influence exerted upon it by the currently reigning choices of their trade partners.

(2) The structure of inter-organizational connections is symmetrical, in that a similar pattern of transactional relations characterizes all of the reference sets.

These homogeneities are, further, assumed to take a rather special form which renders the transmission of influence among the organizations completely deterministic. It is for this reason that we noted, above, that the source of randomness in the model had to do not with whether or not particular firms might be open to the influence of particular neighbors, but, rather with the direction of the re-orientation that such influences would bring about. The concepts and terminology of *percolation theory* provides a precise way of describing these specifications, and showing their relationship to a more general specification of the model.

By the term *percolation process* we refer to the dual of a diffusion process. 'Diffusion', to speak strictly, refers to the random movements of particles through an ordered, non-random medium — as in the case of the diffusion of molecules of salt in water; whereas, the term 'percolation' conjures up the image of droplets of water moving under the deterministic pull of gravity through a disordered, random medium such as filtration tank filled with sand, and pebbles of different sizes. When the water, entering at some source sites, eventually finds its way into enough open channels to pass throughout, wetting the entirety of the interior surfaces, *complete percolation* has occurred — from whence the statistical models of analogous processes take their name.

Adapting the notation of Hammersley and Welsh (1980) to the Markov random field framework, let **G** be a graph in which some, none, or all of the edges may be directed. Thus **G** consists of a set of vertices or *nodes*, **O**, connected by a set of (possibly directed) edges or *connections*, **T**. Thus, we begin again with a set of *node-organizations*, $\mathbf{O} = (o_1, o_2, \ldots, o_n)$, and a set of *transactional lines* connecting them, $\mathbf{T} = (t_1, t_2, \ldots, t_m)$. An *operative path* in **G** from an organization o_1 to an organization o_n is a finite sequence of the form $\{t_{12}\, o_2\, t_{23}\, o_3 \ldots t_{[n-1]n}\, o_n\}$, where t_{ij} denotes a relational line connecting o_i to o_j. The graph **G** is *connected* if for each pair of organizations o_i and o_j there is a path in **G** from o_i to o_j. We now construct a *random maze* on **G**, as follows. Let each organization o of **G** be *open*, or responsive to the influence of any of its neighbors' standards choices, with probability p_o, or *closed* (unresponsive to local network externalities in deciding on its EDI documents standards) with probability $q_o = 1 - p_o$. Similarly, each inter-organizational transactions line t_{ij} attains some minimal or *threshold density* of transactions between i and j (which is sufficient to decisively influence the ith and/or the jth firms' choices of a standard) with probability p_t, or it fails to do so with probability $q_t = 1 - p_t$. Moreover, we shall assume all these events are to occur independently of each other.

An operative path, $\mathbf{d} = \{t_{12}\, o_2\, t_{23}\, o_3 \ldots t_{[n-1]n} o_n\}$ from o_1 to o_n is said to be 'open' if all its transactional links attain the minimum sufficient density and all its organizations are influenced by network externalities in selecting their standard. Thus, the probability that the particular path **d** is 'operational' in that sense is given by $(p_o p_t)^{n-1}$.

Let **R** be some given set of source organizations, from which a particular transactional standard emerges into **G**. The decisions to adopt that standard can flow along any open path from a source organization and will then similarly reorient the other organizations on such a path ('wetting' them, to use the natural percolation metaphor). The *percolation probability* $P(p_o, p_t \mid \mathbf{R}, \mathbf{G})$ is the probability that **R** can thus reorient some infinite set of nodes in **G**. We call the parameters p_o, p_t, respectively the *receptivity* and the *connectivity* of the process. In other words, in a large population, it can be expected that a proportion p_o are receptive to their neighbors' choices of a standard, while a proportion $1 - p_o$ are unreceptive. The transactional lines of **G** connect neighboring pairs of organizations and the model supposes that an 'infected' organization (i.e. already committed into a given standard) has a chance p_t of 'infecting' a neighbor, provided that the latter is receptive. Then $P(p_o, p_t \mid \mathbf{R}, \mathbf{G})$ is the probability that a standard initially established in the source organizations **R** will become adopted universally.

Suppose that **R** and **G** are fixed, that **G** is infinite and that we abbreviate

$$P(p_o, p_t \mid \mathbf{R}, \mathbf{G}) = P(p_o, p_t) = P,$$

then, clearly, the mixed percolation probability P is a non-decreasing function of p_o and p_t, and: $P(0, 0) = P(1, 0) = P(0, 1) = 0$, while $P(1, 1) = 1$. Thus, $P_o(p) = P(p_o, 1)$ and $P_t(p) = P(1, p_t)$ are, respectively, called the *node percolation* and *connection percolation* probabilities of this system.

A fundamental mathematical property of the percolation process is that there exists some critical values of $p_o > p_o^*$ and $p_t > p_t^*$ beyond which there will be a positive probability that percolation occurs, and below which the percolation probability is zero.[29] In other words, the system undergoes a 'phase transition' when these underlying critical probabilities are attained. As Fig. 2 indicates, there are corresponding critical values at which the node-percolation and edge-percolation probabilities, respectively, become positive. These define the endpoints of a region above which a 'mixed-percolation process' (one for which it is not certain that either all nodes or all edges of the graph are open), will have positive probability of achieving complete percolation.

If we now return to the process described by the 'voter model', the limiting outcome — of lock-in to one of the extremal configurations whose identity may be predicted from initial configuration — may be seen to obtain under

[29] See Hammersley and Welsh (1980); and, for a more recent exposition of the theory with numerous physical applications, see Grimmett (1989). The more usual terminology in the mathematical literature associates the vertices and edges of **G** with 'atoms' and 'bonds', respectively, and so refers to the probabilities of atom percolation and bond percolation.

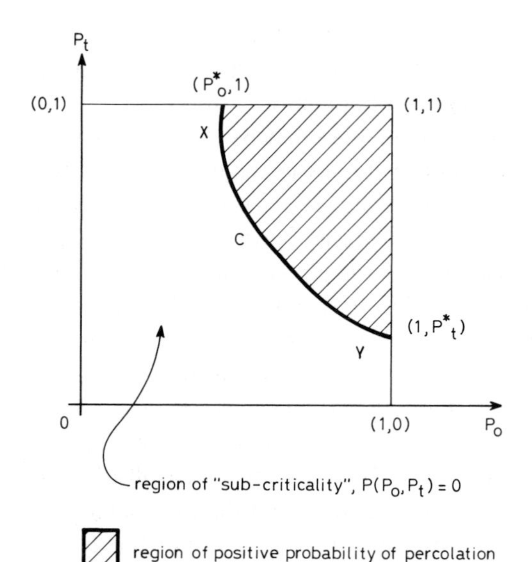

Fig. 2. Critical boundary for percolation in infinite graph, **G**.

the special conditions:

$$P_o(p) = P_t(p) = 1.$$

It is clear, however, that the pair of homogeneity specifications which correspond to those conditions do not correspond well with the realities characterizing populations of EDI-using business organizations.

First, some firms (e.g. those with a very high ratio of internal to external message transactions) are not likely to be highly responsive to the choices their trade partners have made with regard to electronic document formats. Either the internally-oriented EDI users may expect to incur high fixed costs in converting all their data processing to work with some new, externally compatible format, including some disruptions of their internal control and decision-making structures in the switch-over: or they may fear some increases in their vulnerability to external disruptions created by the existence of inter-organizational networks that render their internal operations informationally transparent (as discussed by Hart and Estrin, 1990). These firms, therefore, pay little or no attention to the problems arising from the incompatibilities of the formats in which they exchange information with outside entities. They are ready to bear high per transaction conversion costs in order to maintain their internal data coding and processing procedures in status quo. Under such conditions, some of the organization-nodes of our graph, **G**, will be 'closed', so that for the randomly drawn (representative firm of the system) the node-percolation probability will be inferior to 1. Indeed, it could be so degraded that the system was 'sub-critical' and there would be no possibility whatsoever of the occurrence of a spontaneous percolation of a given standard throughout the population.

Secondly, some transactional lines may be closed, so that the bond percolation probability (the connectivity of the system) may be inferior to 1. Indeed certain local structures (reference sets) are weakly (or not) connected to the rest of the network, in other words, some reference sets are isolated while others are fully 'embedded' in the global transactions system. Moreover, the patterns of interactions within each reference set may be very different. In other words, it appears that structural heterogeneity in subsets of adopters (in terms of internal connectivity between the agents), as well as fragmentation into essentially free-standing sectoral networks of local relationships, are two possible features of the global relational network that are very important in determining the possibility of *de facto* EDI standardization as a dynamic process. This structural heterogeneity at both the inter- and intra-neighborhood levels can be more rigorously formulated within the framework of clique theory.[30]

We use the term 'clique' in a very intuitive way to mean a highly cohesive subgroup of individual agents or organizations. Cohesiveness, however, has two possible dimensions (Alba, 1973): in one dimension, called *completeness* in the literature, a cohesive subgroup would be one in which a high proportion of its pairs possess the appropriate relation (i.e., pairs of points are adjacent). In the other, sometimes referred to as the *centripetal–centrifugal* dimension, a cohesive subgroup would be one in which relations among members of the subgroup are more important or more numerous than relations between members and non-members. Both properties and their associated probability distributions enable one to represent some heterogeneity in describing the structures of connectivity within the users' system.

Completeness: The first property is the degree to which a short path is present from each point to every other point in the subgraph. According to the most common theoretical interpretation of this property, its economic importance stems from the direct communication ties among the individual members. This property and its probability distribution are likely to influence the speed with which a *de facto* standard may 'percolate' within a sub-structure[30].

Discontinuity with the rest of the network: A second property is the degree to which such structures are tied into the total network. That is, it is sometimes

[30] An n-clique is a maximal complete subgraph, where adjacency between every pair of points (and thus completeness) does not require strictly direct relationships but is limited to a maximal distance n (see Alba, 1973).

said that a defining property of cliques is that they be disjoint and/or that they show marked discontinuities with the rest of the network. This positional property and its probability distribution will influence the probability of the existence of these intersecting networks of relationships serving to transmit the effects of decentralized standards decision through the entire structure (David, 1992).

Thus, recognition of the presence or absence of significant relational structures in the adopter system (n-clique organizations) is important. If we return now to the 'voter model' and its assumptions regarding the mixed percolation probabilities (i.e., $P_o(p) = P_t(p) = 1$), it is clear that those assumptions impose considerable restrictions upon the applicability of the analysis: structural heterogeneity in the node- and connection-percolation probabilities certainly plays a key role in the dynamics and the eventual equilibrium outcome of the system. This conclusion gives rise to the following questions: What happens if both node- and connection-percolation probabilities are degraded? Are the results of the 'voter model' (lock-in into an extremal solution and predictability) altered? Are there critical probability levels for node- and connection-percolation, above which the system can take on the properties of a finite state continuous time Markov chain and reach limiting configurations characterized by perfect or very strong correlation among standards choices? We conjecture that both questions can be answered affirmatively, but to our present knowledge, the relevant theorems have not been proved and these propositions remain to be formally demonstrated for the case of a 'voter model' with a mixed-percolation process.[31]

5. *Policy issues*

The modern economic analysis of compatibility standards suggests clear and strong grounds for setting information technology policy the broad goal of encouraging the selection and diffusion of a single/universal EDI standard.[32]

The framework developed here, based on Markov random fields theory and percolation probabilities, points to two specific and important policy issues in this regard, and offers some guidance in the choice among alternative strategies.

---

[31] Our conjectures in this regard for the case in which P_o and P_t are each serially dependent have received support from some preliminary simulation studies, upon which we intend to report in future papers: for the case of serial independence the conjective seems to hold *a fortiori*.

[32] See David and Greenstein (1990) and Payne and Anderson (1991, p. 12), for the conditions under which a universal standard represents a socially better outcome than the persistence of different and incompatible standards.

A first approach is the one that can be referred to as the 'bottom up' strategy of standardization: to create the appropriate percolation conditions that would induce spontaneous emergence of a standard from a sequence of individual decisions. Simply put, such a policy should aim to increase the mixed percolation probabilities (P), so that the EDI user system approaches the conditions that obtained in the case of the original 'voter model'. But, recognizing the impossibility of a universal *de facto* standard emerging spontaneously in the global market where the system of transactions remained subcritical (i.e., where $P < P^* < 1$), a second approach may be suggested. This would be to rely upon some outside agency to formulate and promulgate a flexible and stable document standard that would fit virtually any business requirements, irrespective of firm and sectoral specificities. Seen in this perspective, the EDIFACT effort represents in some ways a unique experiment with the 'top-down' strategy, involving implementation of a standard whose design must meet the requirement of flexibility that is necessary to achieve its acceptance in a subcritical system. Some further implications for each of these strategy approaches follow from consideration of our model.

5.1. Increasing the percolation probability of the system to achieve de facto universal standardization

The mixed percolation framework provides a useful basis for discussing the first policy implication. Indeed, the parameters p_o and p_t (the receptivity and the connectivity of the system, respectively) are not perfectly symmetric in their influence on the percolation probability P. The following general property of P:

$$P_o(p) < P_t(p), \quad \text{whenever} \quad 0 < p < 1,$$

suggests that a standard percolates more easily via an imperfect set of connections between receptive nodes than through a completely connected graph whose nodes (organizations) are imperfectly receptive. A further generalization of this is provided by the theorem due to Hammersley and Welsh (1980):

$$P(p_o\beta, p_t) < P(p_o, p_t\beta), \quad \text{whenever} \quad 0 < p_o < 1,\ 0 < p_t < 1, \quad \text{and} \quad 0 < \beta < 1.$$

This inequality lends itself to the following epidemiological interpretation: in seeking to prevent a pandemic (universal adoption, in our application) it is more effective to reduce the receptivity of members of a population than it is to reduce the probability of infectious contacts among them (the connectivity among organizations). In other words, if one degrades the node probability p_o, the percolation probability is lower than if one equi-proportionally degrades the connection probability p_t. Thus, when the costs of the alternative policy actions are identical at the margin, the proper policy course for raising the percolation probability — so that the EDI user system approaches the Markov random field model's properties — would be one that was 'receptivity-directed' rather than 'connectivity-directed'. The indicated policy would not rely on efforts to increase the density of external transactions, in order to induce firms to attend to the benefits of achieving compatibility in the use of EDI documents. Rather, it would seek to make firms outward-oriented in their information management practices, and so more responsive to external impulses from their trading partners. It could, in particular, open the 'closed nodes' by subsidizing the initial commitment of the reluctant firms (those firms characterized by a high ratio of internal/external information exchanges) to implement the standard which dominates their local structures.

Another policy goal should be to accelerate the migration from local standards to the emerging *de facto* standard (either EDIFACT or X.12). Here, information about the 'robustness' of the different local structures matters. Robustness belongs to the class of the n-clique properties. Robustness is best characterized with reference to the degree to which the structure is vulnerable to the re-orientation of any given individual.[33] For example, any graph of the form $\mathbf{K}_{1,n}$ (i.e., a graph in which one point is connected directly to all others and there are no other lines) is highly 'vulnerable', in the sense that the central node's decision will carry more weight than the satellites' decisions. This property introduces us to another kind of differentiation within the subsets of adopters. For example, in the case of EDI, both the U.S. DOD sub-contracting system and the automobile sub-contracting system could be described as n-cliques having similar degrees of discontinuity with the total network. But, the former is a structure very vulnerable to the removal of a given actor (the DOD), whereas the latter is a more robust one. In other words, a DOD switching-decision should be sufficient to cause the entire migration of the substructure from one standard to another, whereas in the case of the automobile industry such a migration would require collective action by the major firms. Thus, in the latter case, incomplete information and excess of inertia can be expected to play a great role in blocking spontaneous migration to a new standard (Farrell and Saloner, 1985).

[33] This property is not easily identifiable through the notion of n-clique. For example, any graph of the form $\mathbf{K}_{1,n}$ is a 2-clique. Such structures are very vulnerable, their robustness diminishing from maximal to possibly minimal levels in the transition from graphs of diameter 1 to those of diameter 2. Hence the notion of K-plex structure proposed by Seidman and Foster (1978).

5.2. Below the critical level of bond percolation probability: the top-down creation of a new standard — EDIFACT

Given that the bond percolation probability of the EDI system ($0 < p_t < 1$) cannot be readily manipulated without embarking on policy measures designed to alter the internal organization and information management policies of many firms, some disjoint reference groups and local standards are likely to persist. Given such a subcritical system — which is supposed to remain stable, there being no evolution of the intensity of interactions[34] — it is necessary to design a 'universal' standard. Yet, as has already been intimated, the ambiguous significance of universality in the context of language standardization appears to have resulted, unfortunately, in some policy fluctuations affecting the design development process in the case of EDIFACT.

On the one hand, aspiring to universality could mean seeking to supplant any local standards in each sub-structure. The diffusion of a comprehensive document standard, therefore, would require the introduction of the same codes and forms into every subgroups of potential adopters. Further, the standard in question should in principle be made more efficient to use than any local document standard, in order to counterbalance the sunk costs entailed in migration from one standard to another. The policy of the EDIFACT projectors, to the extent that it aims to be a universal standard policy (in this sense), is thus to design a flexible, superior, and stable document standard that can satisfy all potential users. On the other hand, the quest for universality could mean trying to create a standard for setting local standards. At this meta-level, the 'universal' standard is not supposed to be in competition with the local codes and protocols, but rather to support the regular formation of the local standards — much in the manner of the ANSI X.12 document standard. Foray (1991), in a recent detailed review of the internal EDIFACT design policies, has demonstrated that the succession of policies has reflected the continuous oscillation of EDIFACT's projectors between these contradictory goals: competing with the local standards, or providing a general flexible framework to integrate all these disparate local standards.

As a result, the EDIFACT policy has yielded contradictory developments. Thus, EDIFACT messages are not yet stable, very few EDIFACT messages are used, the level and rate of adoption are very low and, optimistic talk of the long-term prospects for international acceptance notwithstanding, the migration from local European standards to EDIFACT is failing to occur with enough momentum to start a bandwagon. These difficulties may represent dif-

---

[34] EDI does not involve changing the state of the interdependence among the agents, at least during the first period of use. Thereafter, increasing interaction may be expected.

ferent aspects of the natural fate of a universal standard failing to percolate throughout a subcritical system.

6. *Conclusion*

Because EDI deals with the transmission of intent through expression, an analogy has been drawn here between the economics of EDI and the economics of language standardization. We have sought to identify analytically the general conditions that could result in the *de facto* emergence of a single, universal EDI document standard through a decentralized, diffusion-like or, more accurately, a 'percolation' process. Decisions to implement EDI with specific document standards are likely to spread in a population of firms through positive feedback effects from an initial core of users; direct transactions cost savings and indirect benefits are realized by marginal (potential) adopters as positive externalities generated by the intra-marginal group of users. Analysis of stochastic processes belonging to the class of Polya urn models, recently used by economists to study competitive technology diffusion under conditions of positive feedbacks, leads to the conclusion that the presence of potential positive network externalities is not sufficient to induce the emergence of a unique global standard. Those models, however, point only to the possibility that 'weak' positive feedback can be consistent with the persisting co-existence of multiple standards, without characterizing the sources of 'weakness' that are relevant in the case of EDI. Further, there are other features of those models that do not conform to the structure of the choice of technique problem facing EDI users in regard to the selection of a document standard.

A different class of stochastic models, based on Markov random field theory, has been proposed as being better suited to the analysis of the behavior of agents embedded in local networks of transactions. Drawing upon results from percolation theory, we have seen that one may account for the observed co-existence of multiple EDI document standards as a manifestation of the 'sub-criticality' of the inter-organizational transactions structure. A particular standard fails to 'percolate' throughout when the connection probabilities between typical nodes in the system remain below some critical level. The formation of finite 'clusters', or sub-groups of firms committed to idiosyncratic document standards (like natural languages and local dialects thereof), is sufficient to bring this about by fragmenting the transactions structure; in graph-theoretic terms this can be described as insufficient connectedness.

The sub-criticality of the system as whole is likely to be the result of the historical process that created antecedent routine of coordination: EDI networks, as a practical matter, must be implemented so as to conform with local coordination structures that have arisen historically from firm and sectoral specificities, geographical neighborhood, and industry conventions in handling the most frequent forms of inter-organizational transactions. Some basic analytical results in percolation theory suggest the general characteristics of policies that would be most efficacious in removing the condition of sub-criticality: such policies would be directed at the internal document generation and data processing practices of firms, rather than at their electronic network connections. But, since these policies are likely to be particularly resisted as intruding into the organization and internal business operations of firms, an alternative strategy approach has been suggested: promoting the spread of EDI by developing EDIFACT so that it can supplant local document standards on efficiency grounds. In other words, like a Universal Language, EDIFACT is an 'a posteriori language' which has to face the competition of pre-existing, locally entrenched linguistic standards. In such competitions, increasing the flexibility of the language and tolerating a degradation of aspirations to rigorous universality may prove to be the strategy most likely to meet with success. Opting for 'flexibility' of the former sort, however, must contend with, and seek to minimize the unfortunate tendency to render the EDIFACT standard less stable, which has so far inhibited it from displacing idiosyncratic European document standards and weakened its attractiveness vis-à-vis ANSI X.12 in the North American market.

Acknowledgements

The authors gratefully acknowledge useful comments on an earlier draft from Chris De Bresson and Edward Steinmueller. The research reported here was supported by grants from the High Technology Impact Program in the Center for Economic Policy Research at Stanford University, Stanford, CA, and an earlier draft circulated as CEPR Publication No. 326 (September 1992). David also wishes to thank All Souls College, Oxford, where he held a Visiting Fellowship during the completion of this work.

References

Alba, R., 1973. A Graph-Theoretic Definition of a Sociometric Clique. *Journal of Mathematical Sociology*, Vol. 1.

Antonelli, C., 1991. The Economics Theory of Information Networks. In: C. Antonelli (ed.), *The Economics of Information Networks* (North-Holland, Amsterdam).

Arthur, W.B., 1989. Competing Technologies, Increasing Returns, and Lock-In by Historical Events. *Economic journal* 99 (394), March.

Arthur, W.B., 1990. Positive Feedbacks in Economics. *Scientific American* 262 (2), February, pp. 92ff.

Beniger, J.R., 1986. *The Control Revolution: Technological and Economic Origins of the Information Society* (Harvard University Press, Cambridge, MA).
Bloomfield, L., 1984. *Language*, 2nd ed. (The University of Chicago Press, Chicago, IL).
Chandler, W.L., 1928. Simplification's boons to purchasing agents. In: *Standards in Industry*. The Annals, The American Academy of Political and Social Science.
Commission for the European Community, 1989. *Survey of EDI in all the Member States*. TEDIS (Trade Electronic Data Interchange Systems) Report (Brussels).
Cowan, R., 1990. Nuclear Power Reactors: a Study in Technological Lock-In. *The Journal of Economic History* 5 (3), September.
Cusamano, M.A., Mylonadis, Y. and Rosenbloom, R.S., 1990. Strategic Maneuvering and Mass Market Dynamics: The Triumph of VHS over BETA *Consortium on Competitiveness and Cooperation Working Paper No. 90-5* (Center for Research in Management, University of California at Berkeley).
Dang-Nguyen, G., 1990. The Value of Cooperation in Telecommunication Related Services. *Paper presented to the Workshop on Special Problems of Network Industries* (T. Nambu, Convenor), Hokkaido University, Sapporo, July 23–25.
David, P.A., 1985. Clio and the Economics of QWERTY. *American Economic Review* 75 (2), May, pp. 332ff.
David, P.A., 1986. Understanding the Economics of QWERTY: The Necessity of History. In: W.N. Parker (ed.).*Economic History and the Modern Economist*, Chapter 4 (Basil Blackwell, Oxford).
David, P.A., 1987. Some New Standards for the Economics of Standardization in the Information Age. In: P. Dasgupta and P. Stoneman (eds.), *Economic Policy and Technological Performance* (Cambridge University Press).
David, P.A., 1988. Path-Dependence: Putting the Past into the Future of Economics. *Institute for Mathematical Studies in the Social Sciences, Stanford University, Technical Report* 533, November.
David, P.A., 1990. The Dynamo and the Computer: An Historical Perspective on the Modern Productivity Paradox. *American Economic Review* 80 (2), May, pp. 355–361.
David, P.A., 1991. Computer and Dynamo: The Modern Productivity Paradox in a Not-Too-Distant Mirror. In: *Technology and Economic Policy: Challenges for the 1990s* (OECD, Paris).
David, P.A., 1992. Path-Dependence and Predictability in Dynamic Systems with Local Network Externalities: a Paradigm for Historical Economics. In: D. Foray and C. Freeman (eds.), *Technology and the Wealth of Nations* (Pinter Publishers, London).
David, P.A. and Bunn, J.A., 1988. The Economics of Gateway Technologies and the Dynamics of Network Evolution: Lessons from Electrical Supply History. *Information Economics and Policy* 5.
David, P.A. and Greenstein, S., 1990. The Economics of Compatibility Standards: An Introduction to Recent Research. *Economics of Innovation and New Technology* 1 (1/2), pp. 1–32.
David, P. and Rothwell, G., 1993. Standardization, Diversity and Learning: a Model for the Nuclear Power Industry. *MERIT Research Memorandum 93-005*, March (University of Limburg, Maastricht).
Erck, A.H., 1928. Standardization of purchase procedure. In: *Standards in Industry*. The Annals, The American Academy of Political and Social Science.
Farrel, J. and Saloner, G., 1985. Standardization, compatibility and innovation. *Rand Journal of Economics* 16.
Foray, D., 1991. *The Economics of Diffusion of EDIFACT*, Summer 1991 (draft).
Greenstein, S., 1991. *Lock-In and the Costs of Switching Mainframe Computer Vendors: What Do Buyers See?*. Working Paper, University of Illinois at Urbana.
Greenstein, S., 1993. Invisible Hands and Visible Advisors: An Economic Interpretation of Standardization. *Journal of the American Society for Information Science*, 43(8), 538–549.
Grimmett, G., 1989. *Percolation* (Springer Verlag, New York).

Grin, F. 1992. Towards a Threshold Theory of Minority Language Survival. *Kyklos*, 45.

Hammersley, J.M. and Welsh, D.J., 1980. Percolation Theory and Its Ramification. *Contemp. Phys.* 21 (6).

Harris, T.E., 1978. Additive Set-Valued Markov Processes and Percolation Methods, *Ann. Probability* 6.

Hart, P. and Estrin, D., 1990. Inter-Organizational Networks in Support of Applications-Specific Integrated Circuits: An Empirical Study. *Paper presented at the 8th International Telecommunications Society Meetings*, Venice, Italy, March 18–21, 1990.

Haugen, H., 1966. *Language Conflict and Language Planning* (Harvard University Press, Cambridge, MA).

Kimberly, P., 1991. *Electronic Data Interchange* (McGraw-Hill, New York).

Kindermann, R. and Snell, J., 1980a. On the relation between Markov Random Fields and Social Networks. *Journal of Mathematical Sociology* 7.

Kindermann, R. and Snell, J.L., 1980b. *Markov Random Fields and their Applications. Contemporary Mathematics, Vol. 1* (American Mathematical Society).

Mansell, R., 1991. Information, Organization and Competitiveness: Networking Strategies in the 1990s. In: C. Antonelli (ed.), *The Economics of Information Networks* (North-Holland, Amsterdam).

Marschak, J., 1964. Economics of Language. *Behavioral Science* 9

Oliver, S., 1990. *RINET: A Case Study in International Network Services. Datapro Report on International Networks and Services* (McGraw Hill, New York).

Payne, J. and Anderson, R., 1991. *Electronic Data Interchange (EDI)*, RAND, R-4030-P&L, p. 1.

Pfeiffer, H., 1990. *The Diffusion of Electronic Data Interchange* (Institut für Wirtschaftsinformatik der Universität Bern).

Ray, P.S., 1963. *Language Standardization* (Mouton, The Hague-Paris).

Rubin, J. and Jernudd, B., 1971. *Can Language Be Planned?* (The University Press of Hawaii, Honolulu).

Schelling, T., 1971. Dynamic Models of Segregation. *Journal of Mathematical Sociology* 1.

Trauth, E.M. and Thomas, R.S., 1992. Electronic Data Interchange in the Post-1992 World: A New Frontier for Global Telecommunications Policy. *Paper presented to 42nd Annual Conference of the International Communication Association*, Miami, Florida, May 21–25.

United Nations Economic Commissions for Europe, Trade Facilitation, 1980. *Introduction to the United Nations Rules for EDI for Administration, Commerce and Transport*, Trade/W.P.4/INF.105 (Geneva).

Yates, Joanne, 1989. *Control Through Communication: The Rise of System in American Management* (The Johns Hopkins Press, Baltimore).

Global Telecommunications Strategies
and Technological Changes
Edited by G. Pogorel

Chapter 8

Economic and political factors in telecommunication standards setting in the U.S. and Japan: the case of BISDN

Terry Curtis [1] and Hajime Oniki [2]

[1] *California State University, Chico Chico, CA 95929-0504, USA*
[2] *Institute of Social and Economic Research, Osaka University, Ibaraki, Osaka, Japan 567*

1. Introduction

Critical political and economic differences exist between Japan and the U.S. with regards to their telecommunication industries. One area in which the differences have been quite marked is in the area of telecommunication standards. The processes by which telecommunication standards are agreed domestically — and are pursued internationally — in the two nations are quite different. These differences will be briefly reviewed here. The developments in BISDN standards will be used as an example.

2. The setting of Japanese telecommunication standards

In the context of rapidly converging telecommunication and information technologies, the first key point to be made in describing the Japanese situation is that there is a division of responsibility between the MPT (Ministry of Posts and Telecommunication) and MITI (Ministry of International Trade and Industry). A gross simplification of the division is that MPT is responsible for telecommunication network standards and MITI is responsible for information systems equipment standards. On each side of this division the ministries are technically and officially responsible for the standards processes and publish the standards which are ultimately agreed. In each ministry the work of negotiating agreement is actually done by multiple private sector and/or government committees whose work is overseen and in some ways sanctioned by the relevant ministry.

The division of responsibility between MPT and MITI is roughly equivalent to the division at the international level between the CCITT and the ISO. However, the formal cooperation and coordination which has developed between the two international bodies is not characteristic of the work of the two Japanese ministries. Such coordination as there is within Japan results from the fact that many of the same firms participate in the committees which do the work for both ministries. This provides for an adequate informal liaison for awareness of each other's activities. Any formal liaison, however, is made very difficult — if not impossible — by the formal respect that Japanese culture dictates that each ministry must show for the other's authority.

2.1. The MPT sector

Within the area of responsibility of MPT, major changes in procedure occurred beginning in 1985. The 'Telecommunications Business Law,' enacted 1 April of 1985, opened the Japanese telecommunications equipment market in some ways and created a need for broader input into telecommunications network standards, which had previously been exclusively left to the two carriers, NTT and KDD. Articles 49 and 50 of the Telecommunications Business Law also gave MPT the official responsibility for specification of terminal equipment standards and for approving specific equipment as conforming to those standards. Prior to that time both specification and conformance testing of terminal equipment had been left to NTT. Responding to the need for broader input into the process, three non-profit committees were established, the Telecommunications Technology Committee (TTCommittee), the Research and Development Center for Radio Systems (RCR), and the Broadcast Technology Association (BTA). The RCR and the BTA are fora for discussion by interested parties of the technical feasibility of radio and broadcast technologies which may be proposed to the MPT to be recognized as standards. The TTCommittee has a larger role.

Participation in the TTCommittee is open to representatives of any firm, including foreign firms, which is interested in the development of telecommunication standards in Japan. The TTCommittee has the capacity to develop standards in three kinds of cases:

(1) Where there is an international standard established by the CCITT or the ISO, TTCommittee will adopt this standard for Japan, possibly adapting it to the Japanese network's requirements.

(2) Where an international standard is under consideration but not yet agreed, TTCommittee may develop its own temporary standard for Japan if this is felt necessary.

(3) Where there is a need for a standard in Japan on a matter which is not under consideration at the international level, TTCommittee may develop a temporary standard.

Note that the TTCommittee primarily reacts to the international standards situation. Except for some efforts to coordinate work on domestic standards with similar work by T1 in the U.S. and ETSI in Europe, TTCommittee is not involved in the setting of international standards, although it does send observers to CCITT meetings.

This leaves open the question of who should represent Japan in the negotiation of international standards which may come back to the TTCommittee for implementation. This role is officially the responsibility of the Communications Policy Bureau of MPT, which accomplishes its standards work through another organization with the acronym 'TTC' — this one the Telecommunication Technology Council. The TTCouncil is actually not so much an institution in its own right as it is a composition of a number of committees which serve in consultative roles to the Bureau. The key committee for international standardization of telecommunications is called the CCITT Committee, an organization of 20 members nominated by the chairman of the TTCouncil. Its consultative work is done by some 200 subcommittee members and around 500 technical researchers, organized into seven subcommittees, and appointed by the Chairman of the CCITT Committee, or the Chairs of the seven subcommittees, from manufacturers, carriers, associations, or universities. Many of the participants are, of course, from the same firms which serve in the TTCommittee, and, for that matter, many are also participants at the CCITT itself, so there are linkages which allow the coordination of standards work throughout this complex set of institutions.

Despite the appearance of pluralist complexity, the political reality is that most telecommunication standards activity in Japan is dominated by NTT. The Communications Policy Bureau of MPT decides long term standards strategy, using the consultation of the TTCouncil's various committees both for input and as a means of communicating strategy to the industry. The TTCommittee serves as a forum for the discussion of national standards issues by the interested parties. KDD, as the primary international interconnect carrier, plays a significant role with regards to international network interconnection issues. But the fact is that the single biggest purchaser of telecommunications equipment in Japan is NTT. The input of NTT in this virtually monopsonistic environment is afforded enormous weight by the manufacturers, large and small. Moreover, historically the research facilities and staff of NTT were used by MPT almost exclusively for input into the development of standards strategy and for representatives to the CCITT. These economic and historical factors have led to NTT's dominance of the process. This dominance has been somewhat mitigated in the last few years as there has been growing participation in the work of the TTCommittee.

Recently there has been a call in Japan to change the process in order to have the TTCommittee, which more closely than any other institution in Japan represents all the interests in private sector telecommunication, involved in the development of Japanese contributions to the CCITT. (The development of contributions to the international level is referred to as the 'upstream' process, while the recognition of international standards as constituting standards in Japan is called the 'downstream' process.) Such a reorganization of responsibilities would be extraordinary, and seems unlikely. Since the liberalizing effects of the 1985 Telecommunications Business Law, the MPT, through the work of the TTCouncil, sets the long term strategy for telecommunication standards in Japan. It is conceived and implemented as a part of an industrial and trade policy for the telecommunication industry. (With regard to trade policy, it should be pointed out that there is, in Japan, no equivalent to the Office of the U.S. Trade Representative, responsible for negotiation and promotion of trade in general. Trade policy in each separate industry is left to the ministry responsible for that industry.) MPT, however, relies heavily on NTT in the process of setting standards to carry out standards strategy.

While it is conceivable that MPT officials may wish to increase their authority over the upstream process it is unlikely that they would view the involvement of the TTCommittee as a good tactic. Membership in the TTCommittee is open and includes both Japanese and foreign firms. MPT can reduce NTT's dominance by appointing representatives of other firms to the CCITT Committee and other committees of the TTCouncil. Such appointments have, in fact, been occurring recently. Since appointments to the committees of the TTCouncil are within the authority of MPT, allowing other voices into the upstream process through this channel allows MPT to assert itself and recognize other firms' concerns gently and selectively, rather than giving authority over to a private sector institution.

It should also be said that there is a desire that the exact division of authority and responsibility between the bureaucrats of MPT, the executives of NTT and KDD, and the rest of the industry be left somewhat ambiguous. The political culture of Japan requires that power relationships which may involve conflict remain ambiguous in order that there be no clear winners and losers. So long as NTT (and, where international interconnection is involved, KDD) use their dominance in the standards process not only to improve their own networks and businesses, but also to maximize the international market opportunities of the major Japanese telecommunication manufacturers, the present system will continue to work. NTT and KDD officials, of course, recognize this. Foreign manufacturers and competing carriers have opportunities to have input — more in the TTCommittee than elsewhere. Their desire to expand their input and power in this arena leads them to support calls for involving the TTCommittee in the 'upstream' process — development of contributions to the CCITT. The present situation, however, is relatively stable. None of the more powerful participants' interests are at risk beyond their capacity to have influence, and even the relatively weak have the capacity to participate and keep themselves informed.

NTT's dominance of the standards process has the effect of allowing NTT to coordinate its contributions and negotiation of standards to serve its long term view of the telecommunication market in Japan. Although MPT now has responsibility for standards strategy and seems to be interested in expanding this role, it, too, views standards work as a means of furthering long term telecommunication industry strategy. As Chalmers Johnson (1982) pointed out in his study of Japanese bureaucracy and economic policy, Japanese bureaucracies perform much more a developmental function than a policing or administrative function. If there is a difference in perspectives between NTT and MPT, it is minor. NTT seeks to plan a network which is optimized for providing the services which fit into its own view of the telecommunication marketplace of the future. MPT has responsibility not only to plan the development of the network to serve future demand, but also to promote the profitability of the telecommunication manufacturers and the balance of trade in telecommunication goods and services. Given that NTT has shown itself to be sensitive to the health of the 'NTT Family' of telecommunication manufacturers, the interests of NTT and MPT are very similar. Their use of standards work to pursue those interests is almost never in conflict and is usually quite consistent.

2.2. The MITI sector

In the MITI sector, the respective roles of the public and private sector institutions are both more clear and more stable. One division of MITI is the Agency of Industrial Science and Technology (AIST), which has a Standards Department. This department is internally divided into industry specific divisions. The Electrical, Electronic and Information Standards Division assigns information industry standards questions to one of eight industry associations, depending on the subject matter and the relative workloads of the associations. The study group of the appointed industry association researches the issue and prepares the text of a draft Japanese Industrial Standard, which is then submitted to the Japanese Industrial Standards Committee (JISC).

JISC is not formally a part of MITI, but it was created at the direction of MITI in 1952 and is administered by the AIST. It consists of 8700 representatives of manufacturers, user groups, consumer groups, and universities, organized into 29 divisional councils and 1010 technical committees. JISC does all long range planning for standards development in the industries for which it is responsible. It has the power of final approval of Japanese Industrial Standards sent to it as drafts from the industrial associations. In addition, JISC is the sole Japanese representative to the private sector international standards organizations, ISO and IEC. Telecommunication equipment standards questions which fall under the responsibility of MITI are therefore subject to a process characterized by substantial public sector planning and policy coordination with broad private sector involvement and representation.

3. The setting of telecommunication standards in the U.S.

Prior to the divestiture of the operating companies by AT&T, telecommunication standards in the U.S. were set by AT&T as a matter of its dominance of the U.S. telecommunication industry. Divestiture created powerful new players in the industry that collectively recognized the need for a standards setting process. The result was the Exchange Carrier Standards Association, an organization of carriers created — even before divestiture took effect — to fill the standards void.

In the U.S., standards are primarily a private sector matter. Committees representing the firms in each industry set their own standards voluntarily, subject only to accreditation of the fairness of their procedures by the American National Standards Institute (ANSI). The procedural requirements of ANSI are openness, due process, balance, and consensus. ECSA's efforts to create a standards committee for telecommunication ran up against the requirement of balance — roughly defined as the opportunity for all parties to participate without dominance by any single interest. The four kinds of interests in telecommunication standards have been defined as (1) the manufacturers; (2) the local exchange carriers; (3) the interexchange carriers; and (4) general interests (e.g., users, consultants, government agencies). To achieve the requisite balance, ECSA proposed Committee T1 and solicited participation. As would be expected, the greatest number of applications for membership came from manufacturers, large and small alike. In a newly and hotly competitive telecommunication equipment industry, most manufacturers would wish to steer standards along lines where they have advantages.

To protect balance, the membership of T1 was limited such that no one of the four interests can make up more than one-half of the total membership. At present the committee has 43 manufacturer members, 18 exchange carrier members, 10 interexchange and reseller members, and 15 general interest members. The work of T1 is done in the working groups of six Technical Subcommittees with specific technical portfolios. Draft standards which are agreed and approved at each level on the basis of consensus — defined as substantial agreement, much more than a simple majority but not necessarily unanimity — are referred to the American National Standards Institute (ANSI). ANSI's Board of Standards Review verifies evidence of consensus and procedural fairness and, if satisfied, recognizes the draft and publishes it as an American National Standard. The entire domestic process occurs in the private sector and with extensive protection of mandatory pluralist procedures.

Representation of U.S. interests in the CCITT is regarded as a diplomatic matter, falling within the responsibility of the State Department. The State Department has only a small staff with subject matter expertise in telecommunication. This role is carried out by the Office of Telecommunications and Information Standards (OTIS), within the Bureau of International Communications and Information Policy. The work of the OTIS is twofold. First, they organize and direct the work of the U.S. National Committees for CCITT and CCIR. Second, they manage the representation of U.S. interests at the CCITT and CCIR.

The U.S. National CCITT Committee is a set of four advisory Study Groups, open to participation by any interested firm. The firms represented in these study groups tend to be the same firms that are members of T1 working groups. The charter and method of functioning of the U.S. National CCITT Study Groups, however, are quite different from those of the related groups within T1. T1 Chairs are elected by the members. Their output has the authority of a recognized and approved process. U.S. CCITT Chairs serve at the pleasure of the State Department. The output of their groups is advisory to the State Department and to its appointed delegations to the proceedings of the CCITT.

Because USCCITT Study Groups do not consider such matters as potential conflicts with U.S. national or international policy, national security, or embarrassment for the U.S. at international meetings, the U.S. National CCITT Committee's conclusions are not binding on the delegations' disposition of contributions to the CCITT. Because of the emphasis on pluralism and participation in the U.S., and because of the advisory nature of the U.S. National CCITT Committee, normally the U.S. CCITT Committee waits for substantial agreement to have been achieved in T1 before attempting to agree on a contribution to the CCITT. The T1 Working Groups are considered the appropriate fora for the consideration of objections and for the debate over the wording of documents. On the other hand, it is also normal for T1 to work closely with the U.S. CCITT Committee and to delay sending any draft standard to ANSI for recognition as an American National Standard until after the work of the CCITT on the issue. This delay is a recognition that deliberations at the CCITT, agreed to by the State Department's delegation, may substantially differ from agreements previously reached at T1 and may therefore require reconsideration of the issue by T1.

The relation between standards agreements at the T1 level and standards agreements at the CCITT level is therefore a complex one. At T1, standards are arrived at by consensus among the members, one half of which are manufacturers in competition with each other and with foreign companies. But the representation of the U.S. position at the CCITT is subject to the authority of the U.S. State Department. Because of the increasing importance of access to foreign telecommunication markets, the CCITT's Recommendations are increasingly treated as authoritative at Committee T1. Management of the standards process and its complex mix of reliance on authority and consent requires that no individual firm's strategy or view of the marketplace be allowed to dominate the process in the U.S. Given the general lack of consensus on long range views of the telecommunication and information service marketplace, standards which directly affect the ability of the network to serve future demand are hard to achieve. The greatest consensus on telecommunication standards in the U.S. is on matters which are viewed either as neutral with regard to what the nature of the market will be or are viewed as suited to serving present, i.e., known, demand.

With regard to the division of authority over standards in the network and standards in the terminal equipment areas, there are other standards setting institutions in the U.S. including the Institute of Electrical and Electronic Engineers, the Electronic Industry Association, and the Telecommunication Industry Association, among others. Moreover, when a particular standard is perceived as moving too slowly or in the wrong direction, a firm may engage in standards 'forum shopping'. Unlike the Japanese case, however, there is considerable formal as well as informal liaison between the T1 Working Groups and each of these other organizations in order to avoid duplication of effort and potential conflicts. Since all of this work falls in the private sector and largely involves the same firms, there is no comparable division of authority between ministerial bureaucrats, nor any justification for a failure to coordinate.

4. Case study: BISDN standards

In using BISDN as an example of the differences in standards processes in the U.S. and in Japan, the proper focus is on the way the two nations have reached domestic agreements and then have participated at the international level in the movement toward international standards. There have been a great many issues with regard to BISDN about which there have been differences along national lines. Two issues stand out as especially interesting:

(1) SDH (SONET) frame format; and
(2) ATM payload type 2 — real time variable bit rate.

With regard to these issues, the Japanese pattern has consistently been that they have come to the forum with a solid position, usually developed almost exclusively by NTT but completely supported. They have, however, been very accommodating and willing to compromise if the product of such compromise promises to be a true international standard. This reflects NTT's dominance of the domestic process, NTT's interest in optimizing the network in Japan to serve its own vision of the Japanese market, and its willingness to compromise wherever that offers an opportunity to maximize the market for Japanese telecommunication manufacturers. This also serves KDD's position as the international interconnect carrier in that a true international interconnection standard makes it easier for KDD to serve its market.

On the U.S. side, the differences in the domestic situation are reflected at the international level. Because consensus in Committee T1 is reached only after a pluralist process with a guarantee of balance, there is a sense that the positions which go forward from T1 should not easily be modified or bartered away at the CCITT. On the other hand, the State Department's Office of Telecommunications and Information Standards makes it quite clear that the conduct of negotiations at the CCITT is the responsibility of the State Department, and that all input from USCCITT Committee members and RPOA and SIO representatives is advisory and must conform to official U.S. positions. On the face of it this provides for an inevitable tension among the company representatives who work at T1 for their companies' interests and then participate at CCITT where the agreements reached at T1 may be given away. Two factors tend to mitigate this tension. The first is that the State Department tries, where possible, to include the individuals most involved in the resolution of issues at T1 in the consideration of those issues at the CCITT. This means that compromise will only be made by individuals who are fully aware of the positions of all of the U.S. players. Second, on most issues one person clearly takes the position of being both a technical and a strategic leader. Whether because of company interests or personal interests, it is frequently the case that one member of a T1 working group or sub-working group will devote such substantial time and energy to an issue that it becomes 'his' or 'hers'.

In either of these cases a peculiar irony results. In the Japanese system NTT dominates the domestic process and has the power both to determine initial positions and then to negotiate them away. But for cultural and structural reasons it behaves with exceptional care for the interests of the manufacturers and others. In the U.S. system the structure not only allows but demands enormous input from a broad range of interested parties. But on the issues of a new technology like BISDN, one representative of one company can come to dominate the U.S. contribution and the negotiation of the U.S. position at the CCITT, creating the risk of a fairly unilateral decision-making process. Again, the causes of such a situation are both cultural and structural.

4.1 SDH (SONET) frame format

At the Hamburg meeting of Study Group XVIII in May of 1987, the U.S. delegation tentatively gave up the U.S. position of 13 rows of 180 octets and agreed to accept an SDH frame format of nine rows of 270 octets. Their stated position, however, was that they had no power to give up on the 13 row agreement which had come out of Committee T1, and would have to send the matter back to T1 for authorization. This frustrated the Japanese and European (especially British) delegations, which have the authority to negotiate for their countries. Two changes in the U.S. participation at CCITT have since come about. First, T1 has begun to delay final consideration of any standard pending CCITT resolution. This recognizes the increasing importance of international standards to U.S. manufacturers. Whereas at one time the U.S. was a telecommunication island with little concern for global standardization, the market is now a global one and both U.S. manufacturers and U.S. carriers want global standards. In addition, U.S. delegation meetings are held every morning at the CCITT and in some cases the attendance at these meetings includes virtually all of the interested and active members of the T1 working group responsible for the issue under consideration. This allows for some certainty that compromises negotiated at CCITT will be assured of getting support at T1.

4.2. ATM payload type 2 — real time variable bit rate

NTT's view of the Japanese market for switched broadband services is heavily oriented towards video applications. Under the marketing label of Visual, Intelligent and Personal Services (VI&P), NTT plans a variety of services which will require the ATM type 2, variable bit rate payload. For this reason, there has been considerable work at NTT on defining this transmission type. In the U.S., however, there are many competing visions of how video images can best be digitally coded and compressed for transmission. While some of the methods involve variable bit rates, many are continuous bit rate services. With little clarity on what will come out of the codec, there is little interest in the U.S. on standardizing what will be carried through the network. Moreover, in the U.S. the focus on broadband services is on the early market, which is seen as interconnection of Local Area Networks, initially by virtual private circuits but eventually on a switched basis. Video services are seen as creating little near term demand and speculative long term value. As a result, even though NTT would prefer to be moving forward on defining this type of ATM payload, there has been almost no discussion at the CCITT. This reflects the unwillingness of the U.S. delegation to pursue the issue and a sense of futility on the part of the Japanese delegation.

A third issue also bears mention here as relevant to the differences between the Japanese and U.S. standards processes, although the outcome is not yet clear. This issue is the generic flow control, or control of multiple access at the ATM layer. NTT has taken a strong lead in developing a position on this issue. They have worked closely with a representative of British Telecom. (Interestingly, the original SDH frame proposal of 9 × 270 was developed by NTT and BT jointly.) This proposal differs markedly from an earlier proposal which had been developed by the Australians and had significant support in the U.S., and elsewhere. Within the U.S. delegation there was some sentiment that frequent accommodation by the Japanese on other issues deserved some payback and that this might easily be a suitable issue for a payback to occur. The U.S. delegation indicated its willingness to give in on the question but others, including the Australians and some of the European countries, would not agree. The issue remains unresolved at this point.

There are, of course, many other interesting and still contentious issues in the negotiation of BISDN standards. Among those with the highest profile would be:

(1) ATM payload types 3- and 5-alternatives for connection oriented fast packet data services.

(2) Congestion control.

(3) Signalling switched connections.

It will be interesting for a number of reasons to watch the resolution of these issues. One of those reasons is that each resolution will in some way depend upon the cultural and structural differences among the nations involved with regard to how they set standards.

Reference

Johnson, C., 1982. *MITI and the Japanese Miracle: The Growth of Industrial Policy* (Stanford University Press, Stanford, CA).

Global Telecommunications Strategies
and Technological Changes
Edited by G. Pogorel

Chapter 9

Centralization versus decentralization of data communications standards

W. Edward Steinmueller

Deputy Director, Center for Economic Policy Research, Stanford University, Stanford, CA, USA

1. Introduction

The rapid evolution of fiber optics and other digital-based telecommunications media promise a dramatic increase in the types and capacities of telecommunication services in this decade. For example, a U.S. government study reported, "Fiber optic systems are being rapidly added to the mix of information-distribution technologies. The single-mode fibers now being installed in the country's telephone plants have communication capacities significantly greater than copper wire or coaxial cable. One fiber can carry the entire Encyclopedia Britannica from Washington to Baltimore in a second or 300 simultaneous television channels within a city. Laboratory results suggest that *this capacity may rise 1,000-fold or more* with improvements in signal transmission and detection equipment that can be used with existing fiber lines." [1] For economists and others interested in the telecommunication industry, these technological developments offer intriguing questions about the organization of production, the evolution of demand, and the formation of markets to bring this capacity into utilization.

This paper continues a line of criticism of commonly held preconceptions about the likely path of development for using this increase in telecommunications capacity developed in David and Steinmueller (1990) and in Steinmueller (1992). David and Steinmueller (1990) argue that ISDN data communications are unlikely to replace the increasingly complex network of high capacity inter-LAN network linkages based on leased lines with a

---

[1] Emphasis added. See U.S. Congress, Office of Technology Assessment (1986, p. 161).

centrally coordinated high capacity data communications network. While a common ISDN network is an attractive engineering simplification of the increasingly complex inter-linkages that are developing in everyday practice, issues of pricing, access, and flexibility complicate achieving the ISDN simplification[2]. Similarly, Steinmueller (1992) criticizes existing end-user data *services* for their emulation of the broadcast media, arguing that the specific advantages of such services are likely to be found in their ability to offer users customized and responsive methods of acquiring information relevant to their individual needs. The common thread in these two papers is their open scepticism of the social utility and private profitability of replicating a centralized system of mass-distribution in which specialization and variety are suppressed in order to create a mass market based on a homogenized product standard.

Two lessons from deregulation of telecommunications services are (1) purchasers are eager to acquire specialized and often mutually incompatible equipment that will meet their specialized telecommunications needs, and (2) suppliers are anxious to define new market services and create new equipment that accelerates the pace of technological change in telecommunications with little regard for universal connectivity. These developments seem certain to block the re-emergence of a single integrated telecommunications system based on centralized decisions about the rate and direction of capacity expansion, technical compatibility standards, and rate-setting. Substituting market competition for centralized management, however, is neither an instantaneous nor costless process. Regulatory authority decisions about access and pricing will have a major impact on the extent of coordination failure and pricing distortions that arise from 'deregulation' and market competition and, in the limit, can reimpose centralized modes of industrial organization by discouraging competitive entry.

This paper focuses on two areas where persuasive arguments in favor of centralized management have been made in the past and will be made in the future, technological standards for data communications and the development of broadband services for end-users. In both areas, arguments for centralized decision-making are critically examined and found to have significant shortcomings. Decentralized, market-based approaches, however, also have their own shortcomings and requirements, suggesting constructive uses of market intervention by regulatory or legislative authority. The aims of this paper are

---

[2] Much of the 'simplification' of ISDN is likely to disappear during implementation when pricing will need to preserve the elements of price discrimination necessary to finance capacity and dynamically respond to competition, where access is a politically sensitive issue, and where flexibility has been significant in user preferences and in the entry of equipment suppliers offering differentiated products.

to direct attention at areas where research is likely to inform policymaking and to identify issues that are currently being neglected.

2. Technological standards for data communications

Few areas of telecommunications offer as marked a contrast in market organization as data communication networks in Europe and the United States. In Europe, public switched packet networks are the predominant method for linking computer networks at different sites. In the U.S., private leased lines utilizing the SNA standard and, to a lesser extent, the Decnet standard (increasingly using packet transmission methods) are the dominant methods of inter-site data communications.

Data communications in Europe and the U.S. have evolved over almost three decade with the outcome that, in the U.S., data communication volumes appear dramatically larger than in Europe. The extent of coincidence between rates of utilization and the relative importance of public packet data switching in the U.S. and Europe requires an explanation. I will begin by considering the causes of the relative prevalence of packet switching and leased line services in the two areas. One possibility is that the needs for integration of European data traffic combined with centralized decision-making of the PTTs smoothed the adoption process, encouraging users to adopt PAD (packet assembly and disassembly) technology at a faster pace than occurred in the U.S. A second explanation is that regulatory prohibitions on AT&T prior to divestiture, combined with the earlier widespread adoption of SNA and DNA, tipped the adoption process in the U.S. toward leased line service and the resulting excess 'adoption momentum' has frustrated the growth of public packet switched data networks.

Since it is generally accepted that packet networks offer technological efficiencies in data communications (see Heldman, 1988, p. 200; Rosner, 1982, chapter 3), these alternative explanations about the prevalence of packet switching have important implications for policy. If the first explanation of Europe's more widespread utilization of packet data networks is accepted, it follows that 'coordination' failure in telecommunications markets in the U.S. may have imposed substantial costs on data communications users realized in higher payments for under-utilized leased lines, in inflexibilities of inter-connection, and in excess costs of using dial-up lines for data communications[3]. If the second explanation is more accurate, any excess costs of the

[3] I adopt the simplifying assumption that the same volume of traffic may be carried at a lower actual cost utilizing public switched packet networks than can be achieved using leased lines. Whether the tariffs on such traffic are set to reflect actual cost is a separate issue. Substantial cross-subsidization among services may exist in both Europe and the U.S. as the result of private aims such as price discrimination or social policy such as subsidization of residential service.

leased line network may be attributed to historical accident and regulatory burden. In addition, however, the size of the social welfare loss arising from coordination or regulatory failure depends on differentials in pricing. It may well be true that social gains available from European packet switching have been reduced by tariffs on these services, particularly across national boundaries, to a level where the U.S. losses from coordination or regulatory failure are small compared to the losses in European welfare from pricing decisions. Some insight into the size and nature of markets is necessary before we can evaluate these alternatives.

As of 1985, AT&T and the Bell Operating Companies accounted for 81% of the $7.5 billion in U.S. data communication revenues (Heldman, 1988, p. 200). By 1990, the total market had grown to $12.1 billion with the carriers accounting for $9 billion or 75% of the market (Mulqueen, 1990, p. 96). A significant factor in the slippage in the market share of carriers has been the growth of specialized service providers such as Teleport Communications Group and Metropolitan Fiber Systems Inc. and bypass carriers such as Locate Inc. and Eastern Microwave Networks Inc. (Mulqueen, 1990, p. 96). The predominant share of revenues realized by U.S. telephone carriers comes from the provision of leased lines and, to a lesser extent, dial-up service. In 1985, 70% of market revenues and traffic passed through leased-lines while 14% of revenues and 28% of traffic came from dial-up lines [4].

A 1985 breakdown of the extent of public switch packet networks in the U.S. shows that only about 8% of the total data communications market is served by public switch packet networks and that the telephone carriers account for less than 1% of the total market (Heldman, 1988, p. 207) [5]. In 1985, U.S. private packet networks outnumbered public networks (provided by VANs and the telephone carriers) by a factor of 2 to 1 (Heldman, 1988, p. 208). In short, public switched packet networks play a minor role in the U.S. market. Publicly switched packet network services are, however, only part of the market for packet data transmission since a leased line may be used for packet data transmission with the appropriate PAD equipment. The U.S. market for X.25 packet equipment (PADs, packet switches, etc.) is larger than in Europe (Vertical Systems Group, 1989). This indicates that while U.S. telecommunications users are not using public switched packet networks, their private networks using leased lines heavily utilize packet transmission methods. Moreover, this observation indicates that SNA (IBM's standard for inter-site data communications) and Decnet (a similar product

[4] Note that the VAN and satellite services received a considerable premium on the 2% of user traffic they carried by earning 16% of industry revenues (Heldman, 1988, p. 207).

[5] Value added network service (VANS) providers such as Tymnet account for the balance.

from Digital Equipment Corporation) have not locked users into proprietary data communications solutions. In fact, while SNA and Decnet were originally based on proprietary standards, both are now compatible with the X.25 packet standard.

In Europe the development of packet switching networks has dominated the growth of private leased line service. As of 1988, there were a total of 465 thousand leased line circuits within the EC compared with approximately 416 thousand inter-exchange leased lines in the U.S. in 1987 and 2.9 million local exchange carrier unswitched (leased) lines in 1986[6]. Although the early adoption of a technologically compatible standard within Europe for public switch packet data networks has been heralded as the basis for a European wide data communications markets, international data communication tariffs within Europe (Ungerer, 1990, pp. 114–116) and, in some nations, utilization of more sophisticated (and therefore more expensive) data terminals[7] as well as other factors have resulted in a modestly sized packet switch data network. In other words, even though public switch packet data networks are predominant in Europe, the overall size of these networks as measured by connected terminals or X.25 nodes is modest. In 1985, the total number of packet switch terminal ports in Europe was estimated at 69,000 (Gabler, 1986, p. 212[8]) and the total number of terminals of all types (including asynchronous terminals) in 1986 was estimated to be 500,000[9]. By comparison, in the U.S. there were approximately 11 million asynchronous terminals in place in 1986 as well as an additional 3 million IBM 3270 type terminals. While it is difficult to compare the number of packet exchanges in the U.S. and Europe, in 1986 there were some 1,932 public VAN nodes in the U.S. (U.S. Department of Justice, Antitrust Division, 1987, p. 5.6) compared with approximately 200–300 X.25 exchanges in Europe[10]. For publicly switched networks the relevant comparison is between AT&T nodes [approximately 100 (U.S. Department of Justice, Antitrust Division, 1987, p. 5.6)] and the 200–300 X.25 exchanges in Europe.

---

[6] The first figure is from Ungerer (1990), p. 52, the second from Vertical Systems Group (1989), p. 52 and the third is from U.S. Department of Justice, Antitrust Division (1987), table PA.7 following p. 5.10.

[7] Some 84% of terminals in Germany that were connected to the packet switch network in 1984 were themselves X.25 standard terminals (Gabler, 1986).

[8] Gabler reports that in April of 1985 the German packet switch network consisted of 35 exchanges in 17 locations.

[9] Ungerer (1990) citing ITU Yearbook of Common Carrier Statistics for reference year 1986.

[10] This latter figure is a rough estimate. As noted earlier, Germany had 35 X.25 exchanges in 17 locations and Gabler (1986) estimated 13,000 terminals in use in Germany. Applying the same terminals per exchange figure Gabler reports to his estimates of terminals in other European countries results in the 200–300 figure reported in the text.

By this standard, the regulated public utility component of the U.S. packet switch market is considerably smaller than that of Europe. At the same time, however, offerings by private companies to the public are enormously larger in the U.S. than in Europe. In short (1) despite common technical standards, Europe's data communication market appears to have been much smaller in the mid-1980s than the U.S., and (2) AT&T's share of the U.S. market appears to have been somewhat smaller than European PTTs and very much smaller than the private U.S. companies offering packet switch services to business.

The small size of AT&T's position in the U.S. market can be attributed to regulatory history. High-speed leased line data communication services began in the late 1970s through AT&T offerings of DDS. The provision of packet data switching services was prohibited to the Bell Operating Companies until May, 1983 when the FCC allowed Basic Packet Switched Service (BPSS), an unenhanced version of X.25. At the same time, however, AT&T was allowed to offer T-1 lines which operated at twenty four times the highest speed available from DDS lines (56 kBps). Larger users began migrating to T-1 leased lines rather than BPSS. This migration was enhanced throughout the 1980s by equipment vendors offerings X.25 packet equipment for leased line service with enhanced features. The FCC has gradually relaxed its position on packet switching, allowing packet protocol conversion in 1985 and, very recently, a much broader range of services. At the same time, however, enhanced leased line services such a frame relay are increasing the attractiveness of leased line services compared with public switched packet networks. Hence, the deregulation of packet switching in the U.S. appears to have been a regulatory case of too little and too late, significantly constricting the growth of public switched packet networks, and suggesting a substantial regulatory burden [11].

The extent of regulatory burden on packet switching is, unfortunately, not limited to the delays in deregulation of public switched packet networks. In pricing, the U.S. and Europe have experienced remarkable divergence in regulatory practice. In the U.S., tariffs for the use of intra-LATA (local access and transport area) switched voice grade lines and of interexchange traffic have subsidized leased and voice grade lines while discouraging the use of packet networks. Tariffs on intra-LATA calls have been held flat regardless of the duration of the call and in most areas unlimited local calling is permitted for a single monthly access charge. Such tariffs dramatically reduce the cost of using dial-up data lines as is indicated by the above estimate that 28% of data communications pass through dial-up lines while generating only 14% of

---

[11] Lest the reader conclude that this outcome was the result of regulatory error it should be noted that the FCC and state regulators have been concerned throughout this history with the possibility that AT&T and later the LECs would utilize their voice customer rate base to subsidize market expansion in data communications.

data communication revenues. State regulatory authorities which in the U.S. have authority over local rates have been reluctant to replace the flat rate system with measured service in which charges would be usage dependent, believing that voice users would experience an increase in rates. Moreover, it is impractical to distinguish which calls are data rather than voice. The result is that it is common to set flat local rates on the assumption that the typical residential phone generates 3 calls per day of 3 minutes, with somewhat higher 'representative' use by business phone service. The effect of this tariff policy is that voice users subsidize data users as the latter occupy a growing share of the capacity utilization of local telephone exchanges. In Europe, higher tariffs for voice grade lines and the absence of interconnection opportunities for data communication suppliers have prevented the arbitrage of subsidized local service by data communications companies.

The problems introduced by flat rate billing in the U.S. are further complicated by access charges for the use of inter-LATA calls. While local exchange carriers must pay approximately one third of their revenues for the use of inter-LATA switches and transmission capacity, companies that bypass the inter-LATA network do not have to pay these access charges. Thus, a voice call and a data call using the same amount of local exchange capacity will pay different interexchange costs, the voice call pays the access charges while the data call may escape them. As Heldman (1988) observes, this is another subsidy of data by voice communications. Nor should it be surprising in light of these details that data calls account for twice as much network usage as they do revenue to local exchange carriers. In Europe, regulatory decisions that limit the extent of inter-exchange bypass opportunities also constrain the market for data communication services.

The above account suggests the following conclusions. The avoidance of public switched packet switching for data communications in the U.S. is not the consequence of technological factors such as SNA and Decnet's early adoption; the market for packet switching equipment in the U.S. while proportionately smaller than that in Europe is, nevertheless, absolutely large. Nor does market coordination failure appear to be a plausible explanation; offerings of public switch packet services have been quickly paralleled by enhancements in leased-line services suggesting that the telephone companies perceive public switched network services as an inferior business strategy for developing data communications at the tariff levels allowed by regulators. While European public switch packet networks have attained a larger market position than AT&T, their size was much smaller than the publicly available packet switching offered by private companies as of 1986, an outcome that appears to be the result of European pricing decisions. While markets have grown rapidly in the past six years, the basic picture of a modest U.S. telephone company participation and a lagging European data communication market appears to hold true to the present. Given the technological superiority of public packet switching for efficient network utilization and the revenue losses from data communications occurring within LATAs in the U.S., the most plausible explanation of telephone company business strategy is that regulatory restrictions have tipped the playing field away from public switch packet networks in the U.S. and pricing and interconnection restrictions are responsible for the lag in the development of the European data communications market.

The larger significance of this controversy is its prospective application to the issues of future telecommunications services such as ISDN or, more generally, broadband network services. Because U.S. policymakers are deeply concerned about issues of international competitiveness, the argument of a technological gap similar to that suggested by European authors such as Jacques Servan-Schrieber (1968) almost twenty five years ago may be advanced in the U.S. Such an argument would proceed from the premise that regulatory intervention in Europe averted a market coordination failure that has been responsible for U.S. failure to build an advanced public infrastructure in data communications. This argument would have much more weight if the European market, with economic integration, and with convergence of data communication trends to the U.S. began to dramatically expand the size and usage of its public data network. The analysis above suggests that this outcome might occur with changes in European pricing and interconnection opportunities [12].

There is, however, an additional possibility that does not directly fit in the above explanation. Despite regulatory burden and price differentials, the U.S. market may have grown more rapidly because the variety of offerings produced using leased line services has introduced a range of solutions that better meet user needs in the U.S. market. If this explanation is correct, than it would be incorrect to conclude that the U.S. *at this stage in its market evolution*, would benefit from the construction of a publicly switched data communications network. Moreover, the possibility that regulators would grant attractive tariffs for investment in broadband capacity to hasten the replacement of the 'obsolete' leased line services with universal interconnectivity might well result in a social welfare loss compared with other paths of future development of the U.S. system. Correspondingly, by the same argument, European regulatory changes involving interconnection might have

[12] The market for value added network services in Europe still lags substantially behind the U.S. In 1988, the U.S. market as $5.96 billion compared with $2.07 billion in Europe (Genschel and Werle, 1992, p. 20).

substantially larger effects than changes in tariff (although revision in inter-country tariffs seem essential to build pan-European economies of scale in data communication services).

While a universal connectivity outcome might occur through market competition in either the U.S. or Europe, it should not be mandated through regulatory fiat without considering a full range of competitive alternatives. For example, regulatory fiat about interconnection could preclude the entry of CATV, cellular, or other local loop service providers offering decentralized hubs that could concentrate traffic and perform switching operations. Correspondingly, regulatory decisions about pricing could tip the playing field in favor of the more uniform and integrated solutions, such as those that local exchange carriers or inter-LATA carriers would likely offer [13]. Neither path can be justified by prior failures of market competition or coordination.

3. Determining standards for end-user broadband services

While the previous section concludes that universal connectivity is not a sinecure of social welfare, this section takes up the issue of uniform technical compatibility standards for data communication services with a similarly sceptical conclusion. The relationship between technical compatibility standards and market structures are complex and subtle. In markets with large numbers of inter-related components the definition of technical compatibility standards may well dictate the possibilities for entry and either complicate or simplify the coordination problems of providing reliable products and services.

One of the most striking illustrations of this principle is the relation among personal computers, operating systems, and application programs. During the past decade, IBM and IBM-compatible personal computers have garnered an enormous market share by using relatively 'open' standards encouraging entry and price competition. Despite its success, this strategy has been criticized for failing to deliver to IBM larger rewards in creating the standard, for retarding technological progress in hardware components such as graphic displays and in software components such operating systems, and for promulgating idiosyncratic and time-consuming variety in user interfaces in application software. Nonetheless, the market defined by the standards associated with IBM and IBM-compatible personal computers has offered many entry opportunities, and the resulting competition has been fierce even if, in certain areas such

---

[13] This is the implicit suggestion of Genschel and Werle (1992) in their prescription for more international political control of the standardization process.

as in Microsoft's operating systems, a dominant competitor has emerged [14]. Moreover, both IBM and Microsoft have (so far) avoided the temptation to use their capability to redesign standards in ways that would complicate the coordination problems of offering complementary hardware and software products [15].

By contrast, Apple Computer's Macintosh series has been predicated on exclusive control of the operating system and, through this control, an exclusive right to produce the hardware in which this operating system is embedded. At the same time, Apple has actively promoted the opportunity that their system offers for entry of software application products. The tight control that Apple has maintained over the operating system has contributed to high returns on investment, to rapid progress in some hardware components such as graphic displays as well as rapid progress in the operating system itself, and to greater uniformity in user interfaces — in short, developments that address all of the perceived shortcomings of the IBM strategy. Despite this, Apple has only achieved moderate success in the overall personal computer market and has had severe conflicts with some of the suppliers of complementary products [16].

While it might be tempting to conclude from this experience that 'open' standards are more likely to bring market success and larger public welfare gains, such a broad conclusion is not possible from any single example. In the case of personal computers, control of the operating system and related hardware has proven to be a less conducive to growth than an 'open' standard approach. This does not prove that 'open' standards are either appropriate or inappropriate for other information technology systems. What is relevant, however, is an examination of the factors explaining controlled versus 'open' standards market success.

Since the development of large information technology systems have often been managed or dominated by a single large firm and each system (e.g main-

---

[14] It is important to note that while Microsoft has come to be regarded as dominating or even, by some accounts, monopolizing the operating system market, its market position has been won largely through market competition. Alternatives to MS-DOS and imitators of MS-DOS (including fully compatible augmentations) have existed almost from the beginning of this market.

[15] Readers who find an exception in the micro-channel bus should note that IBM avoided integrating this standard with the architecture of its PS-2 line allowing producers to use the existing AT and the higher performance EISA busses to create machines that were still fully compatible with the newer iAPX microprocessors and that could claim IBM-compatibility in all respects but the micro-channel bus.

[16] The most dramatic of these conflicts has been the effort to dislodge Adobe from its dominant position as a supplier of typefaces and the Postscript graphic description language for high quality printed output, a position that forced Apple to share its rents on printers, and to a lesser extent, on the Macintosh itself.

frame computers, broadcast network television, national cable television, and the telephone network) have unique features, deriving a well-specified set of factors determining market success that could be tested directly is problematic. Nevertheless, several common factors can be identified as candidates for further exploration.

For users of a new information technology system, the extent of complementary investment that they must undertake in order to make use of the system is a major factor determining the rate of its adoption. The size of investment, holding other factors constant, will bias users toward turnkey systems. Moreover, the extent of installed base of compatible systems will accelerate the rate of adoption. While microcomputer communications programs offered inter-machine text and graphics transfer by the early 1980s, this mode of transmission of data was superseded by the development of facsimile transmission using standalone 'fax' machines and it is only more recently that a large market for personal computer-based facsimile transmission and reception peripheral cards has begun to grow at a rapid rate. Facsimile was able to bypass direct transfer of text and graphics files because it offered common standards and, over time, a larger installed base of compatible terminals. Experience with the competition between facsimile and 'microcomputer to microcomputer' file transfer suggests that markets which generate a large number of incompatible solutions to data transfer problems may be subject to coordination failure.

In the same fashion, it is unclear that standardized approaches to the user interface for accessing information services will prevail simply because they promise large network externalities. Users must find such interfaces useful. The variety of unsuccessful experiments with teletext services suggests that a common standard and a significant installed base are insufficient to generate the volume of use that would make such systems economically viable (see Steinmueller, 1992).

Market creation and definition for high capacity (broad bandwidth) data communications services faces similar problems to the above examples drawn from microcomputer, text and graphic transmission, and teletext markets. Of late there has been considerable discussion of public infrastructure investment in data highways and, in the U.S., legislation to extend the current Internet research network's capacities as a public 'data highway' have raised hopes that an accelerated introduction of broadband data communications services may occur at a more rapid pace than was supposed. There have been solid endorsements of video dial tone as a potential service offering and, in the U.S., we are at the beginnings of the enormous struggle to manage competition and collusion among cable TV and telecommunication carriers.

As a thought experiment, suppose that some nation makes a national commitment to install broadband capacity and the problem now confronting us was how to assure its utilization. Even those of us in the economics community who have reservations about when to install the capacity are determined to make it as cost effective and vital a medium as possible once it has been established.

My basic point is that regulators and businesses will choose one of two fundamentally different approaches for utilizing broadband capacity. The first will be reliance on a central actor to define network services. This is what may be expected, the road most likely to be taken, and, I suspect, the road preferred by telecommunications carriers and regulators. It simplifies the process of defining services, establishing rates, and a host of other thorny issues. The second road, the road I suspect we will not travel, is based on decentralization of service offerings, protocols, and the other technical standards that serve to define uses for broadband networks. Each approach has a plan of attack and an underlying economic rationale. Let's begin with centrally defined network services.

The plan of attack for centrally defined network services is to define a menu of network services. This menu becomes the basis for rates, it also serves to create the value added network services such as information utilities, e(lectronic)-mail, enhanced facsimile service, and ultimately television-grade signals. Defining these services is the first step. The next is offering these outlets to producers and consumers. The effect is not unlike the choice of defining a dial for selecting what service you want to tune in and hence I call this opening of pre-defined service markets a 'standard channel' approach.

The effect of the 'standard channel' approach is to fit demand to supply capability. The great advantage of this approach is that when the enormous capacity of the system comes 'on line' we have lots of information services to transmit over it and a guarantee of technical compatibility among terminals for accessing these services. The drawback is that, except for TV, we don't have a very good idea what 'channels' people will really use or pay for on such a network.

The next, decentralized approach, is far messier. The aim of decentralization is to allow service offerings to percolate upward through a competitive process toward broader utilization. Since it defeats the whole purpose of this approach to define services or access methods, we must instead define the basic elements of connectivity, i.e. we must define a method for defining 'channels'. For example, one approach would be to leave open the higher levels of the OSI model for the definition of service providers, creating data transmissions that could be processed by only a subset of the terminal equipment in use.

By creating a process of setting diverse 'channel' standards and service offerings, the decentralized approach leads to numerous technical standards for connection and a commitment to encourage and stimulate the entry of service suppliers. The outcome would be a process of adapting the supply capacity to market demand. The service and interconnection methods that are chosen through market competition are the ones that meet demand. The advantage of this approach is that it assures that users ultimately *want* what they *get*. And if the suppliers are doing their job, users will *get* what they *want* as well. The drawbacks include:

(1) The possibility of market fragmentation in which, for some class of service, too many firms offer competing services and none make a profit. This is a variant of the destructive competition argument which economists deride. But 'ruinous competition' is a credible possibility in the short run for high fixed cost industries and for industries where network externalities play a critical role as in our example of how facsimile prevailed over text transfer in data communications.

(2) The process of open standards is going to be confusing to users. It will resemble the selection problems induced by the fine print on computer software packages that outline the system requirements for the software's use.

(3) We may not gain very much by allowing this freedom to define interconnection standards and services. In my view, however, the notion that we can accurately pre-define broadband service 'channels' that people will use to pay for (except TV) is excessively optimistic.

An alternative view is offered by Robert Wienski in Heldman (1988): "In order for the residence market to develop on an ubiquitous basis, two fundamental prerequisites are necessary. First, user-friendly low cost terminals need to be made available. Second, value added services must also be made available. Expecting the residential market to develop on an ubiquitous basis is probably dependent upon identifying a 'trigger' where the terminal equipment can be cost-justified by the service provided (based upon cost savings?) rather than by the end-user. This is because in residence markets, '*people spend time to save money*' ... while in business *people spend money to save time*." (p. 213, emphasis, ellipsis, and parenthetical question as in original). While this comment stresses the importance of the installed base of end-user equipment, in my view it leaves open the questions of how to build market demand for this terminal equipment and how to structure the provision of value added services. Much of this argument relies upon creating a cluster of high value applications, perhaps based on television channels, that address individual segments of the market. Thus, the argument for a universal standard amounts to an argument for some 'base load' demand for capacity that can financially carry the system while experimentation with newer services is attempted. Moreover, if we accept that a decentralized system for creating 'channel' standards has substantial value, a possible compromise is to allow some channels to follow a universal standard, while freeing others for the decentralized process.

There are however reasons to limit the extent of such universal standards because they create important incentive effects or 'imperatives'. The traditional approach of a universal standard for network services is based on the economics of fixed costs. Fixed costs create economies of scale. The extent of these economies depend upon the size of fixed costs. Pre-definition of a very ambitious and comprehensive set of network services and access methods assures high fixed costs. Once the problem is defined in this way, all alternative methods of access and services are threats. They bleed demand away and raise unit costs to those remaining. It therefore becomes an imperative to regulate access and assure that customers use the high fixed cost service.

The second imperative of the centrally defined network service approach is the discovery or creation of a common service type to create a mass market. Of course at least two such mass markets, television and telephone service, already exist. Others or variants will have to be discovered or created, e.g. dial a movie. Regulatory decisions that aid in the extension of universal mass distribution services through pricing or interconnection may therefore tip the playing field away from more specialized service offerings.

By contrast, the decentralized approach undertakes at the outset only to provide a competitive alternative to other methods of achieving links e.g. switched or leased lines. The economics of a decentralized approach dictate going for higher value product or service markets first, before defining mass markets, and perhaps as an alternative to mass markets. This approach recognizes that innovative products or services appear in a form that is far inferior to what they eventually become through incremental improvement and interaction with market demand. As Rosenberg has observed, "It seems to be extraordinarily difficult to visualize and anticipate the uses to which an invention will be put" (Rosenberg, 1976). Moreover, the decentralized approach recognizes the common fallacy that new products and services will be conceptualized in familiar terms or, as Rosenberg says, "Even when an invention genuinely contains important elements of novelty, there is a strong tendency to conceptualize it in terms of the traditional or familiar" (Rosenberg, 1976). If successful, the decentralized approach still repays the costs of network creation by filling capacity with a varied product mix.

Finally, it becomes important to reduce the costs of product specification. For example, if you wish to deliver interactive services based on video and data transmission, often called multi-media, it becomes useful to endorse standards (perhaps market created ones) for distributing this type of service. Carriers can influence these outcomes by encouraging or producing their own complementary services that enhance the value or diminish the costs of such services as they arise and demonstrate their market fitness.

The two different approaches highlight two different types of efficiency. The 'standard channel' approach exploits traditional economies of scale to achieve low cost and thus, hopefully, high utilization. The decentralized approach exploits market competition to select those services and access methods that are most valued. Neither approach is guaranteed to be successful. Pre-defining services that no one will buy is just as bad as creating a flourishing jungle of competing and *inconsequential* niche communication services.

The remainder of this paper will explain a specific example of new services based on a decentralized approach to compatibility standards, the idea of shared domain services. First, I will describe in more detail what I mean by shared service domains. Second, I will illustrate what I see as the problems of the 'standard channel' approach which highlight features offered by shared domain services.

There are two basic methods at present of using the telecommunications network for data communications. The first employs switched or leased lines to establish a private link between two parties. Establishing a terminal to service such links is as simple as plugging a modem into the phone line and launching your communications software. High capacity networks will simply extend the bandwidth for transmitting information over such links. Such improvements are conceivably very important, allowing user interaction with more complex graphic images that could be rapidly rendered on the user's screen [17].

Dedicated links are established to accomplish pre-defined types of activities such as inter-LAN networking or file up- and downloading. Again broader bandwidth permits transmission of certain items in real time (e.g. video conferencing) that were previously unattainable. But the immediate implications for other changes in the types of services that will be possible seem modest. One promising locus for change is in the definition of a new class of service which I call 'shared domain'. The defining characteristic of shared domain service is that each party discloses information (often without direct action) that is used as a complement to the provision of services. A rudimentary example of a shared domain service is the use of caller ID disclosure by mail order companies to 'pre-fetch' a customer record during a telephone ordering session.

The main advantage of the exchange of information that occurs during shared domain services is that we can greatly improve network utilization by building systems that actively service users' real needs for information. At present, connection to any service involves either a prior expectation that an

---

[17] The value of GUIs (graphic user interfaces) are now well-established and are rapidly penetrating the IBM and IBM-compatible personal computer market through the adoption of Microsoft Windows and related applications programs. In the Macintosh and workstation markets, GUIs are the predominant user interface method while most teletext systems are based on some sort of GUI.

information need can be filled or an intent to explore the possible information needs that the service can fulfill.

Both users and services are as idiosyncratic as any computer system. Users don't know what they want, and in most cases services can't describe what they have in ways that users can easily comprehend. The result is like going into a store where all the stock is at the warehouse and the sales clerk only speaks a dialect of your language that you would rather not learn. The role of the shared service is create methods of eliciting from both users and services information that will enhance service. It is akin to the good reference librarian who studies available information services and translates your incoherent information request into a precisely defined query.

At present we do not allow service providers to retain some of the information that would be useful for creating shared services. Nor have we opened most terminals to incoming messages that would allow service providers to advertise their services. Both are concomitants of intensively utilizing higher capacity data networks. In order to understand why I believe shared information is so important let's talk about problems with the 'standard channel' model. The dream of the promoter of the 'standard channel' is that all user's needs for information, entertainment, etc. can be served from the same channel since this results in an enormous revenue at prices that are impervious to competitive entry.

The reality is that competition among channel providers divides the audience. Users benefit from the diversity of offerings but lose from the costs of establishing multiple links and even more importantly lose from foregoing particular services that they would have 'tuned in' had they only known about them. The decentralized approach to defining high capacity networks will certainly continue to offer private and dedicated service domains. Flexible interconnection standards and permissive policies regarding new services will make both of these classes of service feasible in the context of higher bandwidth services. In fact part of the problem is the proliferation of new services competing for users. If a new class of shared domain services arises, however, we will see a new method of utilizing these services in a common or shared domain. Such a domain may be 'virtual', i.e. defined by software under the user *or* service providers control, *or*, more likely, it will be real and defined by a service provider, perhaps a telecommunications carrier, who aims to create a superior service environment for users. Assuming that users are willing to share information about their needs and preferences, and later about their experiences with the actions taken on their behalf by the service, we have the elements of a shared domain service. An analogy that is useful for understanding this approach is the relation between a social network of interacting people and the sort of episodic and random social exchanges that occur outside of social networks.

To be more specific, I suspect there is a hierarchy of shared services. At the outset, the shared service offers the basic tools for accessing services and other user domains. At this level we are simply establishing a social etiquette and context for interaction, i.e. what information will be shared, what are the permissible uses of shared information, and how may individuals achieve connection with a given network. The next level of such a service is to aid the user directly in defining preferences and choices. How may individuals shape their appearance and interests, and what is admissible in terms of misrepresentation of individual identity. In other words, the user will be given the option to explore and customize their interface with the shared service to more directly define what is in the shared domain. Finally shared services may progress beyond being the agent of the user or information service to the role of 'anticipating' user needs or business opportunities. The extent of experimentation required to make these transitions seems to be fundamentally incompatible with a 'standard channel' approach.

Retaining user specific information may seem to be a violation of standards of privacy that offers the potential for misuse. The following points should be noted in this connection: (1) we already are known to others including business enterprises and government in ways that were not of our own volition; (2) we need not represent our 'true' self to the shared service in ways that make us uncomfortable; and (3) a concomitant of all services and most business transactions is improving knowledge of the other party. In short, not adopting some variety of shared information is the same as insisting on being treated anonymously.

So I have come part way back to the beginning. Some common approach is useful for fully exploiting high capacity networks. Whether carriers and regulators are prepared to implement or allow such shared services remains to be seen. What differentiates the centralized and decentralized approaches is the mechanism by which services are defined. Pre-definition is hazardous. Competition appears to lead to a need for fundamental changes in the interaction between users and services. A high capacity network can be based on the broadcast model that treats us anonymously. Or a decentralized approach may be employed where some new organizing principle like shared services serves to guide, service, and inform our individual needs.

The conclusion is straightforward. When offered the opportunity to choose centralized coordination of end-user broadband services, the first question to ask is whether such an offer permits flexibility in the definition of user interfaces and service offerings. If the offering restricts flexibility in order to pursue economies of scale and network externalities, a market test should be a pre-requisite before regulatory intervention mandates a common, and perhaps sub-optimal, standard for such service.

Acknowledgements

I am grateful for research support from the Markle Foundation and the Program in Regulatory Policy at the Center for Economic Policy Research, Stanford University.

References

David, P.A. and Steinmueller, W.E., 1990. The ISDN Bandwagon is Coming — Who Will be There to Climb Aboard?: Quandaries in the Economics of Data Communication Networks. *Economics of Innovation and New Technology* I (1/2), 43–62.

Gabler, H., 1986. Packet Switching in the Federal Republic of Germany (with a View to the Situation of CEPT-Countries). In: L. Csaba, K. Tarnay, and T. Szentiványi (eds.), *Computer Network Usage: Recent Experiences* (North-Holland, Amsterdam).

Genschell, P. and Werle, R., 1992. *From National Hierarchies to International Standardization: Historical and Model Changes in the Coordination of Telecommunications*, February (Max-Planck-Institut für Gesellschaftsforschung, Köln).

Heldman, R.K., 1988. *ISDN in the Information Marketplace*, (TAB Professional and Reference Books, Blue Ridge Summit, Pa.).

Mulqueen, J.T., 1990. Data Communications 1991 Market Forecast. *Data Communications*, December (McGraw-Hill).

Rosenberg, N., 1976. Factors Affecting the Diffusion of Technology. In: *Perspectives on Technology*, pp. 189–210 (Cambridge University Press).

Rosner, R.D., 1982. *Packet Switching: Tommorow's Communication Today* (Lifetime Learning Publications, Belmont, Calif.).

Servan-Schrieber, J., 1968. *The American Challenge* (Atheneum, New York).

Steinmueller, W.E., 1992. The Economics of Production and Distribution of User-Specific Information via Digital Networks" In: C. Antonelli (ed.), *The Economics of Information Networks* (North-Holland, Amsterdam).

Ungerer, H., 1990. *Telecommunications in Europe: Free Choice for the User in Europe's 1992 Market, The European Perspectives Series* (Commission of the European Communities, Brussels).

U.S. Congress, Office of Technology Assessment, 1986. *Intellectual Property Rights in an Age of Electronics and Information*, OTA-CIT-302, April (U.S. Government Printing Office, Washington, D.C.).

U.S. Department of Justice, Antitrust Division, 1987. *The Geodesic Network: 1987 Report on Competition in the Telephone Industry* (prepared by Peter Huber), January (U.S. Government Printing Office, Washington, D.C.).

Vertical Systems Group, 1989. *T-1 Network Industry Analysis* (Dedham, Mass.).

Global Telecommunications Strategies and Technological Changes
Edited by G. Pogorel

Chapter 10

Establishing standards for Telepoint: problems of fragmentation and commitment

Peter Grindley [1,2] and Saadet Toker [1]

[1] *London Business School, Sussex Place, London NW1 4SA, UK*
[2] *Haas School of Business, University of California at Berkeley, Berkeley, CA 94720, USA*

1. Introduction

Telepoint is a new public cordless telephone system which uses pocket-sized handsets to access the public phone network. When launched in the UK in late 1989 it promised to be a major step in bringing mobile telecommunications to mass markets with a low cost service, made possible by new digital technology. It also involved an innovative approach to licensing and standards, which brought market forces into the standardization process. Services were licensed to four competing groups, and instead of setting standards in advance they were to be determined by market competition as the technology was developed. The results have been disappointing. The services achieved almost no market success and just two years later the initial three operators withdrew services. The fourth licensee subsequently launched a service which hoped to avoid some of the earlier problems, but this too has not been taken up as rapidly as expected. Although there has been talk of issuing new licenses, it appears that the market opportunity for Telepoint in the UK may have passed. [1]

The difficulties of Telepoint highlight the dangers of overlooking some basic standards strategies for new network services. Service providers did not

[1] This paper focusses mainly on the experience of Telepoint up until the withdrawal of the three initial services (Phonepoint, Zonephone, Callpoint) in late 1991. It considers the experience of the fourth service (Rabbit), from its launch in 1992 until mid-1993, in outline only.

make the full scale commitment essential to establish new standards. They introduced trial services with few base stations and high prices, and made inadequate attempts to promote the services. Rather than cooperating on a single standard they tried to differentiate their services with pricing and other strategies. The net results were tiny installed bases and low system credibility. Potential users were confused and they essentially ignored the system. Not only have the services miscarried but their potential as a showcase for UK manufacturers, aiming at a *de facto* standard for European cordless telephone, has been set back. Although many of these problems may be traced to the licensing terms set out by policy, the strategies used by the services and manufacturers within these ground rules share the responsibility for a missed opportunity.

In this paper we examine the strategies used to introduce new standards by the service providers and, indirectly, the manufacturers. The case demonstrates the over-riding importance of commitment and demand-enhancing strategies. These include full-scale investment in the installed base of access stations and users, establishing credibility for the standard, pricing for market penetration rather than maximizing initial returns, cooperation to achieve a single standard, and the importance of timing. For Telepoint each of these components was given insufficient weight.

The services chose these strategies within a framework of changes in the regulatory system. Telepoint was an experiment in making more use of market forces aiming to use competition to develop standards more quickly than the traditional committee approach. Licenses were issued with only broad guidelines, and detailed standards were to be determined in the market place. The primary aim here is to understand how the market behaved within this framework; lessons for policy are pursued elsewhere [2].

In the paper, section 2 describes Telepoint history and the licensing conditions. Section 3 outlines the role of standards for telecommunications networks and section 4 evaluates the strategies used to establish Telepoint. section 5 discusses future prospects for Telepoint and the European standard, and section 6 draws some conclusions. A list of abbreviations is given at the end of the paper.

---

[2] Policy implications for mixing committees and markets are discussed in Grindley and Toker (1993).

2. History of Telepoint

2.1. Product description

Telepoint is a public cordless telephone system which uses pocket-sized handsets to access the public phone network via short-range radio base stations. It was originally intended as a mass market, low cost, mobile telecommunications service, made possible by new digital technology. The phones are about the size and appearance of a large hand calculator, similar to miniaturized cellular telephone handsets. Each base station, typically a small box on the side of a building, has a range of about 200 meters and can accept up to 40 users at the same time. There would be many thousands of these stations over the country, an estimated 40,000 being needed to cover the UK. The cost of handsets and the service could be relatively low, certainly compared with cellular telephone. It has been called the 'poor man's mobile phone' or 'phonebox in your pocket'.

The technology is a development of the existing portable telephone. The main technical feature is that it uses digital transmission, called CT2 (Cordless Telephone Second generation). Its initial configuration only allowed outgoing calls. It is portable but not truly mobile, as the user has to stay close to the base station and there is no 'handover' from cell to cell on the move, as needed for car phones. The concept is thus somewhere between public payphone and cellular phone. It is used by first dialling a personal identification number (PIN) to validate the user. The telephone number is then dialled as usual. Charges are made to the user's account with the particular Telepoint service with which the user is registered. Apart from Telepoint, CT2 technology may be applied to cordless phones for business and domestic use, such as Cordless PABX and private portable phones, each potentially large markets. In the UK it was decided that the first application would be as a public service. Eventually the same handset could be used for all three areas, public, home and business.

The potential market for Telepoint depends on the choice of product features and price. Its initial configuration in the UK placed it at the lower end of the mobile market. The limitation to only outgoing calls put it in close competition with payphone. This may still have been more convenient than payphone, though it meant carrying a handset and there was no booth to cut down noise. However, with enhancements such as adding a pager Telepoint may be as much as the average user needs in a portable system, provided it is priced competitively. Other possible system enhancements are mentioned below. Even with these, Telepoint would not perhaps compete at the high end of the market with fully mobile two-way cellular systems or others being developed. There is a question whether there is too much overlap between Telepoint and the other systems to leave a substantial market[3]. The available gap gets smaller the longer Telepoint is delayed, as more advanced systems appear, payphones are improved and card payment is extended. It is difficult to estimate the intrinsic demand for Telepoint as an established system. Original market forecasts believed that it would reach millions of users, though it is difficult to know how much reliability to place on these.

2.2. Licensing conditions

Licenses to provide Telepoint services in the UK were awarded to four operators by the Department of Trade and Industry (DTI) in January 1989. These were to Phonepoint (main shareholders BT and STC), Zonephone (Ferranti), Callpoint (Mercury, Shaye) and BYPS (Barclays, Philips, Shell). Three services were launched at the end of 1989, with BYPS delaying introduction. The consortia membership for Telepoint and for other UK mobile phone systems are given in Table 1[4].

The broad product specification and the industry structure were determined by the regulator rather than by the technology. This was part of an unusual approach to standards by the licensing authority, here the Department of Trade and Industry (DTI). It outlined the basic requirements for the system, but did not specify detailed transmission standards ahead of awarding the licenses. The intention was to speed up both introduction and standards setting by allowing the market to determine the standard. Most of the technical details were left for the services to define for themselves. The aim was a simple system which could be developed quickly and would not compete directly with more sophisticated mobile systems being planned. Many of the reasons for promoting a public rather than private service, for the product configuration, for using market standards to speed up standardization and encouraging bids from multinational consortia stemmed from DTI aims to use Telepoint as a base for European standards. The hope was that a working public service would give UK manufacturers a lead in setting European CT2 standards and allow them to participate strongly in the private markets.

Each of four services were allowed to choose their own technology independently of the others. They were incompatible in that handsets could only be used on one service, though initially there were only two technical designs.

---

[3] The size of this gap varies by country. It may be larger in Europe, where cellular telephone is not yet as well established as in the UK, and narrower in the USA and Japan, where payphones are more easily available. Thanks to Edward Steinmueller for noting this.

[4] Table 1 shows the consortia membership at the beginning of 1991, unchanged since the licenses were issued. The BYPS service was sold in February 1991 to Hutchison Telecom. There have also been subsequent changes in ownership of the PCN interests.

Table 1

UK mobile phone consortia membership (1991)

System	Licensed	Service	Membership	%
Cellular	1985	Vodaphone	Racal Telecom	100
		Cellnet	BT	60
			Securicor	40
Telepoint	1989	Phonepoint	BT	49
			STC	25
			Nynex	10
			France Telecom	10
			Bundespost	6
		Callpoint	Mercury	49
			Motorola	25
			Shaye (Nokia)	25
		Zonephone	Ferranti	55
			C&W	5
			Venture capital	40
		BYPS	Barclays	33
			Shell	33
			Phillips	33
PCN	1990	Mercury PCN	Mercury	60
			Motorola	20
			Telefonica	20
		Unitel	STC	30
			US West	30
			Thorn-EMI	25
			Bundespost	15
		BAe	BAe	35
			PacTel	20
			Millicom	14
			Matra	10
			Sony	4
			IBA, Litel	17

The equipment manufacturers and investment for the four systems are shown in Table 2. Phonepoint and Callpoint used a Shaye design, Zonephone used a Ferranti design. BYPS, with no experience in telecommunications, decided to wait until a common standard was developed before selecting a technology. The licenses specified that the operators must accept a common standard, Common Air Interface (CAI), within about a year, by the end of 1990, and convert their systems to this within a further year. Handsets from one system would then be fully usable on the others, with arrangements for billing between services (Inter-System Roaming, ISR) by mid-1991. A common standard greatly increases the number of base stations available to any user, and hence the value of the system as a whole. The rationale given for the two stage

Table 2
Telepoint equipment suppliers and operator investment

Service	Leader	Bases	Handsets	Invest (£ million)*
Phonepoint	BT	STC/Shaye	Shaye	25
Callpoint	Mercury	STC/Shaye	Shaye	20
Zonephone	Ferranti	Ferranti/AB	AB	20
BYPS	BYPS	Philips/GPT	GPT/Orbitel	25

* Estimates.
(Source: Company Press Releases; FinTech).

approach was to exploit proprietary equipment immediately, as the authority 'does not want to restrict the services if they want to use their own equipment' (Department of Trade and Industry, 1989). It was expected to take a year to develop all the technology for the standard interfaces.

The means by which this common standard was to be defined were not clearly specified. It was to be decided by the DTI but only after further consultation and taking into account how the existing systems had performed. It was widely expected that it would be a modification of one of the original designs, and it was on this basis that the early entrants began their services. In fact when the DTI chose the CAI standard in 1990 it was neither of the Shaye or Ferranti designs but a third design, not yet in use, developed by GEC-Plessey Telecommunications (GPT) and Orbitel. Market forces did not appear to have been considered after all; the standard had been chosen on *a priori* technical grounds. Its main feature was that it conformed to Integrated Services Digital Network (ISDN) standards, which the others did not. This was preferred by the DTI to make it compatible with data networks for the business market, also an advantage for becoming a European standard. Since the ISDN interface was put in the handset rather than the base station and the only available chip had high power consumption this had the unfortunate side effect of greatly increasing the size of the handset. The new handsets were no smaller than cellular and four times the weight of the Shaye handset, negating Telepoint's main selling point. The sophistication of ISDN is probably more than the average Telepoint user would need, especially with wide adoption of ISDN still years away (David and Steinmueller, 1990). In fact the operators delayed their phase-in of the new standard, as noted below.

The main product restrictions were that Telepoint could not receive incoming calls, it was to use low powered base stations with limited range, and there was no provision for 'handover'. Several potential enhancements to the system were suggested after the launch, listed in Table 3. Most of these changes have been authorized but the process has been very slow. The main product weakness was one-way calling. The first response was to add a pager to the

Table 3
Potential enhancements to Telepoint

Feature	Status
1. Proprietary handsets and base stations	Current
2. Handsets with built in pagers	Approved 1990
3. Common Air Interface (CAI)	Defined 1990
4. Inter System Roaming	Defined 1991
5. Telepoint Plus	Accepted 1991
6. Two-way Telepoint	Accepted 1991
7. Limited Handover	Feasible 1991

handset, initially billed and operated as a separate service. This was approved by the DTI in 1990. More substantially a means of providing two-way calling was developed whereby the user may 'log in' on the nearest base station to a call routing system, called 'Telepoint Plus'. The handset can then receive calls. This has been approved in principle. Finally, the development of full two-way calling, called 'Neighborhood Telepoint', was allowed in March 1991 [5]. In the meantime, the technical feasibility of limited 'handover' had been demonstrated (by Northern Telecom), which allowed the user to move from one base station to another by automatic re-connection. Although all these enhancements were eventually accepted in principle they were too late to affect the UK system. None of them were implemented by the original services.

Part of the intention in restricting Telepoint features was to avoid conflicts with the DTI's plans for two other new mobile systems and, probably misguidedly as it turned out, to avoid confusing the market. The first of these was Digital Cellular. European standards had already been set for this, called GSM (Groupe Spéciale Mobile). Cellular Telephone services using analog technology had operated in the UK since 1985, and digital systems with greater capacity were planned for 1992, now being introduced. The second was Personal Communications Network (PCN). This is a mobile system similar to cellular, probably with higher capacity but smaller cell sizes (1–5 km radius) and more frequent 'handover' on the move. It is still not clear exactly what form it will take. It will use GSM standards in high frequency bands (1.8 GHz). It could be a high or low price system. It will be independent of the public network, using base stations connected by microwave links. A policy aim was that it should be able to rival the fixed link telecommunications services and was seen as a main plank in extending competition in the UK industry.

[5] Department of Trade and Industry (1991). Hutchison Telecom began trials on Two-way Telepoint in its Hong Kong Telepoint operation in 1993 (*Fintech*, March 1993).

PCN was already being discussed in the late 1980s before Telepoint licenses were issued. Licenses were awarded to three operators in 1990, less than a year after the Telepoint launch. Services were originally planned for 1993, though this schedule has since slipped. As with Telepoint the design of PCN was not decided before issuing licenses and was to be left to the operators. PCN competes closely with Telepoint as a small cell mobile telephone but with the advantage of two-way access and mobility. On one reading of policy Telepoint was intended as a temporary system until PCN was developed. On another it was intended to fill a low price niche between payphone and Cellular/PCN mobile telephone [6]. In fact PCN is unlikely to begin operations as planned. Facing the huge investment in infrastructure of around £1 billion per service for PCN some of the consortia members have dropped out and the original three licensed operators have now merged to one. It is not expected to appear soon.

Policy considerations also aimed to ensure a high level of competition in the Telepoint market. The number of operators was set at four, a large number in telecommunications. British Telecom (BT) and Mercury were allowed to bid for licenses but were required to take minority stakes and set up their interests at arms length, to avoid possible cross subsidy or preferential treatment. Although Racal (a cellular operator) entered the bidding it was not licensed. The DTI favored bids from groups of firms, especially including foreign firms, to equalize market power and to broaden the support for the eventual standards. Several US and European telephone services and equipment manufacturers took small stakes in the consortia, presumably to gain experience in the new services and access the technology. Bidding costs were low as the technology had been developed already by Shaye and Ferranti, though there was an implicit commitment by successful licensees to actively pursue the service.

2.3. System history

The basic Telepoint technology was developed well before services were set up. Ferranti and others had proposed a similar system in the mid-1980s, at about the same time as cellular telephone was being introduced. These

---

[6] The current approach for policy has been to invite industry responses to a discussion document before deciding policy. For example, a discussion document for Personal Communications systems emphasized that the DTI had "no unique insight into the future' and invited 'informed discussion' about what 'appear to be the main trends ... towards digital technology" (Department of Trade and Industry, 1989). Recently a green paper on the fixed link duopoly review (Department of Trade and Industry, 1990) outlined the policy issues and a white paper (Department of Trade and Industry, 1991) stated policy.

Table 4
Installation of Telepoint base stations

Service	Launch date	Initial bases	Actual 1990	Plan 1990	Plan 1995	Close date	Final bases	Final users
Phonepoint	16-8-89	100	1050	5000	25,000	1-10-91	3500	800
Callpoint	7-12-89	150	1000	2000	25,000	1-7-91	1000	<1000
Zonephone	27-10-89	150	1000	7500	25,000	1-7-91	1000	<1000
BYPS *	(1-7-90)	–	–	7000	20,000	–	–	–

* Interest sold to Hutchison February 1991, launched as Rabbit May 1992: 10000 bases, 8000 subscribers by early 1993.
(Source: Company Press Releases; FinTech, May 1990, October 1991, March 1993).

suggestions were not acted on by the DTI until the awarding of licenses in 1989, by which time cellular was safely established.

Telepoint services were begun in late 1989 by three of the groups, Phonepoint, Callpoint and Zonephone, less than a year after licensing. BYPS, waiting for the common standard, announced its launch for summer 1990, a launch which did not take place until much later and under new ownership. Although the services got off to an early start this was on a small scale. The initial number of base stations at the end of 1989 and subsequent history are shown in Table 4, compared with the very large projections being made for the future network. The services were introduced at trial levels, to test the market for the new product. Largely for this reason much of the public were unaware that the services existed, according to private surveys. There was little advertising or distribution effort. There were several press articles when the licenses were issued and services launched, but little more until the PCN announcements.

The situation in 1990, a year after the launch, had not changed much. The introduction rate was far below the rapid build up envisaged in earlier plans. The installation rate of base stations was slow and by May 1990 the three services had only just topped 1,000 stations each[7]. This may be compared with the original projections for market growth in Table 5. Interest in the system by the public was disappointing: in 1990 the three services had together about 3,500 base stations, and admitted having 'less than 4,000' subscribers[8]. Possibly fewer people were aware of the system than at its launch. While original projections were that by 1990 there would be many thousands of stations and hundreds of thousands of users, the reality was only hundreds of each.

---

[7] *Financial Times Business Information*, 24 May 1990.

[8] *ibid.*

Table 5

Telepoint income forecast (1989)

Year	Users (× million)	Connection (£ million)	Subscription (£ million)	Sales (£ million)	Total* (£ million)
1989	0.2	7.0	16.8	30.0	53.8
1990	0.5	10.5	42.0	40.0	92.9
1991	1.0	17.5	84.0	50.0	151.5
1992	2.0	35.0	168.0	50.0	253.0
1993	3.0	35.0	252.0	50.0	337.0
1994	4.0	35.0	336.0	50.0	421.0
1995	5.0	35.0	420.0	50.0	505.0
2000	15.0	70.0	1260.0	100.0	1445.0

* Connection fee £35; Subscription £7 p.m.; Handsets £150 (1989) falling to £50 (1992 onwards). (Source: Retail Business).

Table 6

European Telepoint and mobile market forecast (1990)

Country	Telepoint		Total Mobile			
	Users (× million)	Value (£ million)	Users (× million)		Value (£ million)	
	1995	1995	1989	1995	1989	1995
UK	1.66	496	1.48	5.15	986	2500
France	1.38	368	0.41	3.81	408	2646
Germany	1.10	144	0.37	3.81	476	3675
Other	2.83	576	1.84	9.63	1530	6027

(Source: MZA).

Market forecasts were reduced to about a third of the original estimates, but still projected a large market for Telepoint. A forecast for the UK and other European countries is shown in Table 6. A late hope of the operators was that the business and domestic cordless market would provide the base of CT2 users which will then create demand for public Telepoint. Some hopes were pinned on BYPS, which stated that it would invest heavily in base stations, and only launch its service, called 'Rabbit', once it had 500 stations in place. An added complication was the licensing of PCN at this critical time, in 1990. This was seen as a direct competitor and a number of press articles questioned the need for Telepoint when it would be obsolete in a few years[9].

Seeing the low user acceptance the operators successfully lobbied to modify the license restrictions during 1990–1991. The CAI standard was specified in May 1990, as planned. As this was significantly different from either of the ex-

[9] For example, *Guardian*, 10 November 1989; *New Scientist*, 4 November 1989.

isting designs and was not seen as commercially attractive the operators were reluctant to convert to it and lobbied for the relaxation of the compatibility requirement. As a result in 1991 the operators were allowed to keep their own standards until they felt able to convert.

All this came too late. The services seemed to be marking time until an opportune moment to withdraw from operations. There was a last attempt in mid-1991 to revive demand by dramatically reducing handset prices to £99 (from £180) and call charges by 50%. This went unnoticed. BYPS sold its interest in Telepoint to Hutchison Telecom in February 1991, without launching a service. Callpoint and Zonephone both withdrew services in July 1991 and finally Phonepoint announced its withdrawal in September 1991. It finally became clear how few users there were. The largest, Phonepoint, had installed 3,500 base stations and only had 800 subscribers. Each service had invested around £25 million. The DTI stated that licenses would be available for new operators, on the same terms as before, though no new commercial interest has been evident. The promised business and domestic products had not yet appeared on the market.

The fourth licensee, now owned by Hutchison, finally launched its Rabbit service in May 1992, as a local operation in the city of Manchester. It has also followed an incremental approach, though arguably with more commitment than previous services. It extended coverage to other areas including parts of London by mid-1993, though it has yet to launch a national service. It has taken a more positive approach to installing base stations (with 10,000 by March 1993), has offered a combined handset and pager and has made a significant attempt to advertise the system (including a £4 million campaign in 1992). It also offers a home base station. Even so, Rabbit reports only 8,000 subscribers (against initial forecasts of 50,000 by the end of 1992), despite recent offers cutting equipment prices by half. Some observers believe this shows that the window of opportunity for Telepoint has already passed [10].

3. *The role of standards*

3.1. *Telepoint standards*

For Telepoint the standards define the technical protocols used to transmit messages. These cover the formats used to encode messages, duration of data package transmissions and other specifications, as well as the coding needed for accounting purposes to verify and bill customers. For handsets from one

---

[10] *FinTech (Mobile Communications)*, 11 March 1993.

service to use the base stations from another requires that the standards are compatible. This may be achieved either by redesigning the handset (or base station) to use the same technical format, or by arranging for one or other to use both formats and select the right one when in operation. There must also be accounting systems included which can bill between services.

There were originally two standards, the Shaye and the Ferranti designs, using slightly different transmission techniques. Each agreed to convert to a common CAI standard, by replacement or modification of equipment. When it was finally defined, the CAI standard was neither of the existing standards but a third, with more elaborate protocols than the others. Converting to compatibility was expensive, and probably not justified for a simple portable phone system.

3.2. Standards strategy and network externalities [11]

Our concern here is with compatibility standards, which allow different pieces of equipment to be interconnected in a system. These are distinguished from general quality or safety standards which affect all goods. Compatibility standards are important in products such as telecommunications networks where the value of the service increases the greater the size of the network. They work primarily via the demand side. The more users the network has the more support facilities are provided making the service more convenient to use. There may be economies of scale in production of complementary equipment, such as handsets and base stations [12]. A larger network is more valuable directly by having more users to call or share equipment with [13]. These advantages are called *network externalities*, since new users benefit existing users as well as themselves. Similar network effects apply to standards for a great number of systems products, such as computers or video recorders. The general mechanism is that the larger the *installed base* of equipment on a given standard the cheaper and more various are the complementary goods

---

[11] For discussions of the economics of standards see David (1986, 1987), Farrell and Saloner (1986), Arthur (1987), Farrell (1990) and Grindley (1990, 1992a).

[12] On the production side, incompatibility and fragmented standards may raise costs because: (a) the smaller market for each standard's hardware reduces economies of scale and experience effects and there are fewer units to cover R&D costs, (b) there is less competition between fewer producers, and (c) there is less opportunity for small specialized producers. These effects may be less important if hardware production has reached minimum efficient scale (Gabel and Millington, 1986).

[13] The direct benefits of more Telepoint users are spread over all users of the public fixed link network, not just Telepoint users. With two-way access, all phone users have more people to call. Even if Telepoint has only one-way access, all phone users benefit by receiving more calls and potentially relaying Telepoint messages.

and services provided, such as computer software or pre-recorded video tapes. This in turn makes belonging to the standard more valuable. It usually makes a common standard more attractive than a system fragmented into several incompatible standards, other than cases where variety is important.

The installed base effect leads to some important dynamics in setting up a new standard. As its base gets larger a standard gets cumulatively more attractive, as 'bandwagon' effects start to work. If a standard can establish a significant base early on the externalities start to work in its favor and it has a good chance of success. The *credibility* of a standard in the crucial early stages is also very important, as users make early purchasing decisions based on expectations of whether the standard will ultimately be successful. In a contest between standards an early lead is often decisive, provided the product has reached a basic level of user acceptability, i.e. introduction is not premature. This defines a window of opportunity from the time an acceptable design appears until the bandwagon effects of one or other standard take over the market.

4. Standards strategy

The key requirements of standards strategy are (1) building the *installed base* quickly, and (2) establishing the standard's *credibility* in the early stages. These need heavy investment and promotion. The strategic implications are that strong *commitment* is needed to establish a new standard — a trial approach is usually fatal. There are a number of supporting strategies, including: (3) ensuring the product is developed to an adequate level of user acceptability before launch, (4) pricing for market penetration rather than skimming high margins, (5) coordinating efforts with competitors to agree a common standard, and (6) timing in the window of opportunity. We argue that the strategies used for Telepoint under-emphasized each of these basic elements:

(1) Building the installed base

For Telepoint the installed base is measured by the number of base stations, users and handsets. The main determinant of Telepoint usefulness is the number of base stations. Users must be sure of a base station nearby when they want to make a call. This means that the services have little choice but to invest in a large network before they can attract subscribers. Without this, other means of building the user base will not have much effect. This is a familiar problem that standards involve relatively large initial risks which do not allow the services to be introduced gradually. The risks may be reduced, such as by cooperating on a common standard, but can not be avoided altogether.

The cost of investment needed in base stations is not excessively large. Base stations cost £700–1500 each, so that the full 40,000 for the UK represents say £60 million. The three services each expended about £25 million, but little of this seems to have been on base stations. For example, an estimate of the cost of Phonepoint's 3500 base stations may have amounted to about £5 million. Although these figures are significant, they are tiny compared with the £1 billion or so estimated for PCN systems with fully self-contained communications links.

Instead of this, the initial three services followed a cautious schedule, installing services on a trial basis with limited coverage. There were operational problems in installing access points to the public phone system quickly enough, which contributed to the slow build up. An unfortunate response to this was to erect 'Coming Soon' signs, most of which turned out to be empty promises. This did little for the system's credibility. Base station numbers never really got beyond test levels. The subsequent entry by the fourth service, Rabbit, was accompanied by the installation of a larger number of base stations, but this was still essentially a gradualist approach.

This investment should be taken in parallel with other means of building the user base. These include initially subsidizing handsets and call charges, and strong distribution. As we discuss later, handsets remained expensive, indeed more expensive than cellular, and call charges were kept high until too late. Users wishing to buy handsets had difficulty finding a retailer who stocked them. These all helped to hold back subscribers.

(2) Credibility

The lack of clarity and commitment also reduced the credibility of the system with potential users. Early users make their decisions based on expectations of whether the service is going to succeed. These are determined partly by the proven installed base but also by subjective evaluations of the value of the system and the commitment of the operators.

The definition of Telepoint has never been completely clear. The system started out as an unfinished product, with private base stations and PABX applications to be added later. The terms of the license were left vague and the system had difficulty finding its position in the market. Users were not sure where it fitted in with their needs or with future developments in telecommunications. Users were further confused by the many changes in Telepoint. Specifications have changed twice, to add pagers and then allow incoming calls. To this were added concerns about the number of incompatible systems and how these would be resolved. The common CAI standard expected to clear up these doubts only added to them, as one more complication. It was different from the preceding standards, seemed over elaborate and left the suspicion that it too would soon be changed.

The dedication of the initial operators to the future of Telepoint was not obvious from their trial approach and less than full-scale promotion of the service. They made limited efforts to publicize the system or educate the public. Surveys indicated that few people were aware of Telepoint's existence. If people had noticed the signs at base stations or seen a handset they were often mystified as to what they could all mean. The operators' limited commitment may have been inevitable given the licensing terms. The requirements for four operators and a simple service ensured that margins would be low. This was compounded by the fact that the operators were consortia of large firms, each with parallel interests in other services. It was difficult for Telepoint to be a primary interest under such conditions.

The biggest blow was the announcement of PCN licenses only a few months after the launch, in January 1990. Though it was not fully defined, this could potentially do all that Telepoint could do plus two-way communication and mobility, and possibly could also be a low cost service. This severely damaged what was left of Telepoint's credibility [14]. Even though the final versions of PCN may not overlap the low end of the market as much as feared, the uncertainty was damaging enough. In fact PCN may be closer to cellular than to Telepoint, with full features exploiting digital technology. Had Telepoint already become established within its first year, before the PCN announcement, this would not have been a problem. Telepoint would have had its installed base and could have found a place alongside PCN. Even though PCN services have not been introduced in 1993 as planned, and it is now far from clear when services may appear, the delay this caused to Telepoint was the last straw. Although the effect on Telepoint by PCN was largely a consequence of government policy decisions, the operators also did not make strong efforts to reduce the potential conflicts between the systems.

(3) Product acceptability

A standard has to have reached a basic level of user acceptability before it is launched on the market. Because of credibility problems it is particularly hard with a network product to make major modifications once it is launched, as might be quite usual for other products. Similarly, once a basic level of acceptability is achieved, further technical improvement may be less effective in building the installed base than more direct methods, so that adding product sophistication is usually not a very useful strategy for setting standards. The important step is to get the product acceptable first time.

---

[14] Press comments warned that Telepoint was "set for disaster" (*Guardian*, 10 November 1989) and "already outmoded" (*New Scientist*, 4 November 1989).

As became apparent very soon after launch, Telepoint entered the market below a basic level of acceptability. The initial definition of Telepoint was set primarily by the authority and did not have an obvious market niche. The main problem was the limitation to outgoing calls. This did not make it clearly more attractive than payphone, especially as the price was significantly higher. To make the product acceptable it needed some form of two-way access or at least a built-in pager. Even though the operators successfully lobbied the DTI to allow pagers and later Telepoint Plus to allow limited two-way access, it never recovered from this initial loss of interest. The time taken to make these changes was crucial. It was nearly a year after launch before pagers were allowed and Telepoint Plus was still being discussed when the initial services were withdrawn.

The basic market research had not been done prior to letting the licenses, while the operators with their trial approach believed they had adequate time to make changes later. This was not the case and once the initial interest had been dissipated it did not recover.

(4) Pricing policy

The most valid niche for Telepoint seems to be at the lower end of the mobile market, where it competes most closely with payphone. Market positioning depends on the bundle of features offered and the price. 'Mapping' consumer tastes onto product attributes is hard for new untried products. The relevant characteristics for mobile phones are in five groups:

(1) Access: One-way or two-way calling.
(2) Mobility: (a) Calls made anywhere or only when close to a base station.
(b) Used on the move with 'handover' from cell to cell.
(3) Useability: Lightweight handset, easy to handle and use.
(4) Coverage: National or local.
(5) Service: Data transmission, interference, security, other quality features.

No one of the existing systems is superior on all counts [15]. Telepoint has only one-way access, localized base stations and no handover but has a lightweight handset and national coverage via the public phone network, provided there are enough base stations. It compares poorly with cellular other than on useability. It is closer to payphone, being more mobile but less useable. For the service to survive it has to appeal to groups of users with specialized and limited needs (e.g. who do not need full mobility) and be priced competitively enough to outweigh the limited characteristics. This means that it must be

---

[15] For a comparison of prospective UK mobile services and their pricing policies see Toker (1992).

priced well below cellular and not far above payphone. It also needs wide coverage of base stations.

The implication is that Telepoint should be positioned as a low price, large scale system. This coincides with one of the main recommendations from standards analysis to use low penetration pricing to establish the installed base quickly. In contrast the services have been predominantly high price, small scale, trial introductions. Call charges are closer to cellular prices than payphone. Some representative call charges are given in Table 7. For some time bands Telepoint is actually priced above cellular. In addition there is a connection fee of £20–30 and a monthly subscription charge of £8–10. Handsets cost £170–200. These compare badly even with cellular, which has subsidized handsets to attract users and was sometimes giving them away to new subscribers.

Pricing has been essentially based on the fixed link charges plus the cost of Telepoint access. The service operators claimed that they were limited in their pricing strategy as they must charge for their service on top of those for using the public network. Little attempt seems to have been made to negotiate lower block rates. The prices were too far above payphone to be attractive. The original claims were that Telepoint should cost very little more than payphone.

The reasons for a high price policy may either have been because the operators at first misread the market as a premium product, or because

Table 7
Representative Telepoint call charges

	Peak [a]		Standard		Cheap	
	local	long	local	long	local	long
Home Telephone	13	44	9	31	4	22
Public Payphone	26	87	17	61	10	44
Cellular Phone:						
Cellnet (in M25) [b]	99	99	99	99	30	30
Cellnet (out M25)	75	75	75	75	30	30
Telepoint:						
Phonepoint	39	90	39	90	30	75
Zonephone (in M25)	38	125	25	88	13	63
Zonephone (out M25)	30	100	20	70	10	50

Note: Charges in pence for 3 mins. (including VAT).
[a] Peak 8:00–13:00 hours; Standard 13:00–18:00 hours; Cheap 18:00–8:00 hours; Long = over 56 km.
[b] In M25 = within London area.
(Source: Operator price lists).

they were following a price skimming policy as part of their trial approach, intending to first reach the 'high value' users then gradually to bring down prices to expand the market. It was clear very early on that positioning as a premium service in the initial configuration was not appropriate. The premium service was Cellular Telephone, which continued to attract more users while Telepoint stalled.

A response to the slow take-up of the services could have been to abandon the skimming policy and reduce prices substantially to develop the market and the installed base quickly in a penetration strategy. The services did not do this. Continuing the trial approach they were more interested in taking the product up market, to make it look more like the cellular systems its prices resembled, than bringing prices down to what the product justified. Adding new features, other than a basic way to alleviate one-way calling (i.e. built-in pager or Telepoint Plus), was the wrong focus for strategy. Instead of technical improvements this needed to concentrate on radically expanding demand and the installed base, to start the bandwagon effects. First this meant getting the pricing low. While some of the enhancements, such a pagers, were cheap and necessary, the others required new development which added substantial delays and costs. Also there needed to be broader publicity and distribution. It was not easy to find any handsets to buy for the initial three services, and then they were often sold as luxury items. What had been billed as the 'poor man's mobile phone' was only retailed in London by two luxury department stores, Harrod's and Selfridge's. Making an initial misreading of the market is perhaps understandable, but the response once the error was clear should have been in the opposite direction to the one chosen. The firms seemed to realize the pricing error only at the very end, in mid-1991, when they halved the price of handsets and call charges. This was too late to have any effect [16].

(5) Collaboration vs. differentiation

A major problem with the installed base was that it was fragmented between three, and possibly four, services. To get around this the services could either have collaborated to establish a single standard, or one could have made a determined effort to establish itself as the *de facto* standard by building up a dominant position. Either would have increased the available installed base of each of the services and raised the credibility of the system, at a stroke.

The three initial services did neither. Instead of converging they tried to differentiate their services from one another. This is a strategy which

[16] Rabbit made handsets more broadly available, including distribution through electronics retail stores. It also halved some equipment prices in early 1993, a year after launch, in what some observers saw as a last attempt to expand the market (*Fintech*, March 1993).

might be expected with normal products but which is counter-productive where network externalities are involved. By emphasizing the differences between the standards the effect was to convince users that each standard was separate and could only count on its own small installed base. The services followed intricately different pricing policies for call charges, subscription charges and handset costs. Some of these complexities are indicated in the charge schedules, which used a different blend of fixed and variable charges for each service, shown in Table 8. There were different time bands for peak, standard and cheap rates, shown in Table 9. The different call rates are shown in Table 10. Phonepoint used only two time bands and two distance categories. Callpoint had three time bands but only one distance category. Zonephone had three time bands with four distance categories plus different rates for calls initiated within the London area. It was virtually impossible for potential customers to compare prices between services. These intricacies further confused users who now had to make fine decisions on the features of

Table 8
Telepoint charge structures

	Connection	Subscription	Call	Distance	Area	Peak	Duration
Phonepoint	×	×	×	×	–	×	×
Callpoint	×	×	×	–	–	×	×
Zonephone	×	×	×	×	×	×	×
Payphone	–	–	×	×	×	×	×
Cellular	×	×	×	–	×	×	×
Pagers	×	×	–	–	–	–	–

× = Charge levied.
(Source: Operator Price Lists, in Toker, 1992).

Table 9
Time charge bands (24 hours)

Service	Peak (hours)	Standard (hours)	Cheap (hours)
Telepoint:			
Phonepoint	8:00–20:00	–	20:00–8:00
Callpoint	7.30–13:00	13:00–18:00	18:00–7.30
Zonephone	9:00–13:00	8:00–9:00, 13:00–18:00	18:00–8:00
Payphone:	9:00–13:00	8:00–9:00, 13:00–18:00	18:00–8:00
Cellular:			
Cellnet	8:00–22:00	–	22:00–8:00
Vodaphone	7.30–21.30	–	21.30–7.30

(Source: Operator Price Lists).

Table 10
Comparative call charge schedules

Service	Cheap	Standard	Peak
Phonepoint:			
Local	30	39	39
≤56 km	75	90	90
>56 km (low)	75	90	90
>56 km	75	90	90
Callpoint:			
Local	30	48	60
≤56 km	30	48	60
>56 km (low)	30	48	60
>56 km	30	48	60
Zonephone:			
Local	12.5 (10)	25 (20)	37.5 (30)
≤56 km	25 (10)	62.5 (50)	87.5 (70)
>56 km (low)	37.5 (30)	75 (60)	100 (80)
>56 km	62.5 (50)	87.5 (70)	125 (100)
Home Telephone:			
Local	4.4	8.8	13.2
≤56 km	8.8	22	30.8
>56 km (low)	13.2	26.4	35.2
>56 km	22	30.8	44
Public Payphone:			
Local	10	17.4	26.1
≤56 km	17.4	43.5	60.9
>56 km (low)	26.1	52.2	69.6
>56 km	43.5	60.9	87
Cellular:			
Local	30 (30)	99 (75)	99 (75)
≤56 km	30 (30)	99 (75)	99 (75)
>56 km (low)	30 (30)	99 (75)	99 (75)
>56 km	30 (30)	99 (75)	99 (75)

Note: Charges in pence for 3 mins, 1989 (excluding VAT). Charges outside London M25 in parenthesis.
(Source: Operator price lists).

the individual systems. The pricing structures also made it harder to devise a common standard with an accounting procedure for 'roaming'.

Under the terms of the licenses the operators were required to adopt a common standard after a year. The uncertainty about how this would be decided, whether by authority or by market forces, and the short time allowed may have left the services little incentive to attempt to establish a market standard. With the services making no effort to define a common standard the responsibility returned to the authority. The one devised, though intended to be set in consultation with the services, was neither of the existing standards and probably not one they would have chosen based on market acceptance.

The differentiation strategy shows the dangers of mistaking the competitive objective when standards are involved. Firms may be overly competitive for market share at the expense of market size. Collaboration on standards increases the size of the total market. Although this helps competitors as well as oneself, so that investment in common base stations helps all services equally, this may be the only way to ensure any market exists. Competition may then take place using other variables [17].

(6) Timing

As with most standards there was a narrow time window during which to establish Telepoint, from the time the product was technically possible until competing systems or standards became serious threats. There was also a need to maintain the momentum of the new standard to keep its credibility. Once the standard was seen as a failure it was very hard to regain credibility. As the initial services were launched the time window for public Telepoint was closing rapidly; the opportunity may by now already have passed. Both the size of the market and the time available before Telepoint becomes obsolete were shrinking. The market niche at the low end of the mobile market, between cellular/PCN and pagers/payphone, became smaller as Telepoint took longer to establish itself. At the high end, newer more advanced technologies are on the near horizon, while at the low end of the market, payphones are becoming more numerous and reliable in the UK. The services did not respond adequately to this double threat. The time window had already been narrowed by the issuance of licenses some years after the system was technically feasible, and the planned arrival of PCN gave a limit of about three years to establish Telepoint. The trial approach and high price entry route show that the operators did not fully appreciated that timing had to be well focused.

5. *Telepoint prospects*

It is an open question whether Telepoint can recover in its current form. This seems a pity. Objectively there seems to be a market opening for a simple, low cost portable phone system, to complement sophisticated, high cost cellular

---

[17] Coordination of standards strategies may be achieved via market competition (*de facto*), collaboration through standards committees and regulatory authorities (*de jure*) or by a combination of these (hybrid policies). For assessments of the merits of these, see David and Greenstein (1990), Farrell and Saloner (1988), Grindley (1992b).

services. Telepoint, especially with a pager, could supply this market segment quite adequately. Telepoint is more attractive in the absence of PCN, though this depends partly on how PCN is defined and priced. Initial indications are that given the high infrastructure investment needed with PCN it is likely to be a premium product, leaving the low cost segment open at least for a few years. The introduction of PCN has itself been delayed from what was originally planned. In 1992, as licensees faced the huge investments needed, introductions were postponed, and the three licensed consortia merged down to two.

One possibility is that new operators take over the licenses, as the prospects for PCN and Digital Cellular services become clearer. There is at least a clearer idea now of what Telepoint should look like and can offer the user. It seems that only a major change away from a gradualist approach could make Telepoint services a success. At the time of writing, in 1993, there is little evidence of such a change. Just before closure the initial operators were still talking of waiting 'two or three years' before services would take off. The introduction of the Rabbit service has been pursued with more attention to building an adequate installed base of stations before each local launch, but the overall process has proceeded gradually and by mid-1993, a year after introduction, Rabbit had still not launched a national service.

A remaining route for Telepoint is that an installed base of users will be built up first in the business and domestic portable phone markets, and create a demand for a public service. This was a late hope voiced by the initial operators and part of Rabbit's plan. This is a reversal of the approach originally taken, with Telepoint launched as an 'unfinished' system for public services and followed by domestic and business applications as they were developed. This could revive the system, though it is not clear that the private markets could be large enough to create an adequate base of users. Also these systems are likely to be technically different from current Telepoint, so developing compatible services could require additional investment. The main problem with this approach, and part of the reason the public service route was followed in the original policy, is that this puts Telepoint further back. Before demand could build up in this way, Telepoint could be obsolete.

Looking more widely, the UK manufacturers may still participate in some way in European markets if they can disassociate the technical development of Telepoint from its commercial problems. The technology and standards were developed quickly and UK firms have some lead. The failure of UK services provides a poor example for customers, and negates any benefit from being first. Also the expertise may be with the wrong standard. The UK manufacturers' chances of participating in larger markets are almost back where they started.

6. Conclusion

Telepoint involves two major standards issues: strategies for establishing a winning standard, and the role of official standards bodies. Both have been causal factors in Telepoint's problems. We have concentrated here on the strategies used by the operators. The main strategic problems in introducing Telepoint were in building the installed base and establishing credibility. There appeared to be less than whole-hearted commitment to the service by the initial operators, which showed as insufficient investment in the access network needed to attract subscribers and inadequate promotion of the system. The system specifications needed some modification to make it fully attractive to users and these changes were too slow coming. This places significant responsibility for the problems on the operators. The result is doubly unsatisfactory as UK manufacturers' hopes for defining a European standard have been set back.

Many of these strategic problems had roots in government policy. The intention was to allow the market to determine the standard, but for one reason or another the market was not allowed to work. The system was defined by the licensing authority and the degree of competition imposed may have been too great for the operators to justify major investment. The standardization rules themselves added to the confusion. Other problems were in resolving internal concerns for firms with parallel interests in other mobile services for which Telepoint could become an added complication. Even so, given the experimental nature of the policy and the willingness of the authority to consider operators' demands, it seems that the service could well have been more successful had the operators paid more attention to basic standards strategies.

Acknowledgements

Thanks go to Paul David, Michael Davies, Barry Karlin, Lionel de Maine, Edward Steinmueller, Peter Swann and others for helpful discussions. Responsibility for opinions expressed here and for any errors and omissions resides solely with the authors.

Appendix — Abbreviations

BAe: British Aerospace
BT: British Telecom
CAI: Common Air Interface
CEPT: European Conference on Posts and Telecommunications Administrations

CT-2: Cordless Telephone, Second generation
DECT: Digital European Cordless Telephone
DTI: Department of Trade and Industry
EC: European Commission
ERMES: European Paging Standard
ETSI: European Telecommunications Standards Institute
GPT: GEC-Plessey Telecommunications
GSM: Groupe Spéciale Mobile
ISDN: Integrated Services Digital Network
MoU: Memorandum of Understanding
PCN: Personal Communications Network
PSTN: Public Service Telecommunications Network
PTO: Public Telecommunications Operator
TACS: Total Access Communications System (cellular)

References

Arthur, B., 1987. Self-Reinforcing Mechanisms in Economics. *CEPR Publication* No. 111 (Center for Economic Policy Research, Stanford University).

David, P., 1986. Narrow Windows, Blind Giants and Angry Orphans: The Dynamics of System Rivalries and Dilemmas of Technology Policy. In: F. Arcangel et al. (eds.), *Innovation Diffusion*, Vol. 3 (Oxford University Press, New York).

David, P., 1987. Some New Standards for the Economics of Standardization in the Information Age. In: P. Dasgupta and P. Stoneman (eds.), *Economic Policy and Technological Performance* (Cambridge University Press, Cambridge).

David, P. and Greenstein, S., 1990. The Economics of Compatibility Standards: An Introduction to Recent Research. *Economics of Innovation and New Technology* 1 (1/2), 3–43.

David, P. and Steinmueller, E., 1990. The ISDN Bandwagon is Coming, But Who Will Be There To Climb Aboard. *Economics of Innovation and New Technology* 1 (1/2), 43–62.

Department of Trade and Industry, 1989. *Phones on the Move: Personal Communications in the 1990s — A Discussion Document* (DTI, London).

Department of Trade and Industry, 1990. *Competition and Choice: Telecommunications Policy for the 1990s (Discussion Document)*, House of Commons, Cm. 1303 (HMSO, London).

Department of Trade and Industry, 1991. *Competition and Choice: Telecommunications Policy for the 1990s*, House of Commons, Cm. 1461 (HMSO, London).

Farrell, J., 1990. The Economics of Standardization: A Guide for Non-Economists". In: J. Berg and H. Schumny (eds.), *An Analysis of the Information Technology Standardization Process* (Elsevier, Amsterdam).

Farrell, J. and Saloner, G., 1988. Coordination Through Committees and Markets. *Rand Journal of Economics* 19 (2), 235–252.

Farrell, J. and Saloner, G., 1986. Economic Issues in Standardization. In: J. Miller (ed.), *Telecommunications and Equity* (North-Holland, Amsterdam).

Gabel, L. and Millington, R., 1986. *The Role of Technical Standards in the European Cellular Radio Industry*, Mimeo (INSEAD).

Grindley, P., 1990. Winning Standards Contests: Using Product Standards in Business Strategy. *Business Strategy Review*, 1, 71–84.

Grindley, P., 1992a. Standards, Business Strategy and Policy: A Casebook. *Centre for Business Strategy, London Business School, Report Series*, 216 pp.

Grindley, P., 1992b. Regulation and Standards Policy: Setting Standards by Committees and Markets. *Centre for Business Strategy, London Business School, Working Paper* No. 130, 28 pp.

Grindley, P. and Toker. S., 1993. Regulators, Markets and Standards Coordination: Policy Lessons from Telepoint. *Economics of Innovation and New Technology*, 37 pp. (forthcoming).

Stewart, R., 1988. A Strategic View of CT2. *Mobile Communications Guide 1989*, pp. 63–69 (IBC Technical Services, London).

Toker, S., 1992. Mobile Communications in the 1990s: Opportunities and Pitfalls. *Centre for Business Strategy, London Business School, Report Series*, 106 pp.

Global Telecommunications Strategies
and Technological Changes
Edited by G. Pogorel

Chapter 11

A monopolist's incentive to invite competitors to enter in telecommunications services

Nicholas Economides

Stern School of Business, New York University, New York, NY 10012-1126, USA

1. *Introduction*

In the post-AT&T-divestiture world, telecommunications services are provided by a variety of vendors in diverse combinations of substitute and complementary goods. Typically, consumers demand *composite* goods or services that are comprised of a number of *complementary* goods. It is not uncommon for these complementary goods to be provided by different firms, with each firm well-established in a particular market so that potential entrants may face significant difficulty. These factors would be of little importance, were it not for the complementarity of the components of telecommunications services and the associated *network externalities*.

An important feature of networks is that the size of the network contributes positively to the value of the goods generated by the network. This effect has been called a network externality. To understand this effect, consider a simple star network with a central switch S and n-spikes, SA, SB, SC, etc., as in Fig. 1. If this is a telephone network, the customers are located at A, B, C, etc., and the goods are phonecalls ASB, BSA, ASC, CSA, BSC, CSB, etc. It is clear that the number of phonecalls that can be made in an n-spike star network is $n(n-1)$. Thus, the addition of the nth spike creates $2(n-1)$ new potential phonecalls. Therefore the addition of the nth customer increases the number of goods in the network proportionately to the size of the (pre-existing) network. If the cost of connection is constant per customer and phonecalls are equally desirable, the total value of all goods in the network

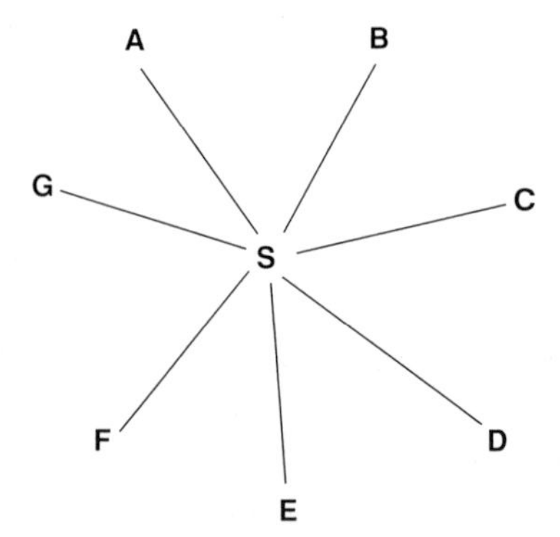

Fig. 1. Diagrammatic representation of a network.

keeps increasing in n as more customers are connected to the switch. This is one aspect of network externalities.

The addition of the nth spike has a positive effect to each pre-existing customer, since he can now make phonecalls to the nth customer. This increase of utility of the old customers through the creation of additional goods is the key feature from which the 'network externality' derives its name [1,2].

It is important to realize that the essential reason for the emergence of network externalities is the complementarity between the components of the good that we named a phonecall in the above discussion. Then such externalities are not confined to physical networks [3], but rather apply to a variety of industrial structures where complementarity is crucial.

In the star network example, a phonecall ASB is composed of two complementary components, AS and SB, the access to the switch of customer A, and the access to the switch of customer B. These two goods look very similar, and typically will have the same industrial classification, and may be seen as substitutes when looked from afar. However, they are complements in the creation of good ASB. And it is this complementarity that creates the network externality: the addition of customer C through the provision of SC allows for good ASC (= AS + SC) to be created thus increasing the utility of customer A. Thus, despite the uniform appearance of the goods 'access to the switch' (AS, BS, CS, etc.), they are complementary to each other, and this is the key to the creation of network externalities.

---

[1] It is an externality because the effect is not mediated by a market.

[2] The network effect is not based on reciprocity, i.e., on the fact the phonecalls can originate in both directions.

[3] A network is a collection of nodes that are connected in a particular fashion.

It is clear now that non-network industries, or industries that only partly utilize networks, can exhibit 'network externalities' as long as the final goods demanded by the consumers are composed of complementary components, and consumers demand diverse varieties. In effect, a 'network externality' appears as an increase in the willingness to pay for good X when the number and the production levels of complements of X (call them Y) increase. If goods Y are produced by a number of competing firms, higher production of Y is typically accompanied by lower prices for Y. This accentuates the network externality because it enhances the value of good X.

In such a setup, a firm would like to influence positively the production of the complementary goods. It could do this directly if it produced both goods X and Y, as, for example, the pre-divestiture AT&T supplied both local and long distance services. However, if a firm cannot easily enter the complementary goods market, or if regulation prohibits a firm from entering such a market, a firm may need to find other ways to create and capture the benefits of the network externality. To create the network externality, a firm has to commit to high production. The problem is that high production is not always credible. In particular, an innovator-monopolist, say the single holder of a new technology, will generally be expected to keep production low, so that he may benefit from the restriction of output and higher prices. In contrast, the innovator may be better off if he invites competitors to enter and compete with him. Their entry automatically implies higher equilibrium production levels, a bigger network effect, and possibly greater profits for the innovator.

We present a model that has two interpretations. In the first interpretation, there is a single good sold, and consumers have expectations of sales that influence positively the demand for the good. At equilibrium, these expectations are fulfilled. In the second interpretation, there are two markets for two complementary goods. In this context, the model describes the interaction of the positive feedbacks across the two markets. Section 2 sets up the basic expectations model. Section 3 describes the market equilibrium with given expectations. Section 4 describes the fulfilled expectations equilibrium. Section 5 determines the incentive of a monopolist to invite entrants. Section 6 discusses the alternative interpretation of the model. Section 7 discusses extensions to licensing and other market structures, and section 8 provides concluding remarks.

2. Setup

We start with a simple model in expectations [4]. Suppose that the expected size of sales in the market is S. Let the *network externality function* $f(S)$ measure the increase in the aggregate willingness to pay because of the existence of the network externality. Thus, the aggregate willingness to pay for quantity Q increases from $P(Q)$ to $P(Q;S) = P(Q) + f(S)$.

We place the following restrictions on $f(S)$.

(1) $f(0) = 0$, so that no expected sales produce no network externality. This is a normalization of the $f(S)$ function and it could have been done at a different level of S.

(2) $f(S)$ is a continuous function of S.

(3) $f'(S) \geq 0$, so that higher expected network sales do not produce a lower externality.

(4) $\lim_{S\to\infty} f'(S) < \theta$, so that eventually, for large expected sales, the marginal network externality, created by an increase in the expected sales by one unit, does not exceed a constant θ. This rules out fulfilled expectations equilibria with infinite sales [5].

3. Cournot equilibrium with given expectations

Suppose that a market is described by inverse demand function [6]

$$P(Q) = A - Q$$

so that with the network externality the inverse market demand is [7]

$$P(Q;S) = A - Q + f(S).$$

Suppose that the innovator has invited $n-1$ entrants, and he competes directly with them as a Cournot oligopolist in market of n participants. Firm i maximizes $\Pi_i = q_i P(Q;S)$ by choosing q_i, where $Q = q_i + \sum_{j\neq i}^{n} q_j$. The first order condition of firm i, $i = 1, \ldots, n$, is

---

[4] Our model is similar to Katz and Shapiro (1985).

[5] Katz and Shapiro (1985) assume that the externality function $f(S)$ is bounded above. Our restriction of the derivative of $f(S)$ has the desirable effects on the properties of equilibria while allowing for a wider class of network externalities functions.

[6] Normalizing the size of units, we set without loss of generality the coefficient of Q to 1.

[7] We assume that the network externality pushes the demand outward without changing its slope, that is, $P'(Q) = \partial P(Q;S)/\partial Q$. This means that the increase in willingness to pay because of the externality is the same for each unit sold. We use this particular functional form to avoid introducing spurious strategic effects from the existence of the externality.

$$\frac{\partial \Pi_i}{\partial q_i} = A + f(S) - 2q_i - \sum_{j \neq i}^{n} q_j = 0,$$

and therefore the market equilibrium is,

$$q_i = \left(\frac{A + f(S)}{n+1}\right), \quad Q = n\left(\frac{A + f(S)}{n+1}\right),$$

$$P = \left(\frac{A + f(S)}{n+1}\right), \quad \Pi_i = \left(\frac{[A + f(S)]^2}{(n+1)^2}\right).$$

Quantities, prices, and profits increase in sales expectations. Given any expectation S, prices, per firm quantity and profits fall in the number of active firms while industry-wide quantity increases[8]. In the next section we restrict expectations to be fulfilled at equilibrium.

4. *Fulfilled expectations equilibrium*

At the full equilibrium, sales expectations are fulfilled. The fulfilled expectations equilibrium is defined by

$$S^* = Q(S^*) \Longleftrightarrow S^* = n\left(\frac{A + f(S^*)}{n+1}\right).$$

(see Fig. 2). $Q(S)$ can be thought of as a mapping of sales expectations into actual sales. Fulfilled expectations then define a fixed point S^* of function $Q(S)$. It is shown in Economides (1992) that equilibrium S^* exists, and that if the network externality function $f(S)$ is weakly concave, $f''(S) \leq 0$, then the fulfilled expectations equilibrium is unique.

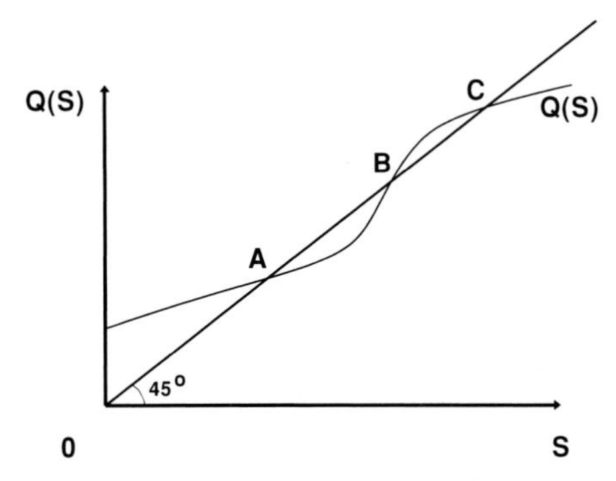

Fig. 2. Fulfilled expectations equilibria.

[8] These are the standard comparative statics as expected since S changes only the intercept of the industry demand.

The equilibrium is *locally stable in expectations* if and only if in the neighborhood of equilibrium S^* the slope of $Q(S)$ is less than 1, which means that if the marginal network externality is not too large. Then,

$$\frac{dQ(S^*)}{dS} < 1, \text{ i.e., } f'(S^*) < \frac{n+1}{n}.$$

For weakly concave network externality functions, $f''(S) \leq 0$, the unique equilibrium is globally stable. In Fig. 2, equilibria A and C fulfill this condition, but equilibrium B does not. Starting with expectations in the neighborhood of an unstable equilibrium but not exactly at the equilibrium value, there will be a tendency to move away from it. Given an unstable equilibrium, such as B, with $Q'(S) > 1$, there always exists another stable equilibrium, such as C at a higher level of sales, $S_C > S_B$. This is because for large S we have $Q'(S) < 1$, and eventually there will be a crossing of $Q(S)$ and the 45° line with $Q'(S) < 1$. Thus, it may not be unreasonable to expect that an unstable equilibrium will be avoided in favor of a stable equilibrium at a higher S.

Intuitively, we expect an increase in market production for any given level of consumers expectations S should support higher fulfilled expectations and therefore higher equilibrium production. This intuition is confirmed for stable equilibria. An increase in the number of firms n increases the quantity produced for any consumers expectations S. That is, an increase in n shifts up the $Q(S)$ function. As a result of the shift, $Q(S)$ intersects the 45° line at a larger S^* if the slope of $Q(S)$ is less than 1 (as in Fig. 3a); conversely, the upward shift of $Q(S)$ results in a smaller S^* if the slope of $Q(S)$ is larger than 1 (as in Fig. 3b). Thus, increases in n lead to increases in S^* if and only if the fulfilled expectations equilibrium is locally stable.

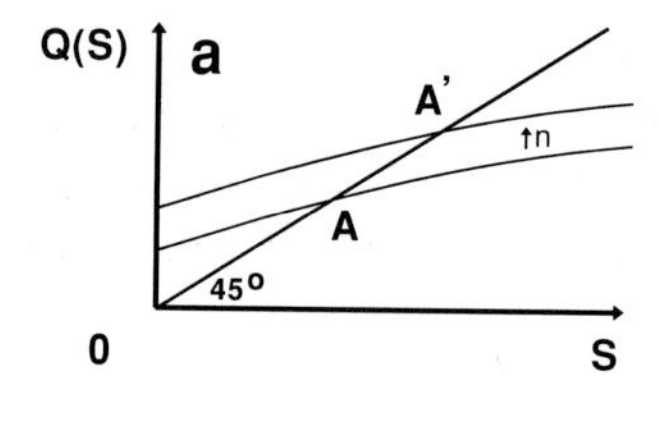

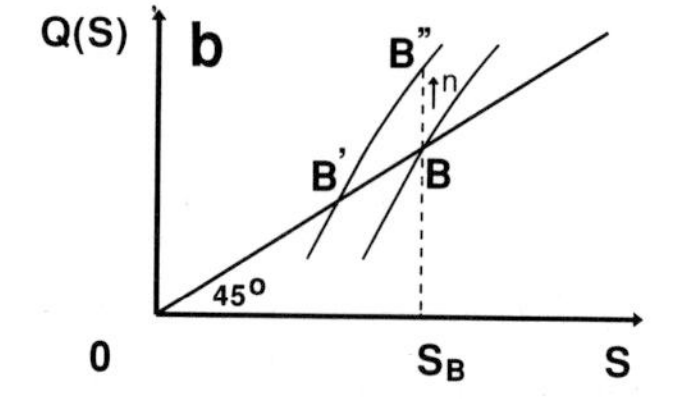

Fig. 3. Upward shifts in $Q(S)$.

Formally, an increase in the number of competitors increases market production (by differentiating the fixed point condition) iff

$$\frac{dS^*}{dn} = \frac{S^*}{\{n[n+1-nf'(S^*)]\}} > 0 \leftrightarrow f'(S^*) < \frac{n+1}{n}.$$

This condition is equivalent to local stability of the fulfilled expectations equilibrium. Now we examine if the increase of sales at the fulfilled expectations equilibrium can also lead to higher profits.

5. *The monopolist's incentive to invite entry*

The equilibrium profits of a firm at an n-firm fulfilled expectations equilibrium are

$$\Pi_i^* = \left(\frac{[A+f(S^*)]^2}{(n+1)^2}\right) = \left(\frac{S^*}{n}\right)^2$$

As the number of firms increases, there are two opposite effects on the innovator-monopolist's profits. First, because the number of competitors increases, his profits fall. This is the *competitive effect*. Second, as the number of competitors increases, the market can support larger expected sales as a fulfilled expectations equilibrium S^*. Increases in expected sales increase the innovator-monopolist's profits because they push up the industry demand through the expansion of the network. This is the *network effect*. These effects can be identified on Π_i^* as follows:

$$\frac{d\Pi_i^*}{dn} = \frac{\partial \Pi_i^*}{\partial n} + \left(\frac{\partial \Pi_i^*}{\partial S}\right)\left(\frac{dS^*}{dn}\right) > 0.$$

The first term, capturing the competitive effect, is negative. The second term, capturing the network effect, is positive since higher expected sales increase profits, and a higher number of firms increases the fulfilled equilibrium (expected and realized) sales. By substitution and simplification we find

$$\frac{d\Pi_i^*}{dn} = S^{*2}\left[\frac{nf' - (n-1)}{n^3[n+1-nf'(S^*)]}\right].$$

Therefore,

$$\frac{d\Pi_i^*}{dn} > 0 \leftrightarrow (n-1)/n < f'(S^*) < (n+1)/n.$$

It follows that an exclusive holder of a technology in a market with strong network externalities at the margin, $f'(S) > (n-1)/n$, has an incentive to invite competitors to enter the industry and compete directly with him.

6. Mutual feedbacks in markets for complementary goods

There is another interesting interpretation and extension of our model in the context of two industries that produce complementary goods. Suppose that one firm is the exclusive holder of the technology in industry 1, but there is free entry in industry 2. The monopolist in industry 1 can invite $n_1 - 1 \geq 0$ competitors. When production levels in industries 1 and 2 are Q_1 and Q_2, the willingness to pay for product 1 is

$$P(Q_1; Q_2) = A - Q_1 + f(Q_2).$$

Here Q_1 plays the role of Q, and Q_2 plays the role of S of our previous discussion. Why is the willingness to pay for product 1 increasing in the production level of product 2? Higher production Q_2 implies a larger number of varieties of product 2 and a lower price for them. Thus, with higher Q_2, the surplus realized by consumers of product 2 is higher. Since products 1 and 2 are complementary, higher surplus generates a higher willingness to pay for product 1. This is captured by $f(Q_2)$.

For example, suppose that there are n_2 locationally differentiated products of zero marginal cost and fixed cost F on a circumference as in Salop (1979). Let consumers be distributed uniformly with density μ according to their most preferred variety and have a reservation price R[9]. The symmetric equilibrium price is $p^*(n_2) = 1/n_2$. Profits are $\Pi(n_2) = \mu/n_2^2 - F$. With free entry, $\Pi(n_2) = 0$, and therefore there will be (approximately) $n_2^* = \sqrt{(\mu/F)}$ active firms. The average benefit of a consumer from the consumption of one unit of product 2 is then

$$R - \left[p^*(n_2^*) + \frac{1}{4n_2^*}\right] = R - \frac{5}{4}\sqrt{\frac{F}{\mu}}.$$

Since all consumers on the circle buy the differentiated product we can interpret μ as Q_2, and we can write the average benefit to a consumer as $f(Q_2) = R - (5/4)\sqrt{(F/Q_2)}$ when output in market 2 is Q_2. Assuming that products 1 and 2 are consumed in 1 : 1 ratio, this average benefit is added to the willingness to pay of consumers for good 1 [10,11].

Firms play an oligopoly game in market 1, taking Q_2 as given. Let the resulting equilibrium output be $Q_1^*(Q_2; n_1)$. This is a direct reinterpretation

---

9 We assume that R is sufficiently large so that all consumers buy a differentiated good.

10 Note that the network externality $f(Q_2)$ (average consumers' surplus) is increasing in Q_2 because a high level of production in industry 2 implies both a larger number of varieties n_2 and a higher degree of competition resulting in lower prices.

11 We could also allow the reservation price R in industry 2 to vary in Q_1 without changing the result.

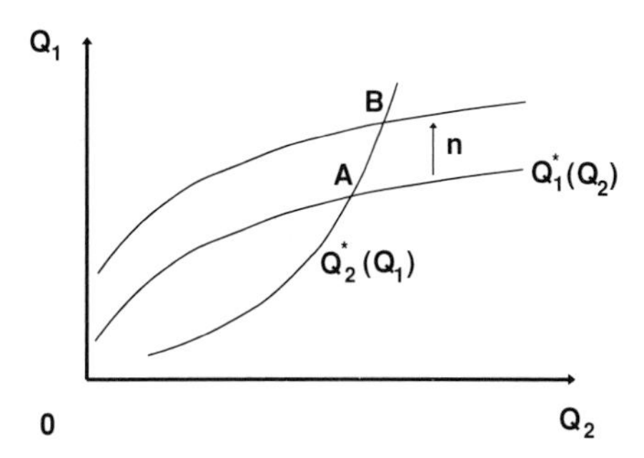

Fig. 4. Complementary markets with feedbacks.

of $Q(S;n)$. Firms in market 2 take Q_1 as given. Let equilibrium output in market 2 be $Q_2^*(Q_1)$. In the expectations model we used $S^*(Q) = Q$, i.e., $Q_2^*(Q_1) = Q_1$, and this applies well in a model where the two types of products are consumed in 1:1 ratio, as in our circumference example. In general, $Q_2^*(Q_1)$ will not be the identity function. Then equilibrium across markets defines Q_1^{**}, Q_2^{**} as the intersection of $Q_1^*(Q_2;n_1)$ and $Q_2^*(Q_1)$ (see Fig. 4). Increasing the number of firms n_1 shifts $Q_1^*(Q_2;n_1)$ to the right. The effect of the increase of n_1 on equilibrium output in industry 1 is $\mathrm{d}Q_1^{**}/\mathrm{d}n_1 = (\mathrm{d}Q_1^*/\mathrm{d}n_1)/[1-(\mathrm{d}Q_1^*/\mathrm{d}Q_2)(\mathrm{d}Q_2^*/\mathrm{d}Q_1)]$, which is positive if the equilibrium is stable $(1 > (\mathrm{d}Q_1^*/\mathrm{d}Q_2)(\mathrm{d}Q_2^*/\mathrm{d}Q_1))$ [12].

The effect of increases in n_1 on the profits of the innovator depends on the particulars of the oligopolistic interaction in markets 1 and 2, as well as on the degree of complementarity between the two markets. For example, if goods 1 and 2 are consumed in 1:1 ratio, this model of interaction across complementary markets is an exact reinterpretation of the expectations model. Thus, all results of the expectations model can be directly reinterpreted for complementary goods model. In particular, when the marginal network externality effect is strong enough, the monopolist-innovator will invite firms to enter.

We have shown in this section that the traditional fulfilled expectations model of network externalities can be re-interpreted and extended to describe markets with positive feedbacks across complementary goods or components. Thus we provided a concrete example of the way in which network effects arise and how and when invitations to enter will be beneficial to the innovator-monopolist.

---

[12] We assume that an increase of the number of active firms in industry 1 results in an increase of its equilibrium output, i.e., $\mathrm{d}Q_1^*/\mathrm{d}n_1 > 0$.

7. Extensions

The oligopoly game played among the firms is not critical for the results. Economides (1992) shows similar results when the monopolist is a quantity leader and the entrants are quantity followers. Extensions to the case of uncertainty about expectations are also discussed in Economides (1992).

Without network externalities, i.e., if $f(S) = 0$ for all S, an innovator-monopolist may want to license competitors. We have shown in earlier sections that, in the presence of strong network externalities, the monopolist would like to invite competitors (charging them zero fees). Now, if he can charge any fees to entrants, he can construct a combination of marginal and lump sum fees that absorbs all profits from the entrants. Essentially with such fees the innovator-monopolist can fully internalize the externality created by entrants. Thus, Economides (1992) shows that in general the monopolist will invite a larger number of competitors to enter if he can charge any licensing (including lump sum) fees.

When lump sum fees are unfeasible, the innovator-monopolist is limited to marginal fees. Economides (1992) shows that the innovator-monopolist charges a positive licensing fee in a market with weak network externalities. Conversely, *in a market with strong network externalities, the innovator-monopolist is willing to give a subsidy to his competitors to encourage higher production*. In the case when the licensing fee chosen by the innovator is positive, the fee also increases with the number of competitors. When the innovator-monopolist chooses to subsidize the followers, his subsidy increases with the number of his competitors.

These results are rather intuitive. In a market with small network externalities, the result is the same as in a market with no externalities, i.e., the innovator-monopolist charges a positive licensing fee. Since in this case the benefit from the externality is small, the innovator-monopolist increases his profit by restricting at the margin the level of output of his competitors through a positive licensing fee. This fee is higher if there are more competitors, to compensate for the higher output. Conversely, when the network externalities are strong, the innovator-monopolist gives a subsidy to his competitors to encourage increased production and greater network effects from which he will benefit. In this case the optimal subsidy increases in the number of competitors to create the strongest externality.

We also find that when the innovator-monopolist uses marginal licensing fees, profits increase in the number of competitors. Thus, the innovator-monopolist has an incentive to invite competitors. Note that it is to the benefit of the innovator-monopolist to invite competitors, both when the license fee is positive and again when it is negative. The intuitive reasons are different in each case. When network externalities are strong, the innovator-monopolist invites competitors, and provides them with a subsidy to enjoy the strong network effects. When the network externalities are relatively weak, the first objective of the innovator-monopolist is to collect the licensing fees; cultivating the network effects is secondary. Of course, in both cases, optimal marginal fee licensing is superior to licensing for free.

8. Concluding remarks

In the fragmented world of today's telecommunications' industry, firms have difficulty capturing or internalizing the underlying network externalities that naturally occur in this industry. Traditionally, network externalities were captured and internalized to a large extent by vertically integrated national monopolies where they appeared as *economies of scope*. The present level of fragmentation in the industry and the inability of many competitors to vertically integrate (because of regulatory or other restraints) limits the ability of firms to exploit network externalities. Our model suggests that in these circumstances it may be desirable to invite competitors to enter, and even subsidize them.

We have shown that in a market with strong network externalities, in the absence of other means of committing to high production (such as vertical integration in the complementary industry or other contractual commitments), it is beneficial for a sole holder of a new technology to invite competitors to enter the industry by giving away his proprietary knowledge. The expansion of output required for the creation of a large network cannot be done in the absence of competitors. The innovator-monopolist cannot credibly commit himself to create a large network (and reap its benefits) because, for any given level of consumers expectations of sales, the monopolist has an incentive to produce a relatively low output. Nevertheless, the innovator can use the fact that a more competitive market will result in a higher output (for any given expectations). By inviting competition, the innovator commits to an expanded amount of market output for any given expectations. Thus, the innovator credibly sustains the expectation of a high production by inviting competition, and thereby creates the desired large network effect.

We also showed that the expectations model is formally equivalent to a model of strategic interaction with mutual feedbacks between two complementary markets. In this framework, the size of sales in industry 2 affects positively the surplus realized by consumers who buy good 2. This in turn affects positively the willingness to pay for the complementary good 1. Because of the formal equivalence, all results can be reinterpreted in the framework of two complementary markets. Thus, in the presence of strong network externalities and strong complementarities, it pays for a monopolist-innovator to invite competitors to enter and compete on equal terms with him.

The innovator does even better and invites more competitors if he can charge lump sum licensing fees. If the innovator can charge only marginal licensing fees, his optimal licensing fee will be positive for markets with weak network externalities, and negative (i.e., a subsidy) when the externalities are strong. For both weak or strong network externalities, the innovator invites entry as well, and has higher profits than when licensing was free.

In conclusion, offering an invitation to enter by an innovator-monopolist is a useful strategy for a firm in markets with strong network externalities arising because of strong complementarities, if vertical integration or other means of commitment to high output are not available.

References

Chou, C. and Shy, O., 1990. Network Effects Without Network Externalities. *International Journal of Industrial Organization*, 8, 259–270.

Church, J. and Gandal, N., 1990. Network Effects, Software Provision and Standardization. *Journal of Industrial Economics*, XL (1), 85–104.

Economides, N., 1988. Equilibrium Coalition Structures. *Columbia University, Department of Economics, Discussion Paper* No. 273.

Economides, N., 1989. Desirability of Compatibility in the Absence of Network Externalities. *American Economic Review*, 78 (1), 108–121.

Economides, N., 1991. Compatibility and the Creation of Shared Networks. In: M. Guerrin-Calvert and S. Wildman (eds.), *Electronic Services Networks: A Business and Public Policy Challenge* (Praeger Publishing Inc., New York).

Economides, N., 1992. Network Externalities, Complementarities, and Invitations to Enter. *Stern School of Business, N.Y.U., Discussion Paper*, EC-92-2.

Economides, N. and Salop, S.C., 1992. Competition and Integration among Complements, and Network Market Structure. *Journal of Industrial Economics*, XL (1), 105–133.

Economides, N. and White, L.J., 1993. One-way Networks, Two-way Networks, Compatibility and Antitrust. *Stern School of Business, New York University, Discussion Paper*, EC-93-14.

Farrell, J. and Gallini, N., 1988. Second-Sourcing as a Commitment: Monopoly Incentives to Attract Competition. *Quarterly Journal of Economics*, 103, 673–694.

Farrell, J. and Saloner, G., 1985. Standardization, Compatibility, and Innovation. *Rand Journal of Economics* 16, 70–83.

Gallini, N. and Wright, B., 1990. Technology Transfer Under Asymmetric Information. *Rand Journal of Economics* 21 (1), 147–160.

Gilbert, R., 1991. On the Delegation of Pricing Authority in Shared ATM Networks. In: M. Guerrin-Calvert and S. Wildman (eds.), *Electronic Services Networks: A Business and Public Policy Challenge* (Praeger Publishing Inc., New York).

Katz, M. and Shapiro, C., 1985. Network Externalities, Competition and Compatibility. *American Economic Review* 75 (3), 424–440.

Oren, S. and Smith, S., 1981. Critical Mass and Tariff Structure in Electronic Communications Markets. *Bell Journal of Economics* 12 (2), 467–487.

Rohlfs, J., 1974. A Theory of Interdependent Demand for a Communications Service. *Bell Journal of Economics* 5, 16–37.

Salop, S.C., 1979. Monopolistic Competition with Outside Goods. *Bell Journal of Economics*, 10, 141–156.

Salop, S.C., 1991. Evaluating Network Pricing Self-Regulation. In: M. Guerrin-Calvert and S. Wildman (eds.), *Electronic Services Networks: A Business and Public Policy Challenge* (Praeger Publishing Inc., New York).

Sharkey, W.W., 1991. *Network Models in Economics*, Mimeo.

Shepard, A., 1987. Licensing to Enhance Demand for New Technologies. *Rand Journal of Economics*, 18 (3), 360–368.

Wilson, R., 1991. *Economic Theories of Price Discrimination and Product Differentiation: A Survey*, Mimeo.

Global Telecommunications Strategies
and Technological Changes
Edited by G. Pogorel

Chapter 12

Reaching compromise in standards setting institutions

Peter Swann

London Business School, Sussex Place, London NW1 4SA, UK

1. Introduction

The vital importance of standards to the telecommunications industry and as a component of telecommunications policy is clear from the papers in this section of the book. It is hardly necessary to reiterate that message to those from the telecommunications industry. Economists too are well aware of the huge strategic and economic significance of standards in the information and communications technologies. Much of the research by economists has been concerned with the process by which markets generate *de facto* standards [1]. More recently, however, increasing attention has been directed at the economic factors influencing the activities of standards institutions [2]. One difficulty facing the economist is that standards institutions perform several rather different activities, and this calls for a variety of analytical models; at present, we do not have an integrated economic theory of the standards institution.

One of these activities is the selection or definition of a standard in a setting where different participants in the standards process have already invested in their own proprietary designs. Indeed, if participation in the

[1] Seminal contributions are those of Katz and Shapiro (1985, 1986), Farrell and Saloner (1985, 1986), David (1985, 1987), Arthur (1989), amongst others. A valuable survey is given by David and Greenstein (1990).

[2] Notably the work of Farrell and Saloner (1988), Farrell and Shapiro (1992), Besen and Farrell (1991), Weiss and Sirbu (1990). See also the studies by Cargill (1989) and OECD (1991), and the papers in Gabel (1987), Berg and Schumny (1990), and Meek (1993).

process is dominated by producers with proprietary interests, then the role of the institution may essentially be to select one of the competing systems, or to achieve a compromise solution between the competing systems. It is to this latter question that the present paper is addressed.

In this setting, the basic economics of the process seems quite straightforward, even if some rather more subtle questions arise in practice. All parties to an agreed standard benefit from that agreement, but some parties may incur additional adaptation costs if the agreed standard lies some way away from their proprietary design. While the group of interested parties as a whole may be better off as a result of the standards agreement, the benefits may not be evenly distributed, raising the possibility that there could be losers (whose adaptation costs exceed their benefits from the agreement) as well as winners. Cast in such terms, it is not difficult to see how some relatively simple economic analysis could be brought to bear on the question of achieving compromise in standards institutions.

Undoubtedly standards institutions conduct other activities which simply do not fit into this model. For example, some observers argue that an analysis based on winners and losers is frequently misleading. Our response would be that it is case-specific. Certainly in the context of many anticipatory standards there are no (substantial) sunk investments in proprietary systems, and conflict is more a result of participants holding incompatible visions of the future for a technology than of participants being unwilling to bear adaptation costs. Even then, there is a growing body of thought that sees the firms technological vision as an asset of considerable value[3]. When the firm is able to see its vision into market reality, the existence and organisational implementation of that vision gives the firm a distinct competitive advantage; but when the vision turns out to be inaccurate, and the technology follows a different trajectory, the value of the vision is eroded, and firms incur substantial costs in adapting to the new trajectory. In short, conflict over diverse technological visions could be a matter of engineers' pride alone; but it could also reflect the competitive interests of their organisations.

The remainder of the paper is as follows. Section 2 presents a simple framework for analyzing the scope for achieving compromise in a standards institution, in a setting where firms have already developed a proprietary technology, and continue to invest in it, but where firms also invest resources in keeping abreast of rivals' technologies. Section 3 uses this framework to explore the scope for compromise at any time, while section 4 uses it to identify windows of opportunity during which compromise is possible.

---

[3] See Swann (1992) for a review of the interaction between technology vision, corporate organisation, technological trajectories and competitive advantage.

Section 5 analyses one simple role for a public agency in this setting, which keeps the windows open for longer, and which can lead to a *better* standard being achieved (in a sense to be explained below), while section 6 uses the framework to explore the question of an optimum date for compromise. Section 7 makes some concluding remarks about the implications of this paper for our understanding of the economics and management of standards institutions in telecommunications and other industries.

It must be stressed that the analysis presented here does not represent a fully analyzed model, but rather an exploratory framework. The simulations presented in the paper should also be taken in that spirit: it is not claimed that they are representative or typical, merely illustrative.

2. A simple triangular framework [4]

The basic framework used here considers the case of three producers or consortia who have invested in their own proprietary systems. These systems are taken to be *horizontally* differentiated — meaning that they represent *different* approaches to a technological solution, but that no one of them is clearly better than the others *ab initio*. It is, however, possible (likely, indeed) that they will become vertically differentiated during the process of competition over the standard — meaning that one *will* be recognised as *better* than the others.

The attraction of limiting the analysis to three producers is that the space for compromise can then be represented graphically in two dimensions. The three proprietary systems can be represented as the corners of an equilateral triangle. A compromise will lie inside the triangle [5], and the distance between the compromise point and a corner of the triangle is a measure of how much that producer has to adjust in making the compromise. If the compromise point is close to a corner, that producer need make little adjustment from his proprietary design, while if the compromise point is far from that corner, the producer will have to make a considerable adjustment from his proprietary system.

While the limit of three producers may seem restrictive, the model can be directly applied to the case of more than three producers if each corner point is treated as the compromise achieved by a consortium of up to three

---

[4] The following is just a very informal account of this framework. Further details are available from the author.

[5] We do not consider outcomes where the institutional standard lies outside this triangle. That is because we only analyze the institution's role in selecting a compromise, and not its role in further developing the standard beyond (or outside) the existing proprietary specifications.

producers. By successive nesting of this sort, it would be possible to apply this framework to cases with many more interested parties[6].

The dynamics of the model is in two parts. The first describes the rate at which the three producers invest in their own proprietary systems, and also the rate in which they invest in assimilating the technologies of their rivals[7]. The second describes the rate at which the (discounted) benefits of joining a standards agreement change over time.

Part 1, the investment behaviour in any period t, is summarised by the matrix **A**:

$$\begin{bmatrix} a_{11} & a_{12} & a_{13} \\ a_{21} & a_{22} & a_{23} \\ a_{31} & a_{32} & a_{33} \end{bmatrix}$$

where a_{ij} is the investment by firm i in producing its own technology (when $i = j$), or the investment by firm i in assimilating the technology of firm j (when $i \neq j$).

The matrix **A** may be a function of t, though in this exploratory study we have kept it fixed. Accordingly, at the end of period t, the cumulative investment by each producer in each technology (so to speak) is given by $t\mathbf{A}$. In this framework, the benefit firm i enjoys from adopting standard j at time t is taken to be proportional to ta_{ij}. Accordingly, if the matrix **A** is fairly diagonal, then firms will much prefer to adopt their own standard, especially as t increases. If however, firm i adopts a compromise standard which reflects θ_1 the wishes of firm 1, θ_2 the wishes of firm 2, and θ_3 the wishes of firm 3 (where $\sum \theta_k = 1$), then the investment available to firm i is assumed to be a straightforward convex combination: $\theta_1 a_{i1} + \theta_2 a_{i2} + \theta_3 a_{i3}$.

There is, however, a bonus for entering a coalition in this framework (part 2). This is a function of the greater network enjoyed by a firm if it adopts the same standard as others. Suppose the network available to one firm operating its own standard is of size n; in the simple illustrative examples, a firm sharing a standard with *one* other firm is assumed to enjoy a network of $2n$, while if all three firms adopt the same standard, the network is $3n$. The bonus is not however a linear function of the network, however, because the framework assumes that beyond a certain size there are diminishing returns to additional network size[8].

---

[6] In practical cases, where n interested parties are to be grouped into successive nests of this sort, a dimension reduction technique like principal components or cluster analysis could be used.

[7] This dichotomy is effectively the same as the 'two faces of R&D' considered by Cohen and Levinthal (1989, 1990): investment in *innovation* and investment in *learning*.

The best strategy, therefore weights the cost of adapting to a compromise standard against the bonus enjoyed from entering a standard coalition. By sticking to its corner, the firm enjoys the fruits of investment in its own technology, but may find that others are reluctant to join it. By giving some ground it may persuade one other firm to join a coalition, and by giving a little more it may persuade both. The next section illustrates the scope for compromise in this framework.

3. *Scope for compromise and stability*

Using the above framework, it is straightforward to compute the payoff for each firm in five different cases.

no coalitions:	three different standards
coalition (1, 2):	3 has separate standard
coalition (1, 3):	2 has separate standard
coalition (2, 3):	1 has separate standard
coalition (1, 2, 3):	all have same standard

In the first case, all we need do is compute the payoffs for each firm at its own corner. For the two-way coalitions (e.g. between 1 and 2), we strictly need only compute the payoffs for compromises along the side 1–2 (the two firms in the coalition). For the three-way coalition, it is necessary to consider the whole space. For simplicity, however, we compute these payoffs at a grid of 10% intervals — that is at those points in the triangle where the θ_j are in multiples of 10% (including 0%).

There will be scope for a three-way coalition at a particular compromise point if no single party is better off deserting the coalition at that point, and reverting to its own standard. If there is such a compromise point from which no one party will desert, then such a universal standard is feasible. However, this alone need not be enough to ensure that it is stable. We also need to check that no *pair* of firms might find it worthwhile to desert the universal standard, and set up a two-way compromise standard.

---

[8] The form of the function relating value to network size has not been analyzed in depth to our knowledge — see Swann and Shurmer (1992) for some very preliminary comments. Assuming a standard diffusion curve for these technologies amongst the user community, discounting of future network growth, and diminishing returns to increases in network size, the bonus curve (for any coalition) follows an upward sloping S-shaped curve over time. Again further details are available from the author. The 3 party coalition curve lies above the 2 party curve, of course.

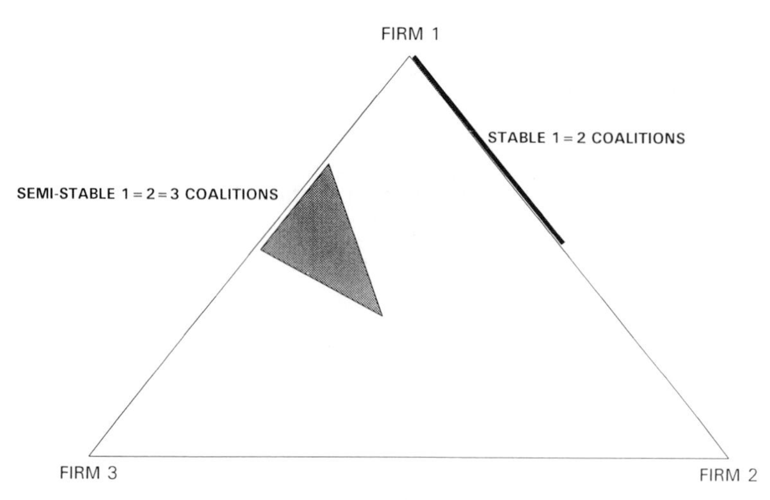

Fig. 1. Possible coalitions: Period 35.

Figure 1 illustrates this for one particular simulation [9]. The lightly shaded area shows what we might call the *semi-stable* coalitions, where 1, 2 and 3 share the same standard. In this area, no *one* firm finds it profitable single-handedly to desert the coalition. But in this simulation, nowhere in the lightly shaded area represents a *fully* stable coalition. Starting from any point in this area, there would be an incentive for firms 1 and 2 to break rank *together* and set up a compromise standard of their own along the side 1–2 (see the darkly shaded area). The coalitions in this latter area are fully stable as firms 1 and 2 are better off than they would have been in any semi-stable three-way coalition.

It is instructive to examine the same question earlier in this particular simulation. Figure 2 shows the equivalent picture for period 25. Here, there are a much wider range of feasible three-way coalitions, and all of them are stable. Not only are they secure against desertion by a single party, but they are also robust against desertion by a pair of firms who share a standard between them. Starting from any point in the lightly shaded area, there is no coalition between 1 and 2 (along the side 1–2) at which *both* firm 1 and firm 2 are better off than in the three-way coalition. One or other may be, but not *both*.

[9] As the full model has not been set out, we do not give the full parameter set here. In descriptive terms, this is a simulation where firms 1 and 2 invest quite a bit in assimilating each others' standards, so that coalitions between 1 & 2 are always easier than those involving 3, whose investment is much more idiosyncratic. Nevertheless, the coalition bonuses are very high in this simulation, so that there is considerable scope for coalitions between all three firms — at least in the middle of the run.

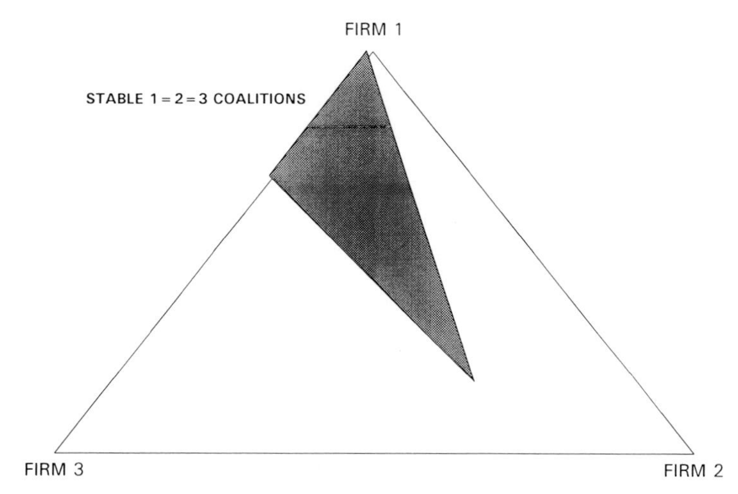

Fig. 2. Possible coalitions: Period 25.

These are the steps required for checking the stability of a three-way coalition. To check the stability of any two-way coalition, similar steps are required. A two-way coalition at a particular point will only be stable if it is the interests of *neither* party to break rank and revert to their own proprietary standard, *and* if neither party finds it profitable to engage the third firm in a three-way coalition.

4. Windows of opportunity

David (1987) analyses the concept of policy *windows* for standards setters, during which standards agencies may be able to influence the outcome of a market process, but outside which, the agency is effectively powerless. Here we discuss a related concept of a *window* of opportunity for compromise. We saw above that while in an earlier period of a simulation, opportunities for a stable three-way coalition were plentiful, they had disappeared by a later period. Figure 3 (derived from the same simulation as in Figs. 1 and 2) shows the number of stable three-way coalitions and the number of stable two-way coalitions between 1 and 2. (There are no stable two-way coalitions between 1 and 3, or between 2 and 3.) For ease of interpretation, these have been expressed as percentages of the triangle area.

In early periods of this simulation, there are relatively few stable coalitions, and those that there are take place between firms 1 and 2 only. But later, the number of stable three-way coalitions grows rapidly, and the number of stable two-way coalitions between 1 and 2 drops to zero. This last result

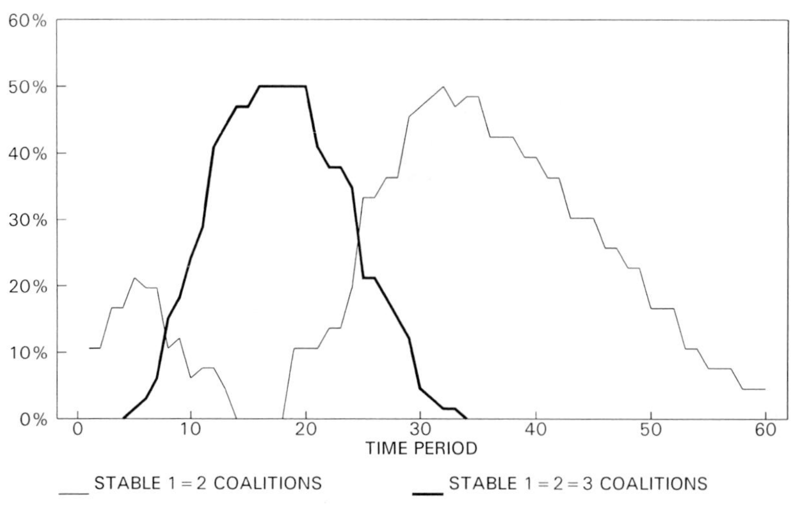

Fig. 3. Probability of stable coalitions.

happens because while many such coalitions are perfectly viable, there is always a three-way coalition that is better for firms 1 and 2. After period 20, however, the scope for three-way coalitions declines rapidly, because firm 3 has continued to invest primarily in its rather idiosyncratic standard: firm 3 now finds that the adaptation costs to any three-way coalition are too great for the coalition bonus — which has started to tail off because of diminishing returns to network size. Now there is an era of two-way coalitions between firms 1 and 2, which continues for quite a while, though the scope tails off steadily after period 35.

Figure 3, in effect, defines the windows of opportunity for reaching compromise. What it says is that if the three parties come to the negotiation table between periods 10 and (about) 25, there is good scope for a three-way compromise. Of course, as Farrell and Saloner (1988) show, reaching consensus will not be instantaneous even when a three-way coalition is feasible. If the negotiation process takes m periods, and all three firms continue the same investment strategies during the period of negotiation, then the window of opportunity must be reduced by m periods. That means that if negotiation starts after (about) period 30-m, then even though a stable coalition was available at the start of the negotiation process, there is little chance of it by the end.

Because we do not model investment strategies explicitly in this illustrative paper, the model is silent about the implications of starting a negotiation process before the window of opportunity. If after an early and (inevitably) unsuccessful negotiation, firms simply agree to re-convene later to see if the prospects have improved, then this framework would suggest that they should be successful at a later attempt. Conversely, in the aftermath of an unsuccessful (and acrimonious?) early negotiation, firms might change their investment strategies: they might concentrate solely on their own systems in anticipation of a full scale battle of the systems in the market. In other words, the matrix A becomes diagonal. If that happens, the windows calculated in Fig. 3 no longer apply. When **A** is diagonal, the windows of opportunity are likely to be very brief (if indeed they exist at all).

5. *Public agencies and compromise*

One stylised function of public agencies in models of this sort is to facilitate the process of reaching compromise. One way in which that can be done is by enabling an informal system of side-payments. *Explicit* side payments are probably rare, and would tend to be viewed with suspicion by anti-trust authorities. Rather, this is achieved by constructing a forum (the standards institution) to encourage informal *give and take* — where one party concedes ground on one issue in (trusting) expectation of being compensated on another issue [10].

Within this framework, the facilitation of informal side-payments achieve three related things. First, it increases the scope for compromise by increasing the area in the compromise triangle that represents stable coalitions. Two firms can compensate the third for moving a long way from his corner because the three of them are collectively better off if the three-way coalition survives.

Second, it provides a mechanism for choosing the *best* compromise from the set of stable coalitions. In Fig. 2, for example, *any* of the points in the highlighted area would be stable three-way coalitions. But the framework as it stood had little to say about *which* of these possible outcomes would be the result. Yet it is unlikely that the collective benefit for all three parties is the same for each of the possible compromises. In particular, a common tendency in this framework is that the compromise that is best from a collective point of view does in fact lie at one of the three corners of the triangle: in such cases, it will be the most developed standard that should be adopted — that is the one in which there has been the greatest collective investment.

---

[10] Immediately we are into a difficult area. In the short term, cheating in such give and take negotiations may be profitable. At the other extreme, economists argue that adverse reputation effects will be a sufficient bar to cheating in repeated games. The reality of standards institutions, where the same parties continue to encounter each other has some of the flavour of a repeated game, but is really much more complex.

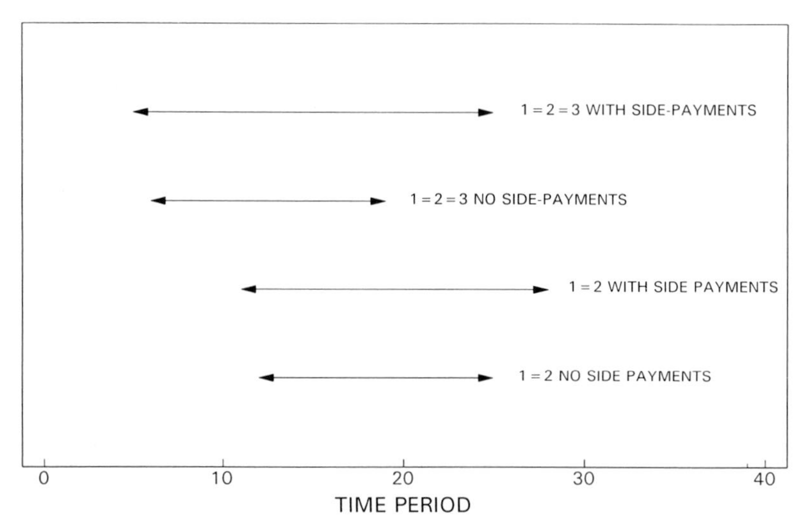

Fig. 4. Windows for compromise.

Third, the facilitation of side-payments keeps the windows of opportunity open for longer. Figure 4 illustrates this [11]. The lines show the scope for three-way and two-way coalitions-corresponding to the non-zero parts of Fig. 3 and comparing the *private* case (no side payments) with the *public* case (side payments allowed). As before, the window for three-way coalitions precedes that for two-way coalitions. Here, the existence of side-payments is particularly effective at extending the window for three-way coalitions, but makes less difference to the window for two-way coalitions. The reason for this is, as noted above, that firms 1 and 2 pursue investment strategies that allow them to assimilate each others' technologies, and hence compromise between 1 and 2 is much easier here than compromise involving 3.

6. *The best time for compromise*

One further insight can be gained from this framework. As noted in the last section, the use of side payments provides a mechanism to ensure that the standard compromise adopted is the one that is collectively best. This enables us to compare the total cumulative investment in standards setting (by the three firms in the model) with the collective benefits realised (when the *best* standards compromise has been adopted).

[11] This is based on a slightly different (and less optimistic) simulation from the previous graphs, because in the earlier simulation, the window never closes when side-payments are possible!

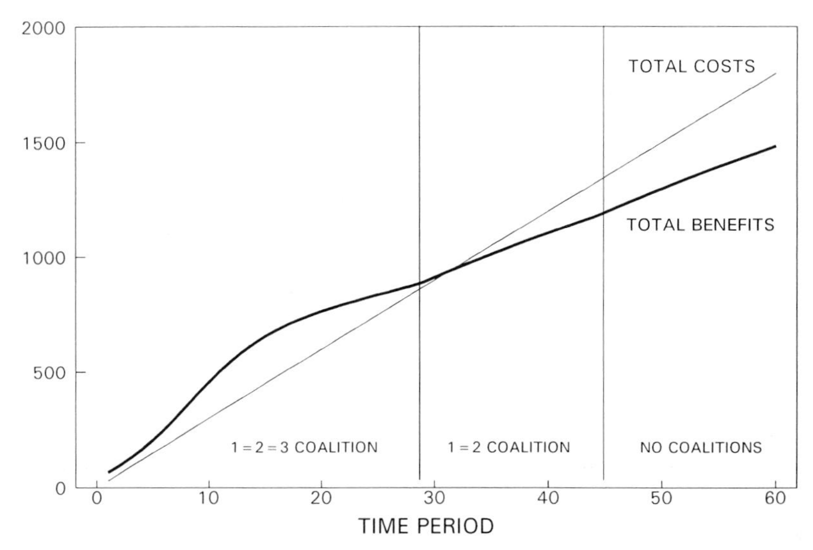

Fig. 5. Total costs and social benefits of investment.

Figure 5 summarises this comparison. Note that, as above, there is an era of three-way coalitions, followed by an era of two-way coalitions (firms 1 and 2), and finally an era of separate standards. The existence of side payments may extend the window of opportunity for coalitions, but not indefinitely. Ultimately, when positions become too entrenched and diminishing returns from further increases in network size have set in, the best solution (from a collective point of view) is to forego the coalition bonuses and market three different standards.

Figure 5 shows that for this simulation, collective benefits exceed total investment up to about period 30, and when we have just switched to the era of two-way coalitions. In terms of achieving the maximum social surplus, the best date for compromise is about period 15, which is in the middle of the era of three-way compromises.

As it stands the model is silent about whether the three firms can expect to achieve compromise at that earlier date. For two reasons, we might expect compromise to be achieved rather later. The first is because of the Farrell and Saloner (1988) negotiation delays mentioned above. The second is that there is an element of *race* to this process, and economic analysis of races frequently predicts over-investment (beyond the optimum). The compromise standard is likely to lie close to the most developed standard, so each firm will want to have the best standard by the date of compromise.

7. Conclusions

This paper has presented a framework for analyzing the scope for reaching compromise in standards institutions. While the paper does not address telecommunications issues directly, it is clear that compromise in standards setting is a most important one in the telecommunications industry. In a national setting, it is usually essential that the standards setting process generates a single standard, but this need not always be the case in an international setting. Under certain circumstances it is reasonable to expect that wider representation in standards institutions can slow down the process [12]. As Besen and Farrell (1991) point out, the desire to reach standards quickly may mean that regional standards organisations (RSOs, such as ETSI) have more success than international institutions (ITU). The narrower representation of the RSO may lead to a greater degree of fragmentation, such as that predicted in this model when one party has made idiosyncratic investments.

The paper showed that to be stable, a coalition of all parties must be robust against one party breaking rank to market its own proprietary standard, but also against sub-sets of firms doing the same. We saw that there was a window of opportunity for reaching a coalition between all parties, although after that window was closed, there would still be opportunities for smaller coalitions to agree on a common standard. The introduction of a public agency to increase trust in *give and take* negotiations was seen to expand the scope for compromise, and lengthen the windows of opportunity. Finally, the framework can calculate a date that is collectively the best for reaching compromise, which may in practice be unrealistically early.

Acknowledgements

This research is part of a project entitled '*How much should be spent on IT standardisation, and how should it be funded?*'. The financial support of the Leverhulme Trust is gratefully acknowledged. I am grateful to Mark Shurmer of Brunel University for his work on this project, and for many helpful discussions. I have also learnt much from discussions of standards issues with Paul David, Ed Steinmueller, Cristiano Antonelli, Robin Cowan, Dominique Foray, Richard Hawkins, Robin Mansell, Martin Cave and Peter Grindley, and from other participants at the ITS Conference, June 1992. None of these are responsible for remaining errors in this paper.

[12] We attempted a simple analysis of this in Swann (1993).

References

Arthur, W.B., 1989. Competing Technologies, Increasing Returns, and Lock-in by Historical Events. *Economic Journal* 99, 116–131.

Berg, J. and Schumny, H., 1990. *An Analysis of the IT Standardization Process* (North-Holland, Amsterdam).

Besen, S.M. and Farrell, J., 1991. The Role of the ITU in Standardization. *Telecommunications Policy* 15, 311–321.

Cargill, C.F., 1989. *Information Technology Standardization: Theory, Process and Organisations* (Digital Press, Bedford, Mass.)

Cohen, W.M. and Levinthal, D.A., 1989. Innovation and Learning: The Two Faces of R&D. *Economic Journal* 99, 569–596.

Cohen, W.M. and Levinthal, D.A., 1990. Absorptive Capacity: A New Perspective on Learning and Innovation. *Administrative Science Quarterly* 35, 128–152.

David, P., 1985. Clio and the Economics of QWERTY. *American Economic Review Proceedings* 75, 332–336.

David, P., 1987. Some New Standards for the Economics of Standardization in the Information Age. In: P. Dasgupta and P. Stoneman (eds.), *Economic Policy and Technological Performance* (Cambridge University Press, Cambridge).

David, P. and Greenstein, S., 1990. The Economics of Compatibility Standards: An Introduction to Recent Research. *Economics of Innovation and New Technology* 1, 3–41.

Farrell, J. and Saloner, G., 1985. Standardization, Compatibility and Innovation. *RAND Journal of Economics* 16, 70–83.

Farrell, J. and Saloner, G., 1986. Installed Base and Compatibility: Innovations, Product Pre-announcements, and Predation. *American Economic Review* 76, 940–955.

Farrell, J. and Saloner, G., 1988. Coordination through Committees and Markets. *RAND Journal of Economics* 19, 235–252.

Farrell, J. and Shapiro, C., 1992. Standard Setting in High Definition Television, Unpublished Paper (University of California at Berkeley).

Gabel, H.L., 1987. *Product Standardization and Competitive Strategy* (North-Holland, Amsterdam).

Katz, M. and Shapiro, C., 1985. Network Externalities, Competition and Compatibility. *American Economic Review* 75, 424–40.

Katz, M. and Shapiro, C., 1986. Technology Adoption in the Presence of Network Externalities. *Journal of Political Economy* 94, 822–841.

Meek, B., 1993. *User Needs in Standards* (Butterworth/Heinemann, London).

OECD, 1991. *Information Technology Standards: The Economic Dimension* (OECD, Paris).

Swann, P., 1992. Rapid Technology Change, 'Visions of the Future', Corporate Organisation and Market Share. *Economics of Innovation and New Technology* 2, 3–25.

Swann, P., 1993. User Needs in Standards: How do we Ensure that Users' Votes are Counted? In: B. Meek (ed.), *User Needs in Standards* (Butterworth/Heinemann, London).

Swann, P. and Shurmer, M., 1992. An Analysis of the Process Generating *De Facto* Standards in the PC Spreadsheet Software Market, Unpublished Paper (London Business School, London).

Weiss, M.B.H. and Sirbu, M., 1990. Technological Choice in Voluntary Standards Committees: An Empirical Analysis. *Economics of Innovation and New Technology* 1, 111–133.

Part 3: Policies and Operators Strategies

Coordinated by François Bar, *University of California at Berkeley*

Global Telecommunications Strategies and Technological Changes
Edited by G. Pogorel

Chapter 13

Inter-firm agreements in telecommunications: elements of an analytical framework

P. Llerena and S. Wolff

Bureau d'Economie Théorique et Appliquée, University of Strasbourg, 38 Boulevard d'Anvers, 67070 Strasbourg Cédex, France

1. Introduction

It is now an admitted fact that, since the 70's, inter-firm agreements have played a increasing role in the development of firms' strategies. From a statistical point of view, the phenomenon seems to be significant (e.g. Chesnais, 1988; Hagedoorn and Scakenraad, 1990). This phenomenon has been interpreted as the effect of growing research costs, of increasing technological and economic uncertainty and of the globalization of economic systems. The major part of the literature focuses on the description and the analysis of the phenomenon.

Our approach is quite different: we take the existence of agreements for granted and we try to analyse why the phenomenon could be a permanent feature of contemporary industrial organization. More precisely, the purpose of the paper is to provide an analytical framework to understand the durability of agreements. As a matter of fact, it is not at all sufficient to explain why, at a point in time, there are more inter-firm agreements, we have also to clarify why such agreements last as stable strategies for competing firms.

The paper is based on field research (Llerena et al., 1991) in telecommunications. The research consisted of case studies and theoretical analysis. The paper presented here will focus on the analytical framework and it will use the following assumption: 'An agreement is a means to cope with a need for increased flexibility due to increasing uncertainties. But the durability of agreements is conditioned by the search for a particular kind of flexibility:

dynamic and pro-active flexibility based on learning processes.'

Interfirm cooperative agreements are broadly defined as explicit interactions between two or more firms, such that the autonomy and identity of the parties are — at least partly-preserved. They thus differ from mergers, acquisitions, and other integration operations. But like the latter, agreements represent a way of achieving external growth. More precisely, their object is generally to promote a new activity, based on the development of a new product or process technology and/or the commercialization on a new market (new geographical area or new type of customers). They can take a wide variety of institutional forms: joint ventures and other minority shares, consortia, commercialization contracts, agreements without equity participation, etc.

According to many data-base analyses [1], several industrial fields — although often to a lesser degree than information technologies — have also been affected by a multiplication of collaborative strategies: automobile, textile, chemicals, biotechnology, aeronautics, etc. The scope and the importance of the phenomenon has puzzled many researchers in the economic and strategic management disciplines, all the more so since it seems to question the prevalence of competition as the only mode of interfirm relationship. Interestingly enough, numerous authors of both disciplines rely heavily on Williamson's transaction-costs economics in order to explain the quantitative development of agreements [2]. Because it focuses on the frontier of the firm in a market/hierarchy perspective, Williamson's approach offers a suggestive conceptual framework for 'hybrid' forms as well. From this point of view, agreements are usually defined as long-term, non-standard contracts, located between market transactions and internal organization. The underlying hypothesis is that they minimize transaction costs, i.e., they are the most efficient institutional alternative, when determinant factors such as asset specificity (or uncertainty) are located at an intermediate level.

In this paper, we shall develop some theoretical elements which analyze the agreement as a means for increasing flexibility. The nature of the flexibility considered will be a key for our analysis.

Section 1.1 will discuss the relevance of transaction-costs analysis, when applied to the particular case of alliances in the telecommunications and connected industries. Telecommunications have been experiencing a major transformation, along with an acceleration of technological change and a growing environmental uncertainty. This turbulent context will lead us to show some limits of the transaction-cost analysis. We will then present (sec-

---

[1] See for instance the data-base analyses of Mariti and Smiley (1983), Jacquemin et al. (1985), Cainarca and al. (1988), D.S.T.I. (1986), L.A.R.E.A. (1986), Camagni (1988), Chesnais (1988).

[2] Transaction costs explanations of agreements may be found for instance in Mariotti and Cainarca (1986), Teece (1986), Zagnoli (1988), Contractor (1990), Joly (1990).

tion 1.2) an alternative approach based on a sequential decision process concerning organizational choices. This approach will make it possible to draw an additional value for agreements, linked to the increasing information and to the flexilibity in terms of future possible choices.

Explaining the stability of agreements in telecommunications requires an explicitly dynamic perspective. In this respect, section 2 will propose an analysis based on a particular notion of 'organizational flexibility'. Two inter-related dimensions of organizational flexibility will be developed: the first one is rather passive and refers to the adaptability provided by agreements in response to unexpected perturbations; the second is pro-active and refers to the emergence of learning between the partners. In this perspective, agreements are better considered as some particular processes of routine creation, which may become organizations themselves. The implications of interactive learning for agreement stability will represent one of the main propositions of the paper.

2. *Flexibilities and unstable agreements*

In this section, an analytical framework will be suggested, where the flexibility of agreements may be considered as a market characteristic.

2.1. *Agreement as an intermediate organizational form*

Indeed, in economic theory, it is usually considered that flexibility is a market characteristic: market mechanisms insure a certain kind of flexibility (through price and/or quantity adjustments), mainly in terms of resource allocation. On the contrary, hierarchies are supposed to be 'rigid'. Rigidities are the result of bureaucratization and centralized decision processes. Transaction cost analysis permits a more appropriate approach to organizational alternatives. Agreements appear then as intermediate forms: more flexible than hierarchies, but less than markets.

2.1.1. *The market/hierarchy alternative*

In his famous precursor work on the nature of the firm, Coase (1937) distinguishes two means for coordinating economic activities: outside the firm, price mechanisms conduct productive activities through a series of decentralized market transactions; inside the firm, the coordination is achieved by the entrepreneur-coordinator, in a rather centralized manner. In this perspective, the firm means much more than a mechanical transformation of inputs into outputs. It has also a coordination function. The reason for the emergence of the firm relates to a kind of market failure: transaction costs. Achieving market transactions implies identifying the potential co-contractor, negotiating adequate contract terms (which may lead up to bargaining), making sure that the terms have been observed, and so on. In other words, it implies some kind of frictions in the market mechanisms, and the economic counterparts of these frictions are defined as transaction costs. When market transaction costs are high, the internalization of the transaction inside the firm represents a more efficient coordination mode than the market.

Williamson (1975) operationalized the concept of transaction costs by determining the particular environmental and behavioural conditions that favour the emergence of an internal organization rather than a market transaction. Economic agents are basically opportunistic, i.e., they are self-interest seeking, which makes allowance for guile, and they are also limited in their rationality, i.e., they know limitations in their cognitive abilities for resolving complex problems. Because of these two human features, market transactions become very costly as soon as the following environmental conditions prevail: the uncertainty surrounding the transaction, and the specificity of the assets required by the transaction [3]. Asset specificity refers to the degree to which an asset can be redeployed to alternative uses and by alternative users without sacrificing productive value (e.g. Williamson, 1989, p. 142). It is related to the notion of sunk cost, and it creates a small number environment that favours strategic bargaining and opportunistic behaviours. There are several types of asset specificity: site specificity, physical asset specificity, and human resources specificity are the most frequently cited.

When strong uncertainty coexists with a high degree of asset specificity, it is impossible to specify ex ante the whole set of contingent clauses required for contract execution (remember the bounded rationality hypothesis). Thus contracts are necessarily incomplete. This in turn generates enforcement problems, because of the opportunism of the agents. As a consequence, long-term incomplete contracting is generally characterized by frequent misunderstandings and conflicts that lead to delays, breakdowns, and other malfunctions — i.e., the transaction costs are very important.

In this case, it is more efficient to internalize the transaction inside the hierarchical structure of the firm, because the latter has a distinctive adaptation property, based on the authority relation. This means that an internal organization has access to particular governance instruments: it enjoys a better access to relevant information and it has better control abilities; it can impose a decision by fiat, which avoids costly haggling; it is able to resolve conflicts more easily, and so on. In short, hierarchy reduces the nasty ef-

[3] Williamson (1975) mentions other factors, such as the frequency of the transaction, information asymmetries, the small number environment and the atmosphere of the transaction. The two concepts cited above are the more frequent and the more developed.

fects of opportunism and bounded rationality resulting from uncertainty and high asset specificity. Hierarchy is the institutional alternative that minimizes transaction costs under those circumstances, despite the fact that it experiences another kind of transaction costs: bureaucratic costs, that are specific to internal organization.

In Williamson's perspective, it is an efficiency criterion — the minimization of transaction costs — that explains the emergence of a specific governance structure, i.e., either the market or the hierarchy. This approach can be extended in order to integrate hybrid forms located between market and hierarchy.

2.1.2. Intermediary forms

Instead of considering only two governance structures (market or hierarchy), it appears possible to consider a spectrum of institutional alternatives going from market to hierarchy: 'Suppose that transactions were to be arrayed in terms of the degree to which parties to the trade maintain autonomy. Discrete transactions would then be located at the one extreme, highly centralized, hierarchical transactions would be at the other, and hybrid transactions (franchising, joint ventures, other forms of non-standard contracting) would be located in between' (Williamson, 1985, p. 83).

From this enlarged point of view, like in the simple market/hierarchy perspective, transaction-costs analysis entails an assessment of the comparative costs of planning, adapting and monitoring task completion under alternative governance structures. The basic efficiency postulate remains unchanged: the governance structure that finally emerges is the one that minimizes the transaction costs. Accordingly, the traditional transaction-costs explanation of agreements supposes that these strategies are selected because they are able to minimize the transaction costs in some particular circumstances (see, for instance, Mariotti and Cainarca, 1986; Joly, 1990; Williamson, 1990).

Generally, the only determinant taken into account is asset specificity (the level of uncertainty is assumed to be fixed at an intermediary level). Consequently, agreements are assumed to minimize transaction costs when the degree of asset specificity is located at an intermediary level. This point is illustrated by Fig. 1, where agreements appear as intermediary forms located between market and hierarchy, and corresponding to an intermediate level of asset specificity. The underlying argument is that agreements combine some of the advantages of the market (high-powered incentives and less bureaucratic costs than hierarchy, but to a lesser degree than the market alternative) on the one hand, and some of the advantages of hierarchy (in terms of control instruments, opportunism reduction and information flows), on the other hand. This combination of advantages would be especially appropriate when asset specificity is not too high.

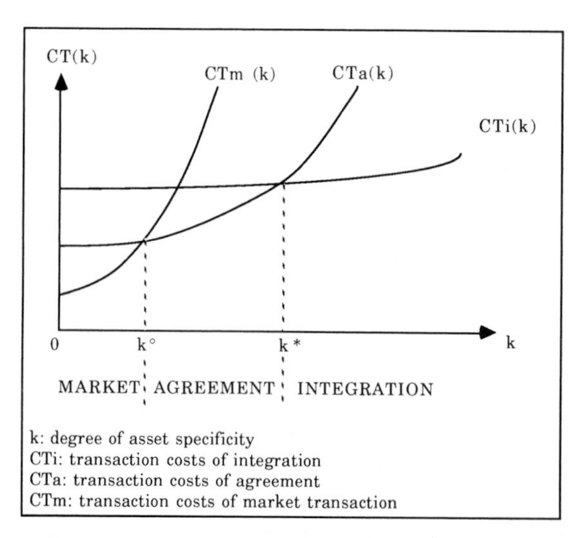

Fig. 1. Asset specificity, market, hierarchy and hybrid forms Source: Llerena et al. (1991).

In fact, the transaction-costs analysis does not seem adequate to explain the proliferation of agreements in the telecommunications industry. Firstly, it will be shown that they do not minimize transaction costs. Secondly, this kind of interpretative framework, although it seems to incorporate a few dynamic elements, suffers from the same limitations as the neoclassical theory: it does not fully capture the dynamic features of technology and of the agreements themselves.

2.1.3. Agreements in telecommunications experience high transaction costs

Empirical observations of alliances in the telecommunications field tend to show high degrees of asset specificity and uncertainty, which leads to the conclusion that transaction costs are potentially important for this type of strategies.

A new, information intensive, economic system is currently at work. It is characterized by an increasing level of demand variety. As a particular case of information technologies, telecommunications play a driving role in the new system: they condition the viability of other firms and organizations; they generate a series of incremental and major innovations in the technical and organizational, as well as the managerial fields, and those changes pervade all other industries.

In turn, the environmental conditions of telecommunications have been strongly influenced by the paradigmatic change (Llerena and Wolff, 1990):

- As a strategic sector, telecommunications suffer a tough, exacerbated, technological competition, which leads to a dramatic acceleration of technical change and rising costs and risks of R&D. Moreover, the internationalization pressures due to deregulation worsen the environmental uncertainty of this industry.

- The variety of demand which characterizes the new paradigm applies to telecommunications also: manufacturers and network carriers have to face a micro-market phenomenon. They must provide a range of 'quasi-customized' goods and services. As a result, relative R&D costs still increase and telecommunications firms have to acquire a deep knowledge of the needs of future users, in order to integrate them early in the innovation process.
- All the changes mentioned above were made possible, and have been amplified, by the internal dynamics of technical progress in the information and communication industries. More precisely, numerous technological opportunities arose from the invention of microprocessors and the subsequent digitalization of networks. The well-known convergence of telecommunications and computers towards the new sector of telematics provides a good illustration of such opportunities. An important consequence for information and communication firms is the necessity to combine different 'upstream' technological knowledge and know-how: micro-electronics, computer, software, network administration, and so on.

In order to sum up roughly the previous discussion, it can be said that agreements between telecommunications firms take place within a very turbulent, complex, unpredictable environment, the components of which are strongly interconnected. The degree of environmental uncertainty is thus very high. Moreover, telecommunications firms are compelled to obtain new competences in order to survive: they need 'upstream' technological knowledge from outside the field on the one hand, and 'downstream' knowledge about users specific requirements on the other.

It follows that agreements in telecommunications have a strong knowledge and innovation dimension. Parties to an agreement are generally seeking access to complementary competences and assets, which may be located upstream or downstream of the production chain. Upstream knowledge refers to technological, R&D and production know-how and information. Downstream knowledge refers to commercial know-how and information about customers' specific requirements.

On the one hand, insofar as technology and know-how are strongly tacit (i.e., firm specific and incorporated in human resources), human asset specificity tends to reach a high level: long-term contracting is thus characterized by important failures and transaction costs. On the other hand, if knowledge is of a more transferable nature, it implies important appropriability problems and information asymmetries; the dangers resulting from an opportunistic behaviour of the partner are potentially high. Hamel et al. (1989) for instance, in their study of Japanese collaborative strategies, show that agreements are sometimes a means for taking over the partner's key competences. Along the same line of ideas, Teece (1986) advocates the integration of the firm holding the complementary assets, particularly when the appropriability regime of innovation is weak (i.e. the corresponding legal protection is inappropriate and/or the degree of tacitness of the technology is low). Here again, it seems that transaction costs are high, because of the strong knowledge content of agreements.

The same type of conclusion emerges if one considers the other determinant of transaction costs: uncertainty. According to Williamson, rising uncertainty leads to the integration of the transaction into the hierarchy, because of its better planning and adaptation ability. However, the telecommunications agreements take place in a highly turbulent, unpredictable environment. Moreover, the object of the agreement is itself often highly uncertain, since the potential results of innovation cannot be precisely specified in advance. Under such conditions, agreements are less efficient than hierarchy as far as transaction-costs minimization is concerned.

Two observations reinforce the idea that transaction costs are important in the case of agreements in telecommunications. Firstly, agreements generally take the form of long-term incomplete contracts, which are precisely those suffering from strategic bargaining, conflicts and other malfunctions, according to Williamson. Exhaustive contracting is too inflexible and generally leads to failures. Secondly, if we admit that the length of negotiations represents an acceptable approximation of the ex ante transaction costs, then it must be recognized that the latter are especially high: duration of the negotiations often exceeds one year [4].

If agreements do not minimize the transaction costs, i.e., if they are less efficient than hierarchy, why do firms not integrate? What is the value of the agreements? How do they compensate for higher transaction costs? The response is related to dynamic elements about which transaction-costs analysis remains too elusive. Section 1.2 will introduce some dynamic arguments, based on a sequential decision process, to show the existence of an additional value for agreements.

2.1.4. Some limits of the transaction-cost analysis: a static approach

Many criticisms of the transaction-costs analysis relate to the efficiency postulate of costs minimization [5]. This one is very difficult to validate empirically,

---

[4] These observations come from the examination of several case studies in Llerena et al. (1991).

[5] See for instance the criticism of Dow (1987).

because the definition of transaction costs is quite ambiguous. As a result, transaction costs are very difficult to measure in concrete terms, and the empirical validation of the efficiency criterion is unsatisfying. But the limitations we want to present here go further than the simple assumption that the postulate is false in the case of agreements in telecommunications. They refer to the elusive treatment of the bureaucratic costs of hierarchy and more generally to the static approach underlying transaction-costs considerations.

Bureaucratic costs are not fully taken into consideration in the previous interpretation of agreements. At best, they play the role of fixed parameters, which are not located on the same level as asset specificity (variable determinants). In other words, the transaction-costs approach eludes them. More precisely, it misses their basically dynamic dimension. Bureaucratic costs are linked to the size and complexity of the hierarchy. They correspond roughly to the rigidity and the inertia of big organizations (persistence of existing, inefficient programmes, and so on). They raise all the more problems since the environment quickly changes and is highly uncertain. From this point of view, it may be said that the analysis of bureaucratic costs requires an explicit dynamic perspective.

However, the transaction-costs approach is essentially a comparative static one. The comparative institutional analysis assumes instantaneous optimization of the institutional form according to a specified set of determinants, like for instance the degree of asset specificity. The process of shifting from one particular form to another is not analyzed. Moreover the technology is considered as given. In his recent work, Williamson (1989) reintegrates the production function of textbooks, asserting that the efficient governance structure is the one that minimizes simultaneously transaction and production costs. The corresponding notion of technology is typically a static, exogenous one. In this perspective, the transaction-costs analysis remains a theory of resource allocation, which can be opposed to an approach in terms of resource creation and innovation [6]. Note however that there is another-implicit-concept of technology in Williamson's theory: asset specificity, and in particular human asset specificity, which refers to tacit knowledge resulting from learning processes. A reinterpretation of the notion of asset specificity should thus allow us to incorporate the dynamic elements of technology and innovation [a more detailed argument may be found in Foray (1990)].

At this stage, we analyzed the limitations of transaction-costs explanation in the case of interfirm agreements in the telecommunications industry. Agreements suffer high transaction costs and, because of its neglect of dynamic aspects, this approach is unable to explain the positive value of such strate-

[6] For a deeper discussion on that point, see Gaffard (1990) and Lundvall (1990).

gies. In section 1.2, we introduce a sequential decision process concerning the choice of organizational forms, in order to show the existence of an additional value to agreements, the option value. Section 2 will finally provide some elements for an explicitly dynamic framework, introducing aspects like organizational flexibility and learning processes.

2.2. *Agreement as a means for future freedom of choice*

This section will formalize the phenomenon by using the option value theory, borrowed from environmental and investment economics (Arrow and Fischer, 1974; Henry, 1974). In this framework, the agreements are considered as 'flexible' decisions, allowing to wait for more information before choosing a definite organizational form. Agreements are temporary situations, used to cope with uncertainty and increasing information.

Suppose a new 'technology', which can be a success with a probability s as well as a failure with a probability $(1 - s)$. To develop it, it is possible to do it internally (integrated solution) or through an agreement. As mentioned in the previous section, agreements induce 'higher' transaction costs. In a very simplified model (two periods, with perfect information in the second period, and two alternatives), it is shown that, by taking into account increasing information and the flexibility of future choices, there is a positive 'option value' in favour of the agreement solution. Here, agreement appears as a kind of waiting position (see Fig. 2 for the detailed model).

This option value can be sufficient to compensate the higher transaction costs of agreements ($c_a > c_i$) in fast growing information situations. But such kind of flexibility, which we shall call 'decisional flexibility', implies unstable agreements by definition [7].

However, we think that there are configurations of agreements which are stable. But in this case, we have to introduce more explicitly some organizational flexibilities: adaptation and learning.

3. *Flexibilities and stable agreements*

Flexibility is an ubiquitous and rather ambiguous concept. The kind of flexibility we are talking about must be clearly distinguished from the allocative, mechanistic adjustment properties generally attributed to market transactions. In this section, we do not argue that agreements are more flexible than

---

[7] Using the same type of approach, Wolff (1992) has developed a model where the sunk cost of different alternatives are considered.

Assumptions:

$ri > ra$; $pi < pa$; $ci < ca$
where r (p) are clash flows in the case of success (failure) respectively for i (integration) and a (agreement)

Sequential decision without increasing information and flexibility:

$$Ri = -ci + s\,ri + (1-s)\,pi \qquad (1)$$
$$Ra = -ca + s\,ra + (1-s)\,pa \qquad (2)$$

$$Ra > Ri \iff ca - ci < s(ra - ri) + (1-s)(pa - pi) \qquad (3)$$

Sequential decision with increasing information and flexibility:

$$R'i = Ri \qquad (4)$$
$$R'a = -ca + s\max\{ri - ci/a\,;\,ra\} + (1-s)\max\{pa\,;\,pi\} \qquad (5)$$
where ci/a are the adjustment costs from agreements towards integration

if $ri - ci/a > ra$ and $pa > pi$

$$R'a = -ca + s(ri - ci/a) + (1-s)\,pa \qquad (5')$$
and
$$R'a > R'i \iff ca - ci < (1-s)(pa - pi) - s\,ci/a \qquad (6)$$

OV (Option Value) is then: $OV = R'a - Ra = s(ri - ci/a - ra) > 0$

Fig. 2. A simplified option value model applied to the agreement/integration alternative.

integration because they have some of the adjustment properties of market transactions, like the ability to change partner and/or dissolve a relationship. Rather, we will focus on a concept of organizational flexibility. Favereau's definition is particularly appropriate for our purpose: flexibility is the property that autonomous systems (such as firms and other organizations) have (or may have) for stimulating and driving their learning abilities, in order to safeguard or increase their scope of action faced with external perturbations (Favereau, 1989a). Two different aspects of flexibility are implicit in this definition. The first refers to the adaptation capability of the firm, its ability to protect itself from external perturbations. It denotes a rather passive behaviour. The second is more pro-active and refers to the learning ability of organizations: we call it 'initiative flexibility', like Gaffard (1989).

3.1. Agreement as an adaptation capability

Agreement is a means to work at the interface of different knowledge bases: between up-stream and down-stream segments of an industry, between different sectors, etc. One of its advantages is that an agreement does not disturb, at least not immediately, the coherence of the existing firms involved in it.

Telecommunications firms have to face three sources of potential perturbations: the turbulence of the environment, the intrinsic uncertainty of innovation activities and the behaviour of the (unknown) firm that holds the complementary assets.

As far as environmental uncertainty is concerned, a traditional response in organization theory relates to the notion of quasi-decomposability of Simon (1962). This notion applies to hierarchical systems in general and to organizations and firms in particular. A system is hierarchical when it is composed of several sub-systems, themselves constituted of sub-units, and so on[8]. A hierarchical system is quasi-decomposable when: (1) the short-run behaviour of each of the component sub-systems is approximately independent from the short-run behaviour of other components; (2) in the long-run, the behaviour of any one of the components depends in only an aggregate way on the behaviour of the other components.

From this point of view, the firm can be seen as a specific, decomposable set of routines aimed at resolving a particular complex problem. This set of routines is decomposable into several more or less independent sub-routines, each of which being in charge of a given part of the global problem. This decomposition offers two advantages. On the one hand, it makes it possible to save on the bounded rationality of the economic agents (cf. their cognitive limitations in information processing), since each sub-system has to deal with only a limited aspect of the problem, independently from the decisions in the other sub-systems. The decision burden is thus distributed and shared among the members of the firm.

On the other hand, it facilitates a progressive adaptation of the whole system in case of external shocks. Actually, each sub-system acts as a buffer vis-à-vis the others. When a sub-system deals with a specific aspect of the environment, it is able to protect the other units from perturbations because of loose coupling. Breakdowns have only a localized impact: at least, they do not spread immediately to the other sub-systems of the same hierarchical level. As a result, the global organizational coherence and the existing routines, which support firms' competences, are better safeguarded.

---

[8] This concept of hierarchy is not necessarily related to the existence of an authority relation. It thus differs from Williamson's definition.

Like Simon's autonomous sub-systems, agreements may be related to a kind of problem decomposition, where each agreement would act as an interface between the parent organization and a particular facet of the environment. More precisely, agreements correspond to the case where there is a strong interdependence between the firm and its environment: in a context of turbulence, all the elements of the environment are strongly interconnected.

Until now, we showed that agreements may represent a solution in the face of environmental uncertainty. The two other sources of uncertainty, innovation and association with an unknown complementary firm, must also be taken into account. Here the interest of agreements turns on the idea that they represent a way to externalize the potential impacts of uncertainty outside the firm and insulate them inside a specific entity. Besides the well-known sharing of the sunk costs and of the risk inherent to innovation activities, an agreement may also act as a fuse between the 'surprises' generated by the novel activity and the existing routines.

The same idea applies to the perturbations initiated by the partner's behaviour. Thus, using agreements avoids the propagation phenomena that lead to questioning — and sometimes to destroying — the whole set of routines and interpretative frameworks in the parent companies[9]. Instead of integrating one framework (a set of routines) into the other by forcing the former to fit into the latter, as is the case of mergers or acquisitions, agreements aim at creating only an intersection between two existing frameworks. 'Cultural', destructive conflicts are thus circumscribed inside the intersection. We do not mean that this is the only advantage of agreements as opposed to total integration[10], but we think that it deserves more attention.

In this sub-section, agreements look less like a particular kind of contract than like a process of routine creation. In fact, they evoke the creation of a new organization and, as such, they are the support of a specific learning process.

3.2. *Agreement as a generator of new specific assets*

In dynamic terms, there is an implicit but strong mechanism of self-reinforcement of agreements as learning processes. One of the possible results is the emergence of new specific assets and, more precisely, assets which are specific to the agreement itself. Durability depends also on the degree of appropriability of this new specific asset. A kind of 'initiative flexibility' appears linked to the ability of creating a new 'state of nature', new resources, etc.

---

[9] The idea of destruction of organizational routines presents strong connections with the notion of 'personality breakdown' described by Loasby (1989).

[10] Other-classical-arguments are also important: cash availability, the proportion of desired competences in relation to the whole size of the firm to be integrated, and so on.

In transaction-costs economics, asset specificity plays an ex ante role: it determines which institutional mode (market, hierarchy, or hybrid form) will be chosen, according to the costs-minimization criterion. It is thus considered as given. Here, we focus on another interpretation of the concept: asset specificity is not given, it is an endogenous, dynamic factor, which evolves and increases during the relationship between firms. More precisely, it is incorporated mostly in human resources and results from a cumulative process of learning by interacting, from informational economies and the progressive specification of a common language and also from the emergence of trust between the parties. The corresponding knowledge is specific to the transaction itself and it grows as the interaction goes on.

Inasmuch as agreements support such a kind of organizational learning, they cannot be reduced to contracts. Like any organization, their main purpose is to create resources through a collective learning process (Favereau, 1989b; Gaffard, 1990). The underlying efficiency criterion has a dynamic nature and relates to firms' survival. This positive potential value of agreements compensates for their high transaction costs. This value presents some strong connections with the notion of relational quasi-rent developed by Aoki (1988).

In his analysis of the Japanese sub-contracting networks, Aoki focuses on the dynamic efficiency generated by the cooperative relation inside the group. Information efficiency due to horizontal coordination, on the one side, and the quasi-permanent association of different competences, on the other, favour learning processes able to create new specific resources (in particular, human resources). In other words, they generate a quasi-rent that is then distributed among the member firms. If the relationships is broken off, most of the benefits from learning are definitely lost.

Note that the idea of agreement as a learning support has some implications for the evolution, the dynamics of the agreement itself. If learning really occurs, and if its outcomes have a strong appropriability regime [11] (i.e., if they are not transferable and not easily imitated), then it may be suggested that an agreement tends to be relatively stable.

We still have to characterize the type of learning that agreements make allowance for. Three different natures of learning appear relevant here: technological and/or commercial innovation, increasing knowledge about the partner (and more generally about the environment), and finally, learning about the management of the agreements themselves.

Argyris and Schön (1978) define three levels of organizational learning: single-loop, double-loop and deutero. Single loop learning refers to successive

[11] According to Teece (1986), the appropriability regime of a given innovation will be strong when the legal protection is efficient and/or the degree of tacitness of the new technology is high.

improvements of routines which do not question the existing framing and norms of performance:

> "... members of the organization respond to changes in the internal and external environments of the organization by detecting errors which they then correct so as to maintain the central features of organizational theory-in-use. ... There is a single back loop which connects detected outcomes of action to organizational strategies and assumptions which are modified so as to keep organizational performances within the range set by organizational norms"

(Argyris and Schön, 1978, pp. 18–19). Double loop learning refers to changes of routines and organizational norms. They imply at least partial unlearning and erasing of the organizational 'memory'.

According to us, the first two natures of learning previously mentioned (technological and commercial, knowledge about the partner) are particular cases of double-loop learning. Finally, deutero learning refers to the ability to learning to learn, i.e. it is the highest level. Learning about how to negotiate and manage agreements in general is a special case of deutero learning. It may crystallize, for instance, in an 'agreements division' of the parent companies.

There are often conflicts between single-loop learning and the two higher levels of learning. The 'success' of a low-level routine generates a reinforcement process: the more the existing rules are perceived as efficient, the more they are utilized and thus the more they are improved, and so on. This kind of lock-in phenomenon may lead to the inertia or, in other words, to the rigidity of existing routines. Bureaucratic costs, such as the persistence of inadequate programmes and other resistances to change, emerge as a consequence. Agreements offer a solution in the face of such organizational failures, because they make it possible to externalize and promote learning outside the existing, bureaucratic structure of the existing company. As autonomous entities, they are better able to promote an inner environment that favours innovation [12] and 'radical' learning (Ciborra, 1990).

Another related, important aspect of innovation through agreements is the potential creativity resulting from the confrontation of several points of view and the combination of different competences. Confrontation induces new problems which often trigger solving activity processes. In cognitive sciences, it is an accepted fact that learning occurs through inter-individual interactions. This idea applies to firms too. The significance of interaction in learning, is that it introduces possibilities for new combinations of different kinds of knowledge. It is a source of unexpected novelty, and, thus, a necessary part of the innovation process.

---

[12] Cf. the concept of 'organic system' by Burns and Stalker (1961).

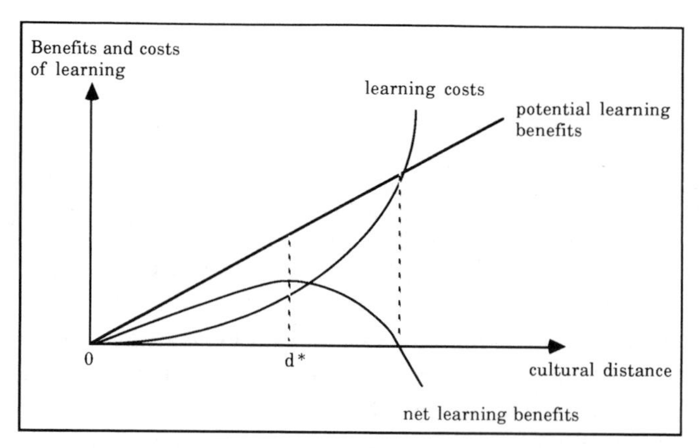

Fig. 3. Cultural distance and the costs and benefits from learning. Source: Lundvall (1991, p. 11).

Lundvall (1991) puts forward an interesting proposition about the costs and benefits of interactive learning: both depend on the 'cultural' distance between the two entities. On the one hand, it is assumed that the more different their knowledge base, the greater their learning potential. On the other hand, interactive learning presupposes access to shared information channels and codes of information. Thus, an initial learning process, consisting in the creation of a common language, must be established. This will be all the more costly as the cultural distance is great. Figure 3 provides a heuristic illustration of these ideas. In order to promote the net benefits of higher learning, cultural distances between the partners should be at an intermediary level.

Section 2 was devoted to specific concepts of organizational flexibility. Firstly, we suggested that an agreement could increase the adaptation capability of the parent company by acting as a fuse between different kinds of external perturbations and the existing set of routines of the parent organization. Secondly, we proposed that an agreement could also be the support of a radical, interactive learning process.

Organizational flexibility, especially initiative flexibility, has some critical implications for agreement stability. If interactive learning effectively occurs, then it generates a new specific asset, which is progressively encoded in the organizational routines of the agreement. Such a process of routine creation, because it is associated with self-reinforcement effects, usually favours the stability of the relationship between the partners. The agreement will be stable insofar as the appropriability regime of the newly created asset is strong.

4. Conclusion

In this paper, we have provided some important conceptual elements to explain inter-firm cooperative agreements in telecommunications and connected industries. More precisely, we defined different types of flexibility, and we derived important propositions about the evolution of an agreement.

We first analyzed agreements as unstable strategies. The traditional transaction-costs approach of hybrid forms was presented and discussed. We showed that this conceptual framework was not fully appropriate for our purposes: agreements in telecommunications did not minimize the transaction costs. Rather, their value lay in dynamic aspects such as flexibility, which the transaction-costs explanation failed to analyze. Option value theory made it possible to specify the value of the flexibility of an agreement, when the latter was considered as a kind of waiting position in an uncertain environment. In this perspective, the specific value of agreements resulted both from increasing information and from the flexibility in terms of future organizational choices.

In order to explain agreements as stable strategies, we had to develop a particular notion of organizational flexibility: the pro-active, dynamic flexibility based on the activation of learning between the partners. Interactive learning was presented as one of the main factors of agreement stability. Inasmuch as such a kind of learning was effective, agreements could be considered as routine creation processes, leading to the irreversible emergence of a highly specific human asset. Moreover, we suggested that learning through agreements was of a rather radical nature, and that it was better analyzed by distinguishing two steps: an initial learning process concerning the partner, followed by a second process which would consist in the creation of new competences. According to us, interactive learning deserves much attention. Going deeper into the study of learning mechanisms represents a very promising track of future research about inter-firm agreements.

References

Aoki, M., 1988. *Information, Incentives and Bargaining in the Japanese Economy* (Cambridge University Press).

Argyris, C. and Schön, D.A., 1987. *Organizational Learning* (Addison-Wesley, Reading, Mass.).

Arrow, K.J. and Fischer, A.C., 1974. Environmental Preservation, Uncertainty and Irreversibility. *Quaterly Journal of Economics* 88, May, pp. 312–319.

Burns, T. and Stalker, G.M., 1961. *The Management of Innovation* (Tavistock Publications, London).

Cainarca, G.C., Colombo, M.G. and Mariotti, S., 1988. Accordi tra imprese nel sistema industriale dell' informazione e della comunicazione. *Rapporto interno No. 88-039* (Politecnico di Milano).

Camagni, R. and Gambarotto, F., 1988. Cooperation Agreements and New Forms for External Development of Companies. *Séminaire Européen Accords de Coopération Technologiques entre firmes*, November (Paris X, Nanterre).

Chesnais, F., 1988. Les accords de coopération technique entre firmes indépendantes. *STI Revue*, December. No. 4, pp. 55–131.

Ciborra, C.U., 1990. Alliances as Learning Experiments: Cooperation, Competition and Change in the High-Tech Industries. *8th ITS Conference*, Venice, March.

Coase, R.H., 1937. The Nature of the Firm. *Economica* 4, 386–405.

Contractor, F.J., 1990. Contractual and Cooperative Modes of International Business: Towards a Unified Theory of Modal Choice. *Management International Review* 30, No. 1.

Dow, G.K., 1987. The function of authority in transaction-cost economics. *Journal of Economic Behavior and Organization* 8, 13–38.

D.S.T.I., 1986. Technical Co-operation Agreements Between Firms: Some Initial Data and Analysis. *Document OCDE*, DSTI/SPR/86.20

Favereau, O., 1989a. Valeur d'option et flexibilité: de la rationalité substantielle à la rationalité procédurale. In: Cohendet and Llerena (eds.), *Flexibilité, Information et Décision. Economica.*

Favereau, O., 1989b. Organisation et marché. *Revue Française d'Economie* 4 (1), 65–96.

Foray, D., 1990. *Coopération industrielle et équilibre organisationnel de la firme innovatrice.* Discussion paper, March (OIT, Genève).

Gaffard, J.L., 1989. Marchés et organisations dans les stratégies des firmes industrielles. *Revue d'Economie Industrielle* 48, 2nd trim.

Gaffard, J.L., 1990. *Economie industrielle et de l'innovation* (Dalloz).

Hagedoorn, J. and Schakenraad, J., 1990. Strategic Partening and Technological Co-operation. In: Dankbar and al. (eds.), *Perspectives in Industrial Organization* (Kluwer Academic Publishers, Dordrecht).

Hamel, G., Doz, Y.L. and Prahalad, C.K., 1989. S'associer avec la concurrence: comment en sortir gagnant? *Harvard l'Expansion*, Autumn.

Henry, C., 1974. Investment Decision Under Uncertainty: The Irreversibility Effect. *American Economic Review* 64, 1006–1012.

Jacquemin A., Lammerant, M. and Spinoy, B., 1985. Compétition Européenne et coopération entre entreprises en matière de R&D. *CCE Document de travail*, No. 80.

Joly, P.B., 1990. Eléments d'analyse des systèmes d'innovation dans le domaine biovégétal: flexibilité et coûts de transaction. *Revue d'Economie Industrielle* 51, 1st trim.

L.A.R.E.A./C.E.R.E.M., 1986. *Les stratégies d'accords des groupes de la CEE: intégration ou éclatement de l'espace industriel européen.* Programme 'Europe Industrielle et Technologique' du Commissariat Général au Plan, October.

Llerena P. and Wolff, S., 1990. R&D in Telecommunication and the New Techno-Economic Paradigm. *Eurotelecom Conference*, 5–7 June, Madrid.

Llerena P., Umbhauer, G., Wolff, S. and Yildizoglu, M., 1991. Accords de cooperation inter-entreprises dans le secteur des télécommunications. *Rapport final, BETA-ULP* (Strasbourg).

Loasby, B., 1989. Organization, Competition, and Growth of Knowledge. In: Langlois (ed.), *Economics as a Process* (Cambridge University Press).

Lundvall, B.A., 1990. Explaining Inter-Firm Cooperation and Innovation-Limits of the Transaction Cost Approach. *Workshop on the 'Socio-Economics of Inter-Firm Cooperation', Wissenschaftszentrum, Berlin*, June 11–13.

Lundvall, B.A., 1991. *Closing the Institutional Gap* (unpublished).

Mariotti, S. and Cainarca, G.C., 1986. The Evolution of Transaction Governance In The Textile-clothing Industry. *Journal of Economic Behavior and Organization* 7, 351–374.

Mariti, M. and Smiley, R.H., 1983. Co-operative Agreements and the Organization of Industry. *Journal of Industrial Economics* 31, No. 4, June, pp. 437–451.

Sachwald, F., 1988. Accords inter-entreprises: éléments d'interprétation. In: G.E.R.P.I.S.A., *Les accords entre constructeurs automobiles. Actes du GERPISA*, No. 3.

Simon, H.A., 1962. The Architecture of Complexity. *Proceedings of the American Philosophical Society* 106 (6), December, pp. 941–973.
Simon, H.A., 1982. *Models of Bounded Rationality, Behavioral Economics and Business Organizations, Vol. 2* (MIT Press, Cambridge, Mass.).
Teece, D.J., 1986. Profiting from Technological Innovation: Implications for Integration, Collaboration, Licensing and Public Policy. *Research Policy* 15, 285–305.
Williamson, O.E., 1975. *Markets and Hierarchies: Analysis and Antitrust Implications* (Free Press, New York).
Williamson, O.E., 1985. *The Economic Institutions of Capitalism* (Free Press, New York).
Williamson, O.E., 1989. Transaction Cost Economics. In: Schmalensee and Willig (eds.), *Handbook of Industrial Organization, Vol. 1*, pp. 135–182 (Elsevier, Amsterdam).
Williamson, O.E., 1990. Comparative Economic Organization: the Analysis of Discrete Structural Alternatives. *Economic Analysis & Policy, Working Paper*, EAP-36.
Wolff, S., 1992. Les accords technologiques inter-entreprises dans le secteur des technologies de l'information et de la communication: une approche en termes de flexibilité décisionnelle. *Communication to the 9th Journées de Microéconomie Appliquée*, Strasbourg, 4–5 June.
Zagnoli, P., 1988. *Inter-Firm High Technology Agreements: A Transaction Cost Explanation*, May (unpublished).

Global Telecommunications Strategies
and Technological Changes
Edited by G. Pogorel

Chapter 14

AT&T, BT, and NTT: vision, strategy, corporate competence, path-dependence, and the role of R&D

Martin Fransman

Institute for Japanese–European Technology Studies, University of Edinburgh, Edinburgh, Scotland

1. Introduction

This paper is concerned with an apparent puzzle. While the major global telecommunications equipment companies, such as Northern Telecom, Siemens, Alcatel, Ericsson, and NEC, spend a similar amount on R&D (as a proportion of sales), there is a significant difference in the allocation of resources to R&D (by the same measure) by AT&T, BT, and NTT. Why is this the case, given that all three of the latter companies are increasingly having to compete in the same global selection environment? To what extent is this difference to be explained by the unique historical background of each company (its 'path-dependence'), to what extent by the strategic choices they have made? In so far as the difference is a reflection of strategic choice, can anything be said regarding the rationale for the company's choice and the factors that are likely to determine the longer run success of this choice? The present paper will analyse these questions.

The paper begins with a description of the differences between AT&T, BT, and NTT in terms of size, profitability, and R&D. The major changes that have occurred in the global selection environment which confronts these three companies are then examined. The question is then posed and answered regarding whether the strategic choices that have been made by the companies are a result of their past history, that is their path-dependence. An overview is then presented of the strategic choices that they have made regarding how they are to acquire the competences they need in order to offer competitive telecommunications services. It is shown that while BT has chosen to use the market to acquire 'network elements' (broadly, telecommunications equipment), NTT has selected cooperative development with a small number of suppliers, and AT&T has opted for vertical integration. In the following section a detailed analysis is provided of the differing visions of the three companies which underlie their varying strategic choices and the reasons for the differences are explained. A more general analysis is then undertaken of the advantages and disadvantages of the market, cooperation, and internal development as alternative modes of coordinating complementary competences. Finally, attention is turned to a major focus of the paper, namely the role of research and development in the three companies. This role is examined in terms of the way in which these companies have dealt with four major questions related to research and development: What research does the firm need now? (an information problem): How to prevent 'irrelevant' research? (a control problem): What research is needed for the future? (an uncertainty problem): and Should the required research be undertaken in-house or ex-house? (an assignment problem). In examining these four questions particular attention is paid to the role of the internal market for research and development. It is shown, however, that while such a market can be crucial, it is also necessary to go beyond the internal market in providing longer term, radical innovation.

2. *AT&T, BT, and NTT: size, profitability and, R&D*

Data on the size and profitability of AT&T, BT, and NTT are provided in Table 1.

As can be seen, AT&T and NTT are of a similar size in terms of sales. While in 1991 the sales of the former were $45.07 billion, for NTT they were $42.22 billion. BT was a little over half the size of the other two companies with sales of $24.31 billion in the same year. In terms of market value, however, the picture was significantly different. As a result of the different values that at the time ruled on the Tokyo Stock Exchange, NTT's market value was significantly different from the other two companies. Accordingly, NTT's market value of $103.00 billion compared with $40.43 billion for AT&T and $40.03 billion for BT. This difference in valuation is reflected in the price/earnings ratios for the three companies. While for NTT this ratio was 60, for AT&T it was 15, while for BT it was 11.

The valuation difference also shows up in one of the measures of profitability, namely the return on equity. While in 1991 this was 22.3 percent for BT, and 19.7 percent for AT&T, for NTT the figure was only 5.9 percent. Another measure of profitability, namely profit as a percentage of sales, yielded similar figures. For BT this latter ratio was 22.4 percent, while for NTT the figure was

Table 1
Size and profitability of AT&T, BT, and NTT

Indicator	AT&T	BT	NTT
Sales (1991)	$45.07 bn *	$24.31 bn	$42.22 bn
Market value (1991) (by week July 15, 1991)	$40.43 bn	$40.03 bn	$103.00 bn
Price/book value ratio (ratio of May 1991 closing price to net worth per share or common stockholder's equity investment)	2.9	2.5	3.6
Price/earnings ratio	15	11	60
Return on equity (earnings per share at May 1991 as percent of most recent book value per share)	19.7%	22.3%	5.9%
Profit as% sales		22.4%	8.8%
Operating return on net property, plant and equipment, 1991 [a]		22.6%	7.3%
Operating margin (before interest and tax), 1991 [a]		26.5%	11.4%
Net income per line, 1991 [a]		$139	$36
Cash surplus/(deficit) per line, 1991 [a]		$33	$6
Lines per employee, 1991 [a]		112	204

* AT&T sales figure consolidates NCRP
Source: *Business Week*, December 2, 1991, January 20, 1992; [a] *Financial Times*, November 1, 1991, p. 21.

8.8 percent. Other measures of profitability also put BT significantly ahead of NTT. In 1991 operating return on net property, plant and equipment was 22.6 percent for BT and 7.3 percent for NTT; operating margin (before interest and tax) was 26.5 percent for BT and 11.4 percent for NTT; net income per line $139 for BT and $36 for NTT; and cash surplus/(deficit) per line $33 for BT and $6 for NTT. In terms of one (debatable) measure of productivity, however, namely number of lines per employee, the figure for BT was 112 while for NTT it was 204. (It is, however, worth noting that these measures of BT's relative profitability and their appropriateness were hotly debated in the pages of the *Financial Times*.)

The R&D intensity of telecommunications equipment companies and operating companies is shown in Table 2.

The top part of Table 2 gives figures for a number of equipment companies for R&D as a percentage of sales for 1987. These show that R&D as a

Table 2
R&D as percent of sales, 1987

Company	R&D as % sales
NEC[a]	13.7%
Siemens[a]	12.8%
Northern Telecom[a]	12.3%
Alcatel[a]	9.8%
Ericsson[a]	9.1%
AT&T[b]	7.3%
NTT[b]	3.8%
BT[b]	2.1%

[a] R&D as percentage of telecommunications sales (estimated). Calculated from Grupp and Schnoring (1992, p. 58, table 4).
[b] Total R&D as percentage of total sales. From Grupp and Schnoring (1992, p. 53, table 2).

proportion of sales for NEC, with the highest R&D intensity, was 1.5 times that for Ericsson, which had the lowest intensity. On the other hand the bottom part of the table provides similar figures for the operating companies. These figures show that for AT&T, the company with the highest R&D intensity, the ratio of R&D to sales was 3.5 times that of BT, with the lowest intensity. Clearly, therefore, there was a greater difference in R&D behaviour between the operating companies as compared to the equipment companies. This difference is explained later in this paper.

Further information on R&D in AT&T, BT, and NTT is provided in Table 3.

As Table 3 shows, while R&D as a proportion of sales increased in NTT from 3.8 percent in 1987 to 4.1 percent in 1991 the corresponding figure for BT *decreased* from 2.1 percent to 1.9 percent in these two years. Additional information has been obtained for NTT. This shows that in 1992/3 the R&D ratio increased to 4.7 percent. As far as the present author is aware, there is no intention to increase the R&D ratio in BT.

Table 3
R&D in AT&T, BT, and NTT

Indicator	AT&T	BT	NTT
Sales, 1991[a]	$45.07 bn*	$24.31 bn	$42.22 bn
R&D, 1991[a]		$ 0.45 bn	$ 1.74 bn
R&D % sales, 1991[a]		1.9%	4.1%
R&D % sales, 1987[b]	7.3%	2.1%	3.8%
R&D % sales, 1992/93[c]			4.7%

* AT&T sales figure consolidates NCR.
[a] *Business Week*, December 2, 1991; [b] Grupp and Schnoring (1992); [c] *Nikkei Weekly*, March 21, 1992.

Despite this significant difference in the R&D intensity of AT&T, BT, and NTT, these companies have confronted similar changes in their global selection environment. This is examined in greater detail in the following section.

3. Changes in the global selection environment facing the major telecommunications operating companies

Since the mid-1980s there have been a number of important changes in the global selection environment within which AT&T, BT, and NTT have had to operate. Amongst the changes have been the following:

(1) Introduction of greater competition in telecommunications services on their domestic markets, leading, all other things equal, to a tendency for profits to be squeezed. (A *threat* to the companies.)

This tendency, however, has affected the companies in different ways. AT&T in 1991 managed to increase its net profit margin to 7.4 percent, up from 7.1 percent in 1990, after a period of severe cost-cutting. While BT has also successfully counteracted the squeeze on profits with its own cost-cutting measures, NTT has experienced a fall in profitability. For the fiscal year from April 1 1992, for example, NTT's pre-tax profits are expected to decline by 6.4 percent to Yen 351 billion. NTT's planned increase in R&D as a proportion of sales, therefore, is taking place against a background of falling profitability, which provides an indication of the company's strong degree of commitment to R&D[1].

(2) Gradual liberalisation of foreign service markets, leading to new possible sources of business. (An *opportunity* for the companies.)

In particular, it is now becoming possible for the operating companies to offer global telecommunications services to global multinational users. This has led to the emergence of several strategic corporate alliances as operating companies have attempted to stitch together alliances to offer 'end-to-end services' covering the world. At the same time this has created an 'outsourcing market' as the possibility has arisen for multinational users to contract out (vertically disintegrate) their telecommunications management activities. We shall return to this point later in an analysis of the corporate competences that AT&T, BT, and NTT believe they require in order to become and remain competitive in these emerging markets.

(3) Gradual liberalisation of domestic and international telecommunications equipment markets, creating the possibility of multi-vendor purchasing

---

[1] See *Business Week*, January 20, 1992, p. 35 and *Nikkei Weekly*, March 21, 1992.

environments for operating companies. (This is an *opportunity* for the operating companies, although it may simultaneously be perceived as a *threat* by those equipment producing parts of these operating companies which now face the possibility of greater competition from potential outside suppliers.)

As we shall see later, a crucial difference between AT&T, BT, and NTT has resulted from the strategic stance they have taken regarding the role of multi-vendor purchases.

(4) Maturation of the 'plain old telephone' market, both domestically and internationally, leading to the necessity to introduce new telecommunications services. (This poses both a *threat* and an *opportunity* for the operating companies.)

For example, while spoken telephone calls constitute about 90 percent of the world's telecommunications traffic, their growth rate is only about 7 percent per annum. Other telecommunications services are growing far more rapidly, such as data communications which is increasing at about 25 to 30 percent per annum[2].

(5) New technologies and standardisation which at one and the same time create the potential for innovative new services while increasing the likelihood of increased competitive pressure through falling costs. (This also is both a *threat* and an *opportunity* for the operating companies.)

It is now widely accepted that in the future the new switching and transmissions technologies imply that the cost and price of telephone calls will be determined by the quantity of information sent (bit-rate tariff), rather than the distance over which it is sent. However, costs are likely to fall where demand is growing most rapidly, namely in long distance telephone and data calls. This will put increasing pressure on the operating companies.

In view of these changes in the global selection environment which affected all the major telephone operating companies, how is their significantly different allocation of resources to R&D, documented above, to be explained?

4. *Strategic choice of corporate competences and the explanation of R&D allocations*

4.1. Path dependence

There are important historical differences between AT&T, BT, and NTT, quite apart from their size and profitability. Perhaps most notable, from the point of view of the concerns of the present paper with research and development,

---

[2] See *The Economist*, March 10, 1990, p. 12.

is AT&T's presence as both a major manufacturer of telecommunications equipment and a telecommunications operator. This reflects the incorporation of the former Western Electric into AT&T at the time of divestiture. Conversely, neither BT nor NTT were involved significantly in manufacturing activities prior to their privatisation and part-privatisation respectively. However, although both BT and NTT were at this prior stage closely involved with several domestic equipment manufacturers in the research, design and development of complex telecommunications equipment, there were important differences between these two companies in terms of their involvement. While for NTT the involvement with their 'family' of suppliers, primarily NEC, Fujitsu, Hitachi, and Oki, was on the whole satisfactory, it is probably fair to conclude that for BT the experience was far more ambiguous. This was particularly the case with the company's most ambitious project, namely the joint research and development of the System X digital switch. Furthermore, BT witnessed a significant change in the circumstances of its suppliers as GEC and Plessey first merged their telecommunications activities in the new firm GPT, which itself was subsequently taken over by Siemens and GEC, and STC, another major supplier, withdrew from switching. (NTT's relationship with its suppliers is analysed in Fransman (1992a, b), while BT's experience is recounted in Molina (1990).) AT&T, BT, and NTT, therefore, each had significantly different experiences regarding the research, design, development, and manufacture of telecommunications equipment.

Are these contrasting experiences sufficient to explain the different role in the three companies of research, design, development and manufacture (differences that will shortly be documented in detail)? The short answer is that such an explanation, couched in terms of path-dependence, is inadequate. The reason is that since their divestiture, privatisation, and part-privatisation, the three companies have had ample opportunity to *change* their activities in these fields. And, indeed, they have made significant changes. As we shall shortly see in greater detail, AT&T has *deepened* its involvement in the research, design, development and manufacture of telecommunications-related equipment with its acquisition of NCR, the computer manufacturing company. At the other extreme, BT has significantly *reduced* its involvement in these areas by pulling out almost entirely from the *joint* research, design, and development of telecommunications equipment. Of the three companies it is NTT which has exhibited the greatest degree of *continuity*, retaining its joint activities with its former suppliers, while adding new suppliers, both Western and Japanese, and simultaneously moving to a multi-vendor supply situation. Clearly, it was both in principle and in practice possible for each of the three companies to have made different choices. Thus AT&T could have opted to reduce its commitment to the manufacture of telecommunications and telecommunications-related equipment; BT could have followed NTT in developing closer relationships with international equipment suppliers involving joint research and development; and NTT could have distanced itself from its suppliers, deciding to specify and purchase equipment rather than jointly researching and developing them.

Accordingly, it must be concluded that it is strategic choice rather than path-dependence which explains the role of research, design, development, and manufacture in the three companies. The point, it should be noted, is not that 'history did not matter', but rather that in the light of their historical experience, and their understanding of their changing global environment, decision-makers in the three companies made very different choices regarding these activities, choices which are reflected in the resources that they have allocated to R&D as documented earlier. It is, therefore, to a deeper analysis of their strategic choices that we now turn.

4.2. Strategic choice of corporate competence: an overview

At a recent conference at The Royal Society, London, Dr. J.S. Mayo, President of AT&T Bell Laboratories, suggested that there is a widespread consensus regarding the telecommunications services that will be provided over the next decade [3]. These are voice, data, and image based services, and a combination of all three, available at any time. The provision of these services will require competence in three core areas: software, electronics, and photonics.

In Fig. 1 a more detailed 'map' is provided of the three main sets of competences that are required to provide these telecommunications services. These competences are in the area of 'network elements' (i.e. switching, transmission, computing, and devices); network design, operation, management, and

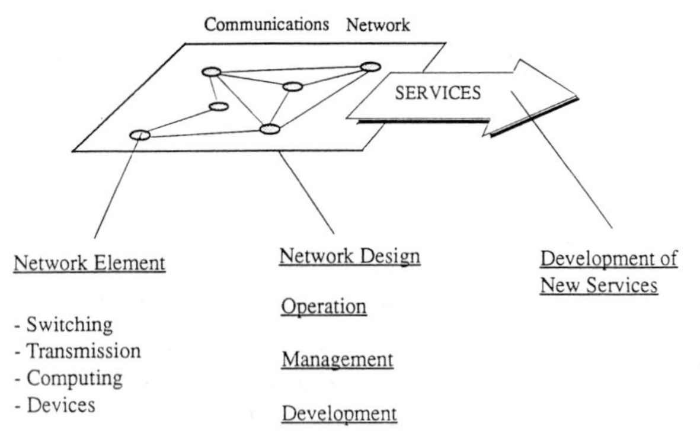

Fig. 1. A map of telecoms competences.

[3] Conference on 'Communications after 2000 AD', The Royal Society, London, March 18 and 19, 1992.

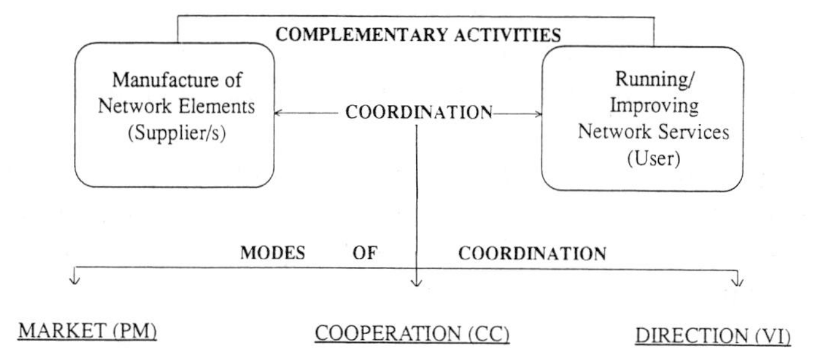

Fig. 2. Modes of coordinating user-supplier activities.

development; and the development of new services. These competences may be divided into two complementary sets of activities, the first set involving running and improving network services (undertaken by the users of network elements), while the second set involves designing, developing, and manufacturing network elements (undertaken by the suppliers). These two sets of activities are shown in Fig. 2.

Clearly, the activities of the users and suppliers need to be coordinated in order to provide telecommunications services. However, as Richardson (1972) has pointed out in a criticism of the dualistic conception of industrial organisation in terms of firms as islands operating in a sea of markets, there are in fact three broad alternative modes of coordination: through the market, through interfirm cooperation, and through direction or development within the firm.

What strategic choices have been made by AT&T, BT, and NTT regarding the mode of coordinating these two sets of activities? Their choices are shown in Table 4 where it can be seen that while all three companies have chosen to develop in-house competences in the areas of network operation and

Table 4

Strategic choice regarding acquisition of competence

Strategic options	Competences		Network elements			
	Network operation, development	New services	Switching	Transmission	Computing	Devices
Buy (market)			BT	BT	BT	BT
Joint development (cooperation)			NTT	NTT	NTT	NTT
Make/do (internal development)	BT NTT AT&T	BT NTT AT&T	AT&T	AT&T	AT&T	AT&T

development, and the provision of new services, they have made very different choices regarding the acquisition of network elements. More specifically, while each firm is a significant purchaser of telecommunications equipment on the market, BT has opted for greater use of the market as a mode of acquiring network elements, NTT has chosen inter-firm cooperation to a more significant extent, while AT&T has selected internal development.

Why have these different strategic choices been made? This question is examined further in the following section.

5. *BT — Using the market*

5.1. *BT's vision of the future*

Operating under conditions of complexity, imperfect information, and uncertainty, corporate decision-makers have no alternative but to construct 'images' of the future in deciding on the moves they should make. These images are not simply the result of processing the information which the company has obtained from its environment, as some would have us believe. While image construction does involve information processing, it is largely a process of interpretation of information based on belief. It is these images which are embodied in the company's vision of the future, a vision which allows the company to envisage the future and therefore to decide which present actions are best to undertake. [4]

BT's vision of the future embodies the following elements:

(1) The key force driving the future competitiveness of telecommunications operating companies is the capability to cater to the needs of customers, particularly large multinational companies.

(2) The essential element in this capability is the ability to tailor the characteristics of telecommunications services to the requirements of specific customers.

(3) In turn, this ability is largely dependent on competences in the area of software and system engineering. These competences are so important that it is essential that they are developed largely by BT itself rather than relying on outside suppliers of software and system engineering. (Software and system engineering relating directly to the provision of competitive telecommunications services, therefore, are seen as core competences which are mainly to be developed within the company. Other software, such as packaged computer software, may be purchased from outside suppliers).

---

[4] For a more detailed discussion of the concept of vision, and the related concept of bounded vision, see Fransman (1990, 1991).

(4) The ability to manufacture telecommunications-related equipment is seen as being largely irrelevant insofar as competitiveness in telecommunications services is concerned. Accordingly, BT will purchase the equipment that it requires on the market. Neither are there competitive advantages in engaging in the costly activity of jointly researching and developing telecommunications equipment.

(5) Although many analysts have referred to the alleged 'convergence' of communication and computing technologies, BT does not believe that the realisation of synergies between these two areas will be an important determinant of competitiveness in the provision of telecommunications services. Accordingly the company has decided not to become a producer and supplier of computers and computer services.

(6) Similarly, BT has decided that it does not for competitive reasons require substantial in-house competences in the design and manufacture of devices, including electronic and optical devices. While these devices are important components of telecommunications systems and the services which they provide, BT can rely on the market to satisfy its requirements. Besides, the escalating research and development costs of devices is constantly adding to the cost of competence in this area, without there being compensating benefits in terms of increased competitiveness.

(7) The implications of all this for R&D are that BT can largely leave research and development in the areas of manufacturing telecommunications equipment, computing, and devices to the supplying companies. BT will rather concentrate its R&D in areas such as software and system engineering which directly and strongly affect its competitiveness. One of the roles of research, however, is to ensure that BT has sufficient knowledge of new technologies to be able to rapidly and efficiently assimilate them. Furthermore, research also has the aim of 'guarding against surprise' and this justifies a certain amount of long term research.

While the question of R&D is examined in more detail in a later section, we will now examine some of the evidence on which the above account of BT's vision rests.

According to Iain Vallance, Chairman of BT, "what we know how to do well" is the "running of telecommunications networks and the provision of services across them." Vallance identified three levels of competence that he regarded as particularly important for BT: software competences in order to customise services and provide them rapidly and efficiently to customers; competences in operating telecommunications networks themselves; and, "most importantly", competences in marketing, sales, and customer service. Technology per se he regarded as less important than customer services: "We have to offer, not technological solutions, but services that our customers need." [5]

[5] Iain Vallance, quoted from Royal Society conference address. See earlier reference. See also Vallance (1990, pp. 84–87).

In line with its emphasis on customer services rather than 'technological solutions', BT has decided that it does not need in-house manufacturing capabilities. Accordingly it has decided, in retrospect, that its investment in Mitel Corp., the Canadian telecommunications equipment manufacturer in which BT owned a 51 percent share, was misjudged. As Richard Marriott, Director of BT's corporate strategy and formerly from IBM, put it: "The one [overseas investment made by BT] that really hasn't worked out as well as we hoped is Mitel... At the time, it seemed like a good idea to be in the hardware business. Our vision is now entirely based on being in network services. That, of course, in turn has meant a major shift in our research and development — moving from hardware into software and systems to give us an edge in meeting customers' increasingly complex needs. Mitel doesn't fit our new strategy."[6]

If technological solutions are not a priority, neither is in-house competence in computer hardware and software. As Iain Vallance put it: "What about 'convergence'? Ten years ago that was the 'in' word. AT&T and IBM were going to clash in the battle of the titans. On a lesser scale convergence was the rationale for STC's acquisition of ICL [the British computer company] five years ago. It never happened. ICL now belongs to Fujitsu, and STC will shortly be part of Northern Telecom [and now is]. And I was going to wonder aloud whether AT&T had lost as much money in computers as IBM had in telecommunications, but in the light of recent announcements by Bob Allen [Chairman and CEO of AT&T] that seems inappropriate..."[7]

A further reason for BT avoiding manufacturing equipment and getting involved with computers was to 'economise on bounded rationality', namely to reduce complexity by focusing on a narrower range of issues. As Richard Marriott put it: "Our international strategy is perhaps the most focused of all of the major global players. We have no distractions, such as providing service to Third World organisations. Nor are we concentrating on manufacturing telecom equipment or computers — we're getting out of manufacturing. Our strategy is very, very focused."[8] Underscoring this point, Iain Vallance said, "there is enough complexity to deal with. We do not want to take on these other activities."[9]

---

[6] Interview with Richard Marriott, 'Can BT Snare the Global Market? Using Connections to Become a One-Stop Telecom Shop', *Information Week*, February 3, 1992, p. 26.

[7] Remarks by Mr. Iain Vallance, Chairman, British Telecom plc to The Common Carrier Summit, Tokyo, 17 December, 1990, p. 11.

[8] Richard Marriott, *op. cit.*

[9] Iain Vallance in answer to a question by the present author at the Royal Society conference, *op. cit.*

6. AT&T — Reaping synergies through vertical integration

6.1. AT&T's vision of the future

AT&T's vision of the future contains the following elements:

(1) The key driving force behind the company's future competitiveness is the synergy that can be realised from its internal competences in software, switching and transmissions, computing, and devices. This synergy will allow AT&T to provide customers with better telecommunications services more rapidly than its competitors.

(2) AT&T's distinctiveness lies in its in-house possession of competences in all of these areas. None of the other major telecommunications operating companies has such a wide range of competences. Neither do the equipment supplying companies, which have also begun to compete in some telecommunications service-related markets. While the latter may also have competences in the areas of hardware and related software, they do not have AT&T's competence in network management and services. They are therefore unable to provide the customer with the same package of services that AT&T can.

(3) By implication, AT&T would not be able to realise the synergies on which its competitiveness rests by bringing together by agreement its own competences with those of other distinct companies. Hence, for example, rather than fostering close ties with one or more computer company, it was necessary to either develop the necessary computer competences internally or acquire a computer company, as eventually happened with NCR after the internal development strategy failed.

(4) The implication of this vision for R&D is that AT&T must undertake research and development activities in all the areas in which it has chosen to develop internal competences. Furthermore, basic research has a role to play in providing the seeds for future improvements in services. Nevertheless, as will be seen later, it is necessary to establish organisational mechanisms that will ensure that research and development activities contribute effectively and profitably to commercial competitiveness.

Let us now briefly examine some of the evidence on which this account of AT&T's vision is based.

In a recent interview with *Business Week*, Bob Allen, AT&T's Chairman and CEO, stated that "There is no other company like AT&T. We are unique, and therefore we have a unique opportunity."[10] *Business Week* elaborated on Allen's vision: "Allen is convinced that, even after years of frustrated efforts, AT&T can be a global information powerhouse by putting together a set

---

[10] See *Business Week*, January 20, 1992, p. 36.

of resources that no other company can match: a sophisticated worldwide network to carry voice and data, plus the equipment to run it, plus the devices that hook up to it... Put them under one roof, the Allen theory goes, and you can build complex, networked information systems that surpass anything available from a computer manufacturer, long-distance company, or phone-switch maker." [11]

For AT&T the crucial point is that as a result of synergies, these competences *in combination* will be capable of yielding greater competitiveness and returns than they would deliver separately. Taken separately, the 'products and systems' part of AT&T has not been nearly as profitable as the 'communications services' part. (In 1991 products and systems — which includes computers, network switches and transmission equipment, office telephone systems, microelectronics, telephones, and answering machines — produced a net profit margin of only 3.0 percent. This compared with 13.8 percent for communications services, which includes long distance telephone and dedicated, high-speed data lines for business. [12])

Although by and large the hoped-for synergies still remain to be realised, AT&T does have a number of examples to illustrate the potential gains. These examples include the intelligent networks that AT&T have developed for the Italian telephone operating company, SIP. According to Claudio Carrelli, head of SIP's research and development, AT&T's competitive advantage stemmed from its combined competences in switching and network operating: "All switchmakers have the same technology but AT&T has the major operating experience. From that point of view, AT&T is far ahead of everyone else." [13] Another example, combining communications networks and computing, is the computer network for after-hours trading at the New York Mercantile Exchange which will be installed by AT&T and NCR. [14] According to Gilbert P. Williamson, Chairman of NCR, synergies such as these could not have been achieved if the two companies had an arm's-length market-based relationship: "If you really want to do something big, it gets very difficult to do it on an arm's-length basis." [15]

---

[11] See *Business Week*, January 20, 1992, p. 35.

[12] See *Business Week*, January 20, 1992, pp. 34–35.

[13] See *Business Week*, January 20, 1992, p. 38.

[14] See *Business Week*, January 20, 1992, p. 39.

[15] See *Business Week*, January 20, 1992, p. 37.

7. NTT — Competing through cooperation

7.1. NTT's vision of the future

NTT's vision consists of the following elements:

(1) In order to remain competitive domestically against the new common carriers and become increasingly competitive internationally, NTT must be a leader in the provision of new telecommunications services.

(2) Leadership in new services requires that NTT plays a leading role in research and development. This means that NTT must be actively developing new services (and the technologies which underpin them) *before* they are developed by the telecommunications equipment manufacturers. This advanced positioning on the part of NTT is necessary since once new services are developed by the manufacturers, they will soon be made available to all the major telecommunications operating companies, depriving any one of them of a competitive lead. It is for this reason that NTT has decided to devote an *increasing* proportion of its sales to R&D.

(3) For historical reasons NTT has not acquired manufacturing competences. In view of the costs of accumulating such competences, however, NTT has decided to leave manufacturing to the manufacturers rather than attempt to compete with them.

(4) Nevertheless, NTT believes that it can benefit from cooperating closely with a number of equipment manufacturers and jointly developing telecommunications equipment and services that are not already available in satisfactory form and price on the market. In this way NTT will be able to 'marry' its competences with those of its suppliers in order to achieve its objective of service leadership through technological leadership.

(5) NTT also believes, as the terms of its part-privatisation legally require it to believe, that it must continue to play an important national and global role (in addition to its private role) as a supporter and advancer of research and development in telecommunications-related technologies.

Since its part-privatisation in 1985, NTT has come under substantial pressure from the new common carriers on the Japanese market. By financial year 1990 the new common carriers accounted for 12 percent of the Japanese long-distance telephone services market and this increased to 16 percent by 1991. The new common carriers controlled 40 percent of the voluminous Tokyo–Nagoya–Osaka market in 1990, rising to 49 percent in 1991.[16] In part the increasing market share of the new common carriers was facilitated by

---

[16] Talk given by Mr. Teruaki Ohara, Senior Vice President, Fair Competition Promotion Office, NTT, to the Anglo-Japanese High Technology Industry Forum, Tokyo, June, 1991, and talk given by Dr. Iwao Toda, Executive Vice President, NTT, to the Royal Society conference, *op. cit.*

the lower prices that they charged relative to NTT, which in turn were made possible by the low access costs that they had to bear for access to NTT's network. Although by 1994 access charges are expected to reach an economic rate, the substantial share of the Japanese market controlled by the new common carriers, together with NTT's recent fall in profitability referred to earlier, put great pressure on NTT to ensure its longer term competitiveness.

How does NTT plan to increase its competitiveness? The answer according to Dr. Iwao Toda, Executive Vice President of NTT, is that NTT intends to improve its services and offer new services through concentrating on research and development. By 1994 NTT expects that new services will account for a substantial 30 percent of total annual revenue. According to Dr. Toda, "We think a substantial portion of [NTT's] earnings are attributable to research and development. For this reason, NTT's management believes R&D is the foundation of NTT's business activities." [17] It is for this reason that, as is shown in Table 3, NTT's expenditures on R&D as a proportion of sales are expected to increase from 4.1 percent in 1991, to 4.7 percent in 1993.

Dr. Toda has explained that NTT has three objectives for its R&D. The first is to develop new telecommunications services, the second is to improve the planning, administration and management of the telecommunications network, while the third is to innovate in the area of 'network elements' (which include switching, transmission systems, devices, and computers). Dr. Toda elaborates on the rationale behind NTT's emphasis on network elements:

> "The technological progress in telecommunication fields is so rapid that the quicker we can introduce newer and more cost effective network elements, the more we can reduce network costs. This is why NTT develops key network elements by itself. Its aim is to produce the element much *earlier than outside manufacturers*. Usually we initiate our own R&D work for network elements which rely on rapidly advancing technologies and, furthermore, for which we can expect more than two years lead over outside organizations." [18]

This does not mean, however, that NTT does not cooperate closely with telecommunications equipment manufacturers. Again Dr. Toda elaborates:

> "NTT has no manufacturing capability. Therefore, NTT has to collaborate with outside manufacturers to develop a network element. That is, we have to ask outside manufacturers to build a prototype of a network element *according to our design*. These collaborating manufacturers are called development partners." [19]

---

[17] Interview with Dr. Iwao Toda, NTT.

[18] *ibid*, p. 5, emphasis added.

[19] *ibid*, p. 8, emphasis added.

But close collaboration with a selected group of manufacturers, however, does not mean that NTT forgoes the benefits of competition. Dr. Toda explains the advantages of the NTT approach, which the present author has called '*controlled competition*' (Fransman, 1992b):

> "The development partners for a particular network element are ... selected by openly inviting tenders [from] the manufacturers world wide. This open and fair selection process encourages competition among manufacturers. We usually select more than two partners for development of a network element. The selected partners compete with each other during the development process."[20]

Furthermore, NTT also benefits from the competitive pressure which non-partner manufacturers bring to bear through market competition with the partner companies that cooperate with NTT:

> "It should be noted that NTT's competition with non-partner manufacturers is most intensive. We believe such competition [between partner companies and between them and non-partner companies] contributes to better product development in a shorter period."[21]

Dr. Toda summarizes the advantages that accrue to NTT despite its lack of competences in the area of manufacturing: "first NTT as a big user of network elements can innovate technologies, and second, NTT can always utilize the best manufacturing technology in the world." However, he does acknowledge that "a disadvantage is difficulty in accumulating the manufacturing know-how necessary for better design."[22] This disadvantage NTT attempts to rectify through close cooperation with its partner manufacturing companies.

Examples given by Dr. Toda, where network elements were developed more rapidly through cooperative development under conditions of controlled competition, include low-loss optical fiber cables and large capacity transmission systems. More recent examples include the ATM (asynchronous transfer mode) broadband digital switching system, where the partners in addition to Japanese companies include Northern Telecom, and the MIA (multivendor integration architecture) project to define interface specifications for NTT's future computer acquisitions which includes IBM (Japan) and DEC.[23]

---

[20] Dr. Iwao Toda, *op. cit.*

[21] *ibid.*

[22] *ibid*, p. 9.

[23] For a detailed account of the development of optical fiber technology in Japan and for the ATM and MIA projects, see Fransman (1992b).

8. Contradictions among the visions? The market versus cooperation versus internal development

8.1. Vision, bounded vision, and vision failure

Returning to Table 4, it would appear that there is a significant contradiction between the visions of AT&T, BT, and NTT as reflected in their different strategic choices of the market, cooperation, and internal development in acquiring the network elements which provide the basis for the telecommunications services which they sell. As Table 4 shows, while BT has opted for the market as the means of acquiring network elements, NTT has chosen cooperation, while AT&T has selected internal development.

How is this different choice to be explained? Does it reflect inconsistent calculations regarding the costs and benefits of transactions involving the market, cooperation, and internal development?

The answer to these questions hinges on an understanding of the way in which corporate visions are formed. In short, it is proposed that the differences in the vision of the three companies is to be explained by the way in which the past experience of the company is interpreted by its decision-makers, and the beliefs of these decision-makers regarding the alternatives that exist and the consequences of these alternatives. Accordingly, it may be suggested that in the case of BT, the past history of joint research and development of telecommunications equipment was on the whole not interpreted very favourably. Seen in this light, the alternative of market procurement seemed to be relatively attractive. For NTT, on the other hand, the experience of joint research and development with a stable group of suppliers was interpreted far more positively. It is therefore perhaps not surprising that NTT has opted for a continuing close relationship with selected groups of suppliers in its attempt to take the lead in researching and developing advanced network elements while drawing on the manufacturing competence of these suppliers. Similarly, AT&T, with its internationally favourable track record in developing network elements internally (through close interaction between AT&T's operating units and its subsidiary Western Electric), emerged after divestiture with a positive interpretation of internal development as a suitable mode for obtaining such elements.[24]

According to this account, what a company 'sees' depends on its knowledge. But this 'knowledge', in turn, comprises the interpretations and beliefs of the company's decision makers. These interpretations and beliefs are influenced by (without being fully determined by) the past experience of the decision

---

[24] For a comparative international analysis of the performance of NTT and AT&T in the area of switching, see Fransman (1992a).

makers. This, however, raises the likelihood of 'bounded vision', namely where the company's vision is bounded by its interpretations, beliefs, and past experience. In some cases bounded vision may result in 'vision failure', where the company's vision results in a failure to correctly anticipate events with the result that inappropriate decisions are made. Thus the copper cable manufacturer may fail to anticipate the significance of optical fiber, the analog switch maker and buyer may fail to appreciate the speed with which digital switches will be introduced, the maker of large computers may underestimate the growth of demand for small computers, the producer of CISC microprocessors may not see the coming importance of RISC, etc.[25] Similarly, it is also possible that telephone operating companies may overestimate, or perhaps underestimate, the advantages and disadvantages of markets, cooperation, and internal development.

8.2. The market versus cooperation versus internal development

We have seen that BT, NTT, and AT&T have made very different strategic choices with respect to the acquisition of 'network elements'. More specifically, BT has tended to rely to a greater extent on markets, while NTT, although also resorting to markets, has opted more than the other companies for interfirm cooperative development. Finally, AT&T, also using markets, has selected internal development (or vertical integration) as its preferred mode of coordination. Is it possible to make any analytical generalisations regarding the advantages and disadvantages of these alternative modes of coordination?

The first step in attempting to make such a generalisation is to construct stylised 'ideal types' of the strategic options each company has selected and then to proceed to analyse the advantages and disadvantages of the modes of organisation they have chosen. In so doing, however, it is necessary to stress that these are ideal types and may not accurately reflect all of the interactions between the telephone operating company and its suppliers of network elements.

The three alternative modes of organisation, or strategic options for the procurement of network elements, chosen by BT, NTT, and AT&T are illustrated in Fig. 3.

BT has tended to opt for the 'pure market' (PM) option typified by a potentially large number of suppliers (at least in the pre-contract stage) all of whom have an arm's length relationship with BT. NTT, by contrast, has chosen 'controlled competition' (CC) for developing those network elements which it

---

[25] Each of these examples either could be documented, or has been documented (in some cases by the present author), as a case of vision failure.

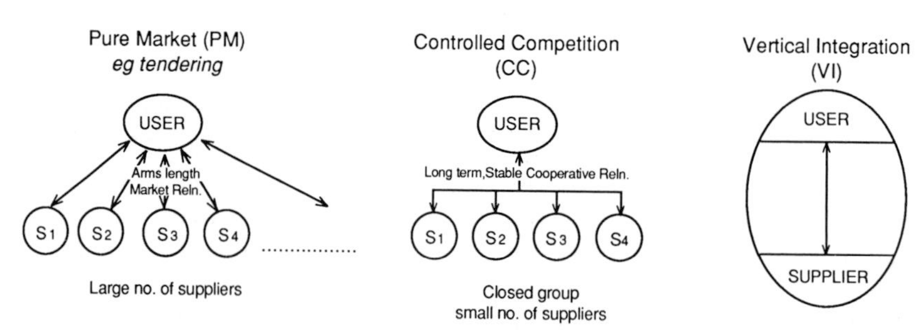

Fig 3. Strategic options for procurement.

believes will give it a competitive advantage and which are not available on the market. Controlled competition has involved a relatively long term and stable cooperative relationship with a selected group consisting of a small number of suppliers. Recently these suppliers have included non-Japanese companies such as Northern Telecom, DEC, IBM (Japan), and AT&T, and Japanese companies that before 1985 did not supply NTT, such as Toshiba, Matsushita and Mitsubishi Electric. Finally, AT&T has selected 'vertical integration' (VI) or internal development for those network elements which it does not wish to purchase on the market and which it would like to develop in a synergistic way.

What are the advantages and disadvantages of these three ideal type alternative modes of organisation? Some of the main advantages and disadvantages are shown in Table 5.

8.3. Advantages and disadvantages of 'pure market' (PM)

One of the main advantages is that the carrier/buyer can choose the best supplier from amongst all those available in the market (and therefore maximise 'static private allocative efficiency'). Furthermore, under PM innovation is largely the responsibility of the supplier, and the carrier, therefore, does not have to commit as much resource to innovation and thus risk becoming 'locked in' to specific technological paths which it has chosen. In addition, insofar as there is competition between potential suppliers, there is also likely to be competition-induced innovation which may speed and improve the quality of new technologies offered by the suppliers.

PM, on the other hand, may also result in a higher degree of competition-induced uncertainty (relative to CC and VI) since suppliers, even those who have won orders, will be unsure from one round of tendering to the next whether they will win future orders. In addition, and following on from the point about increased uncertainty, there will under PM be limited incentive for potential suppliers to invest in 'transaction-specific assets', that is assets

Table 5

Advantages and disadvantages of PM, CC, and VI

PM	CC	VI
Advantage for carrier	*Advantage*	*Advantage*
1. Maximise static efficiency (choose best supplier)	1. Better coordination (compared to PM)/ better information flow	1. Coordination through direction may be more efficient (e.g. synergies)
2. Innovation responsibility of supplier/competition-induced innovation	2. Limited competition possible	2. Knowledge-leakage minimized
	3. Greater incentive to invest in transaction-specific assets	3. Greatest incentive to invest in transaction-specific assets
Disadvantages	*Disadvantages*	*Disadvantages*
1. Competition-induced uncertainty	1. Limited ability in short run to switch suppliers (→ static inefficiency)	1. Innovation responsibility of user (→ high R&D)
2. Limited incentive to invest in transaction-specific assets	2. Possibly high coordination costs (autonomous firms)	2. Absence of competition-induced innovation
	3. Some competition-induced uncertainty	3. No switching from internal to external supplier (→ static inefficiency)
	4. Possible knowledge-leakage	

such as machinery and equipment and human skills which are specific to a supplier's transaction with a particular buyer and which are worth substantially less in alternative uses. The reason is that the supplier is uncertain regarding whether it will win orders and therefore be able to recoup reasonable returns on its investment in transaction-specific assets. Under these conditions suppliers are less likely to invest in such assets compared to where they are more sure of winning future orders. [26]

[26] The problems posed for market transactions by transaction-specific assets were first examined by Coase (1937). Williamson (1985) provides a more detailed elaboration on the importance of these assets. As we shall shortly see, these problems are reduced under CC and even more so, as Coase originally noted, under VI. It is precisely because of problems such as transaction-specific assets and competition-induced uncertainty which arise under pure market conditions, and which can negatively affect the user-supplier relationship, which have led companies to attempt to modify these conditions while retaining the competitive option of switching suppliers. Thus, for example, BT in its publicity material intended for potential suppliers, has stressed that

8.4. *Advantages and disadvantages of 'controlled competition' (CC)*

One advantage that may follow from CC is more effective coordination between supplier and carrier compared to PM. While PM will provide for coordination through the market mechanism, specifically through the price mechanism and competitive process, it may be less effective in facilitating non-market-mediated flows of information between suppliers and carriers who have an arm's-length relationship to one another. Better flows of information between carriers and suppliers who enjoy a longer term, relatively stable relationship may lead not only to better coordination but also to more effective user-oriented innovation. Furthermore, although there is less competition than under PM since there is at any point in time a closed group of suppliers consisting of a small number of firms, there remains the opportunity to benefit from some competition between the members of this group. Less competition compared to PM also means that there is less competition-induced uncertainty and hence a greater incentive, again compared to PM, to invest in transaction-specific assets.

Against these advantages, however, must be set the limited ability in the short run to switch suppliers, which may lead to 'static private allocative inefficiency' relative to PM. It is also possible that there will be relatively high coordination costs under CC as the carrier attempts to share information and get agreement between several autonomous firms. Furthermore, although competition-induced uncertainty is likely to be lower than under PM (since there is controlled competition), to the extent that competition remains there will be some such uncertainty. In addition, there is, from the supplier's point of view, the danger that commercially sensitive knowledge will leak to the

---

"it is vital that we safeguard our position and that of our customers by buying only those goods and services which represent value for money." This, however, requires that high standards of quality are attained by suppliers particularly of complex equipment that directly affects the services which BT sells to its customers. The attainment of these quality standards, however, requires that BT engage in quality-enhancing activities such as "joint quality improvement projects", "comprehensive monitoring of [post-purchase] performance", and the collection of "information on whole life costs." (p. 4) But such activities are not always compatible with pure market practices where the user/purchaser may switch current suppliers in order to take advantage of more favourable price/performance offered by new suppliers. It is for this reason that Brian Rigby, BT's Director of Group Procurement Services, expresses the hope in the introduction to this document that "if we choose to work together we can develop a long term, mutually beneficial relationship." (p. 1) Later it is reiterated that "a close relationship with suppliers is essential. We need a partnership founded on trust, co-operation, support and, above all, continuous improvement."(p. 4) It is precisely these relational characteristics which the supporters of controlled competition claim necessitates the significant departure from a market relationship which controlled competition represents. (BT, 'Selling to BT. A Winning Partnership', undated.)

other competing suppliers who are also involved in the cooperative research and development.

8.5. Advantages and disadvantages of 'vertical integration' (VI)

One possible advantage of VI may be more effective coordination between complementary activities (since the activities are under the same 'command structure' in the same firm) and therefore a greater realisation of synergies between these activities. In addition, under VI there will be the greatest incentive to invest in transaction-specific assets (relative to PM and CC) since the costs of diverting these assets to lower-earning alternative uses will have to be borne by the firm itself (a point that Coase made in his original 1937 article). Furthermore, the chance of knowledge-leakage to competing firms is minimized and there is also a minimisation of competition-induced uncertainty, since the transactions occur within the same firm.

A major disadvantage of VI, however, is that innovation is the responsibility of the carrier and this is likely to lead to relatively high R&D expenses. Furthermore, there is the danger that, having invested in particular forms of innovation, the firm may become 'locked in' and therefore unable to as easily take advantage of alternative innovations that have been generated outside the firm. In addition, since by definition there is only one internal supplier, there is the absence under 'pure' VI of competition-induced innovation. Similarly, again under 'pure' VI, since there is by definition no opportunity to switch from an internal to an external supplier, there may well be greater 'static private allocative inefficiency' than under either PM or CC. (It is, of course, to avoid these consequences that vertically integrated firms usually leave some option for users in the firm to purchase from outside firms.)

8.6. Conclusion

In view of the complexity of the possible effects of PM, CC, and VI it is not possible to reach a definitive conclusion regarding the relative efficiency of these strategic options. It is no doubt for this reason that BT, NTT, and AT&T have been able to make different choices regarding these options and remain convinced that they have made the optimal choice. To a large extent the efficiency of the choice depends on the company's competences including the extent to which it is able, organisationally and managerially, to exploit the advantages which have been analysed, while dealing effectively with the disadvantages. Furthermore, as mentioned at the outset, the three strategic options analysed in this section are ideal types and although the three companies have made broadly different choices with regard to these options, they have also modified the ideal typical modes of coordination in order to increase the net advantages, as is indicated in the last footnote with reference to BT's concept of 'quality suppliers'.

However, the strategic choices that have been made regarding the acquisition of network elements must be sharply distinguished from the strategic choices that the companies have made in the area of research and development. It is accordingly to an analysis of research and development that we now turn.

9. *The role of research and development*

9.1. *Introduction*

The first corporate research laboratories were established in the German chemical industry in the late Nineteenth Century and not long after in American companies such as AT&T, General Electric, and Du Pont.[27] While these laboratories allowed their companies to apply scientific principles and practices to industrial production and in some cases to increase the company's ability to appropriate returns from its investments by securing commanding patent rights, the laboratories also posed new and formidable organisational problems. These stemmed largely from the greater *fragmentation of knowledge* that accompanied the increased division of labour implied by the establishment of an industrial laboratory. Now a cadre of *researchers*, one step removed from the company's routine process of production and with a knowledge base significantly different from those involved in this production process, had to be integrated into the firm's activities in a value-enhancing way. It was not, and still is not, always clear to the management of the company concerned how this integration could be effectively achieved.

This organisational problem became increasingly pressing for the major telecommunications operating companies from the mid-1980s, faced as they were with increasing competition in their major markets. While hitherto their monopoly status in their domestic market afforded them a degree of protection from the necessity to generate immediate value from research, in the tougher world of short run competition the 'luxury' of research relatively uncoupled from direct value-creation could no longer be sustained. The result was organisational reform in these companies designed to integrate research more closely with the immediate value-creating process.

[27] For accounts of the early laboratories see Reich (1985) and Hounshell and Smith (1988).

9.2. The four major research problems

Any company undertaking research has to resolve four major research-related problems (see Fransman, 1991):

(1) What research does the firm need now? This is an *information* problem.

(2) How to prevent 'irrelevant' research? This is a *control* problem.

(3) What research is needed for the future? This is an *uncertainty* problem.

(4) Should the required research be undertaken in-house or ex-house? This is an *assignment* problem.

How have the telecommunications operating companies dealt with these problems?

9.2.1. The information and control problems

> "To ensure that... [research] is well coupled to the needs of the business, the strategy demands that at least two-thirds of the work performed by the central laboratories should be sponsored and financed directly by the customers in the operating divisions [of BT]." (Alan Rudge, 1990, pp. 124–125.)

> "If researchers in the laboratories received payment from the operating divisions in NTT the benefit is that researchers would become more customer/business oriented. However, the cost is that they will become short-term oriented. The business people [in the operating divisions] cannot see the future. So right now we do not use a financial link [between the operating divisions and the central research laboratory]. Head Office funds most of the research in the laboratories." (Author's interview with senior NTT officer, 1991)

(1) Using an internal market for research and development. In an important article, Alan Rudge (1990), Director of Group Technology and Development at BT, has carefully explained how BT since privatisation has attempted to deal with the information and control problems relating to research. The central role in coping with these problems has been played by what Rudge calls "the customer-supplier principle" (p. 127), or what is referred to here as the internal market for research and development.

In short, in order to provide information regarding the research and development priorities in BT and to ensure that the company's researchers allocate their time and other resources in accordance with these priorities, BT has established an internal research and development market. In this internal market, BT's Operating Divisions demand the research and development that is supplied by the company's 4,000 scientists, engineers, technicians, and support staff in the central research laboratories. This demand is expressed in the form of an explicit contract for the particular project, which is defined in terms of milestones and deliverables and stipulates a fixed time period and price. Every task undertaken in the central research laboratories is defined as a project in this way (Rudge, 1990, p. 121). Overall, the central research laboratories are highly dependent on the internal market for the revenue which they need to employ their staff and cover their costs. Thus Rudge reports that in 1989 "75% of work of the central laboratories is funded, and therefore directed, by customers either in the Operating Divisions [who provide the bulk of the revenue] or externally." (p. 126)

Further organisational steps have been taken to "couple" the central research laboratories, which Rudge sees as BT's "engine of change" (p. 127), with their customers in the Operating Divisions. One of these steps has involved the appointment of Programme Office Directors whose main job is to coordinate all of the projects being undertaken in the laboratories for a particular customer in one of the Operating Divisions. While the Programme Office Director plays the entrepreneurial role of "trying to anticipate the client's needs and his likely direction in the future", he is also a "line manager", performing a significant part of his work for the same major customer. (p. 123) In its attempt to make the boundary between the Operating Divisions and the central research laboratories a "managed interface", BT has made use of what Rudge calls the "flow management principle". This provides a "means of measuring the effectiveness of the company's R&D by monitoring the flow of technology, techniques, products and service ideas across the boundaries." (p. 128) Rudge argues that "This kind of flow is measurable and can be quantified. It offers sensible advantages over measures based upon the numbers of publications, which are often totally inappropriate in commercial organizations." (p. 129)

AT&T has also attempted to make use of an internal market in order to coordinate the activities of its laboratories and operating divisions. In 1988 Robert E. Allen, AT&T's Chairman, reorganised the company into some twenty separate business units with profit-and-loss responsibility.[28] The following year, Arno A Penzias, Vice President of Research at Bell Laboratories, initiated a number of corresponding changes in the laboratories. One of the main objectives of these changes was to tighten the link between the laboratories and the business units. The changes included the consolidation of research projects into 15 laboratories divided among 4 divisions, requiring almost half of the 1,200 research staff to support projects in the business units, and assigning all of the 19 directors in the laboratories the task of working with one of the business units partly by establishing cooperative projects with them.[29]

---

[28] See *Business Week*, January 20, 1992, p. 36.

[29] See *Scientific American*, 'Rethinking Research', December, 1991, pp. 92–93.

Through changes such as these it is hoped that AT&T would be able to increase the value that it reaps from research and thus meet the criticism of critics, like A. Michael Noll, a former AT&T marketing executive and now Professor of Communications at the University of Southern California, Los Angeles, who have argued that AT&T's return on investment in research has been too low. According to Noll (1991), in "1984 [Bell] Labs employees were 5.3% of the total number of AT&T employees; in 1989 this doubled to 10.2%." However, Noll argues, AT&T has not derived sufficient benefit from these employees. In support of his argument Noll has published the following rather remarkable claim:

> "One reason AT&T agreed to divestiture was to be free to develop the many new business opportunities that supposedly were available from the technological storehouse of Bell Labs. This hypothesis was not investigated before divestiture. After divestiture AT&T Bell Labs did an internal study of all its ideas for new products, services and businesses. While at AT&T I had the opportunity to review the results of this study, a multi-volumed document. Of the hundreds of ideas submitted, fewer than half a dozen, as I recall, were realistic, and most of them were already under development in some fashion. The simple fact was that the Labs' cupboard was bare. The jewels of the Bell System, the operating telephone companies, had been given away in return for nothing!" (p. 103)

NTT has also, after its part-privatisation in 1985, reorganised in order to strengthen the connection between research and development and the operating divisions. In 1987, under the influence of the new head of NTT, Dr. Shinto, the 4 research locations of NTT's Electrical Communications Laboratories (ECL), based in Musashino, Yokosuka, Atsugi, and Ibaraki, were divided into 11 research laboratories.[30] In order to meet the development needs of the operating divisions, the remaining scientists, engineers, and technicians, who had been part of ECL but were not included in the laboratories, were moved to 2 new development centers, the Network Systems Development Center and the Software Engineering Center, which were closely tied to the activities of the divisions. In 1991 some 3,200 of NTT's approximately 8,200 scientists, engineers, and technicians were allocated to the 11 laboratories while the remaining 5,000 were based in the centers and in the development sections

[30] Of the eleven laboratories, three dealt with network and human interface issues (the Telecommunications Networks Laboratories, Communications and Information Processing Laboratories, and Human Interface Laboratories), three with systems (the Communication Switching Laboratories, Transmission Systems Laboratories, and Radio Communication Systems Laboratories), four with key technologies (the Software Laboratories, LSI Laboratories, Optoelectronics Laboratories, and Applied Electronics Laboratories), and one with basic research (the Basic Research Laboratories). In 1991 a new laboratory, the Communications Science Laboratory, was added, based in the Kansai area. At around the same time a new development center was announced, dealing with development of the optical subscriber loop.

Table 6
R&D in BT and NTT

	BT	NTT
% R&D in central research labs (CRL)	66%	60%
% of CRL R&D funded by Headquarters rather than Operating Divisions	18%	95%
CRL staff BT, 1989	4,000 *	
CRL staff NTT, 1991		3,200 **
% of total R&D 0–6 year horizon	65%	60%
7–10 year horizon	25%	30%
11-20 year horizon	10%	10%
'Basic' research as% of total R&D	NA	5%

* Includes total research-related staff.
** Includes only scientists, researchers, and technicians.
Sources: Rudge (1990); NTT Annual Report, 1991; author's estimate for NTT.

of the operating divisions. In the same year, while about 40 percent of NTT's total R&D budget went to the centers and development sections, 60 percent was allocated to the 11 laboratories. [31]

When Dr. Shinto's reforms were first mooted, they caused a good deal of anxiety on the part of researchers and others high in the NTT hierarchy who feared that they may indicate a tendential swing against NTT's research-led approach to telecommunications. A top delegation of NTT's researchers accordingly went to discuss the matter with Dr. Shinto who reassured them that as a former engineer he understood the importance of research. He had, he said, no intention of undermining the role of research in the company, although he felt that some reorganisation was necessary to increase the effectiveness of NTT's development activities. It was for this reason that he had formed the two new development-oriented centers and linked them more closely with the operating divisions [32]

(2) The internal market and beyond. Although, as documented, use is made of an internal market for research and development to deal with the information and control problems referred to, the internal market mechanism is of only limited value in taking care of a company's research needs. The limitations have been well expressed by Rudge (1990) who has observed that "while research provides the in-depth knowledge, it is noteworthy that the majority

[31] Author's estimate and Table 6.

[32] Interview with senior NTT research leader.

of our customers [in BT] are not interested in buying research as such." (p. 117) However, "If the central laboratories are to provide expertise on demand they must have in place a well-directed research programme to generate enabling technology and 'know how' *in advance of the customer's need.*" (p. 124, emphasis added) This creates something of a dilemma: the central laboratories must generate knowledge *before* the customer has a need for this knowledge, and therefore before the customer will be willing to pay for the creation of this knowledge. This raises the question of who will pay for the creation of this knowledge, that is if it is to be created at all. The answer in most cases given to this question is that it is the company's headquarters which will have to extend what is in effect a grant to the central research laboratories to finance the research that will result in output which the operating divisions will later be willing to pay for. This requires going beyond the internal market for research and development.

However, even if it is accepted that headquarters should play this role, further difficult questions are raised. For example, what proportion of the budget of the central research laboratories should be funded from headquarters? How should the company ensure that research funded in this way is relevant to the needs of the company? In dealing with questions such as these, significant differences have emerged among the companies being studied here. This takes us on to the third research problem referred to earlier, the uncertainty problem, namely: What research is needed for the future?

9.2.2. The uncertainty problem

There are significant differences between BT and NTT regarding the solutions that have been devised for dealing with the uncertainty problem. Reminiscent of the differences between these two companies with respect to the acquisition of network elements, which were analysed earlier, BT has tended to rely to a greater extent than NTT on market forces, albeit in this case internal market forces. This emerges clearly in Table 6.

From Table 6 it can be seen that in the two companies a similar proportion of total R&D is undertaken in the central research laboratories as opposed to elsewhere, namely 60 percent in the case of NTT and 66 percent in the case of BT. The big difference between BT and NTT, however, relates to the proportion of the central research laboratories' R&D which is funded from, headquarters rather than the operating divisions, about 18 percent in the case of BT compared to about 95 percent for NTT (second row of Table 6).

This substantial difference is indicative of a fundamentally different philosophy (or, more accurately, vision) in the two companies regarding both the role of research and the balance of market and non-market forces required to deal with the uncertainty problem. The differing philosophies emerge in the two quotations presented above (under the heading 'The Information

And Control Problems') which are sufficiently illuminating to justify repeating here:

> "To ensure that... [research] is well coupled to the needs of the business, the strategy demands that at least two-thirds of the work performed by the central laboratories should be sponsored and financed directly by the customers in the operating divisions [of BT]." (Alan Rudge, 1990, pp. 124–125)

> "If researchers in the laboratories received payment from the operating divisions in NTT the benefit is that researchers would become more customer/business oriented. However, the cost is that they will become short-term oriented. The business people [in the operating divisions] cannot see the future. So right now we do not use a financial link [between the operating divisions and the central research laboratory]. Head Office funds most of the research in the laboratories." (Author's interview with senior NTT officer, 1991)

The difference in philosophy between the two companies may be expressed in the following way: While both companies agree that a combination of 'internal market pull' and 'science-technology push' (funded from Headquarters) is required in order to deal with the uncertainty problem, they differ regarding the *balance* that is necessary between these two forces; BT ultimately gives greater weight to internal market pull, while NTT tends to favour science-technology push.

According to the NTT view, the vision of the staff in the operating divisions (including the development engineers working there) is bounded by the activities currently undertaken in the divisions. These activities in turn are determined by the existing markets and technologies of the operating divisions. This situation is appropriate as long as only incremental changes in markets and/or technologies are needed. If more radical changes are required, for example new services are introduced based on fundamentally different technologies, then the bounded vision of the operating divisions becomes a binding constraint. In the latter case the lead must come from the research laboratories where knowledge of the new technologies and their applications has been created.[33]

NTT's preference is also apparent in the choices that have been made regarding the appropriate location for the creation of knowledge required for the future and the way in which such knowledge, once created, is transferred to the operating divisions where it is to be used. Before Dr. Shinto's reforms introduced in 1987 referred to earlier, both radical changes in knowledge and applications of this knowledge were created in the central research laboratories and were then transferred to the operating divisions. Under the reforms, an attempt was made to move the site of such application creation to the

[33] Interview with senior NTT research leader.

development centers and operating divisions. However, it was soon discovered that the engineers in the development centers and operating divisions frequently were insufficiently familiar with the new technologies. Accordingly, it was decided to revert to some of the previous practices and locate the creation of applications based on radical improvements in technology in the central research laboratories. In order to transfer these applications to the operating divisions, engineers by and large go *from* the divisions *to* the laboratories where knowledge regarding the new applications is located, rather than the other way round. They then return to their divisions to complete the applications according to the division's requirements. In those cases where researchers from the central laboratories move to the operating divisions to assist the technology transfer process, they generally return to the laboratories rather than, as happens in many Japanese electronics companies, remaining permanently thereafter in the divisions.[34] In this way, the 'causal direction' of innovation is clearly from the central research laboratories to the operating divisions rather than the other way round.

With regard to the proportion of the budget of the central research laboratories funded from headquarters, practice in NTT differs significantly from that in the major Japanese industrial electronics companies such as NEC, Fujitsu, and Hitachi. The central laboratories in the latter companies generally receive about 50 percent of their budgets from corporate headquarters.[35] However, in a recent and important departure from previous practice, several of these companies have established advanced research laboratories which are intended to undertake long-term 'oriented basic research' and which receive all of their funding from corporate headquarters. These laboratories include Hitachi's Advanced Research Laboratory and NEC's Princeton Laboratory.[36] In funding terms these laboratories resemble NTT's central research laboratories (the Electrical Communications Laboratories) more closely, although unlike ECL their mission is entirely longer term oriented with the result that they have significantly less interaction with the operating divisions in the company.

In BT, rather than an autonomous science/technology push from the central laboratories being given responsibility, as it were, for initiating the innovation process, as is the case in NTT, an attempt is made to have the research process itself pulled by the users of the research output. This is true *even within the Central Research Laboratories* as Rudge (1990) makes clear: "One way of en-

---

[34] Interview with senior NTT research leader.

[35] See Fransman, 'Explaining the Performance of Japanese Large Companies' *op. cit.*, for further details.

[36] *ibid.*

hancing [the flow of technology, techniques, products and service ideas across the boundaries between the laboratories and the operating divisions and between the research and development departments of the central laboratories] is to ensure that a significant number of projects on one side of the boundary are being funded, and hence directed, by a manager on the other side. In BT this process has been established not only between the central laboratories and their customers in the Operating Divisions, *but within the Central Research Organization*. While it is difficult for a manager of a corporately-funded research project to guarantee success of his work in the ultimate market place, he can often ensure that his output is sufficiently attractive to his customer *in the development laboratories*, that they will want to take it up, either by adopting his output or by funding extension work with him." (pp. 128–129, emphasis added) As Rudge makes clear, for BT the 'causal direction' of innovation is largely, although not completely, from the users of knowledge to the creators of knowledge, rather than the other way round.

Comments on the difference between BT and NTT. Can the differences between BT and NTT just referred to be reconciled? One way of doing so is to suggest that in the case of intra-paradigmatic technological change (where the changes of technologies occur within a given technological regime) a causal direction from the users of knowledge to the creators of knowledge is indicated. Not only does this assist the knowledge creator by providing more information about the circumstances and needs of the user, it also helps the subsequent flow of knowledge from the creator to the user. However, in the case of inter-paradigmatic technological change (where there is a change in technological regime) a different causal direction is needed, namely from the creator to the user of knowledge. The reason, to give some examples, is that the users of electromechanical switches, or copper cable transmission systems, or records, are able to provide knowledge which is only of limited use to the creators of digital switches, or optical transmission systems, or optical disks. In these cases, therefore, it is necessary to go beyond the market, even the internal market, in order to ensure the creation and implementation of radical new technologies. However, how to ensure in this case that the knowledge created remains relevant for the users, and is therefore value-enhancing for the company, when the users themselves play only a limited role in the knowledge-creation process, is the organisational and managerial problem that remains.

Basic research and long term research. This raises the issues of basic research and long term research. Are there any differences regarding the role of these kinds of research in BT and NTT?

As shown in Table 4, both BT and NTT allocate about 10 percent of their total R&D budget to projects with a commercial time horizon of 10 to 20 years. Nevertheless, as will become apparent, there are important differences between the companies regarding the roles of both long term and basic research.

Rudge (1990) elaborates on the role of long term research in BT:

> "The most difficult part of the R&D portfolio to direct effectively is the final 10% which is devoted to identifying the technological threats and opportunities which may effect (sic) the company's business activities in the longer term. Looking to the longer term there are an increasing number of scientific developments which could be of interest. The approach adopted [in BT] has been to provide a thin screen of relatively small projects which guard against surprises, combined with a lesser number of larger projects focused on the areas which are considered to be key to the Group's strategic development. To strengthen the longer term coverage, a managed programme of research with selected universities and other external research organizations has been put in place." (pp. 125–126)

However, even in the case of long term research, BT has applied its 'customer supplier principle':

> "To provide a customer–supplier framework for the management of the Corporate Research Programme [i.e. corporate funded research with primarily a 7 to 20 year commercial time horizon], a 'Programme Office' structure has been established [analogous to the Programme Office Directors referred to earlier who coordinate the link between the central research laboratories and the operating divisions] led by the Chief Engineering Advisor supported by a small team of specialist staff. Their task is to establish a portfolio of research projects within the agreed corporate budget and to audit the projects on a regular basis. The corporate research portfolio is managed to comply with the broad strategic directives agreed by the Research and Technology Board and the portfolio is reviewed by the Board and an advisory committee of senior technical managers selected from business units across the group. Additional inputs to the Programme Office include commercial support from the Business Operations Division and advice from 30 specially identified technical specialists ... chosen for their eminence in their field of expertise." (pp. 126–127)

As Rudge's account implies, there is little if any 'basic research' being undertaken in BT's laboratories, at least insofar as this refers to research undertaken without any practical or commercial objective in mind. According to this definition it may also be the case that there is little basic research undertaken in NTT. Nevertheless, there is a difference between the two companies. This becomes apparent in the role of NTT's Basic Research Laboratories which, together with so-called basic research in other parts of the Electrical Communications Laboratories, consumes about 6 percent of NTT's total R&D budget and which has been protected as a priority area from budget cuts.[37] Although

---

[37] Author's estimate.

the Basic Research Laboratories is perhaps best characterised as undertaking longer term 'oriented basic research', it has differentiated its activities from the other laboratories in ECL also concerned with oriented basic research by concentrating largely on 'scientific' research, albeit research probably still undertaken with long term telecommunications needs in mind. According to a former director of this laboratory, the publications of this laboratory are now largely in scientific, as opposed to engineering, journals.[38]

Furthermore, in line with NTT's 'science/technology push' approach, neither the other ECL laboratories, nor the operating divisions, are given any formal role in influencing the agenda of the Basic Research Laboratories, although informally of course there is the possibility of influence.[39]

In part this difference in emphasis is the result of the different obligation in BT and NTT to do research in the 'national interest'. Rudge (1990) is clear that this obligation, important pre-privatisation, is now non-existent: "For research and technology [in BT], the first priority must be the BT Operating Divisions and the Corporate Headquarters. Recognizing BT's public sector background it has been emphasized that this first priority does not include 'British Industry', the 'UK Government' or even the 'National Good' except where BT's interests coincide." (p. 117)

NTT's obligations are significantly different. Clause 2 of the NTT Company Law of 1984 (Nippon Denshin Denwa Kabushiki Kaisha Law), which provided for the part-privatisation of NTT, details NTT's responsibilities in connection with research:

> "... NTT should make efforts to contribute to the improvement of telecommunications and increase the level of social welfare of the people by promoting basic-like and applied research in the area of telecommunications and also by diffusing the results of research, since telecommunications plays an important role for social and economic development of the future."

But although there clearly are important legal differences between BT and NTT regarding their obligations to do research in the public interest, it would be incorrect to reduce the differences between the two companies in the areas of basic and long term research to legal obligations. Enough has been said in this paper about the differences in vision and managerial style between BT and NTT to support the conclusion that, even in the absence of legal obligations to do research in the national interest, a significant distinction between the two companies would still remain regarding the role of basic and long term research and the way in which this research is organised.

---

[38] Interview with a former head of the NTT Basic Research Laboratories.

[39] *ibid.*

9.2.3. The assignment problem

Finally, we turn to the assignment problem, namely the question of whether R&D should be done in-house or ex-house.

Earlier we noted a significant difference between AT&T, BT, and NTT in terms of the strategic choices they have made regarding the acquisition of competences in network elements (broadly, telecommunications equipment). More specifically while AT&T opted for the internal development of these competences, BT chose to use the market, while NTT decided on the course of cooperative development with a small selected group of supplying companies.

It is necessary to emphasise, however, that these strategic choices do not simultaneously determine solutions to the assignment problem being discussed here. To take an example, even if a company decides to purchase switches or transmission systems on the market, it does not necessarily follow that it should do no research on switching or transmission. Similarly, if a company decides to internally or cooperatively develop switches or transmission systems it does not therefore follow that it should not buy-in some, even a significant part, of the research and development required. Accordingly, the strategic choice regarding the overall acquisition of competences does not determine whether R&D is required and, if so, whether it should be done in-house or ex-house.

To take this a little further, a company that buys its switches on the market may still require knowledge in the area of switching for at least two reasons. In the first place, the company will require switching knowledge in order to make an effective choice between the switching systems that are available on the market, and in order to install and efficiently implement the switches it purchases. Secondly, the company will also need knowledge to be able to anticipate future trends in the area of switching and to 'tap into' the networks that will provide information regarding trends. Such knowledge will help the company to make suitable decisions.

The question still remains, however, regarding whether the company should accumulate this knowledge in-house via research or whether it should rely on external sources for this knowledge. Choice of the latter alternative would have to confront the following conundrum: in order to acquire complex knowledge, an individual or company needs to already possess some of that knowledge. Furthermore, much of the knowledge that is needed is tacit with the result that it is difficult and costly to acquire, and perhaps in some cases even impossible. These kinds of difficulties may increase the incentive to accumulate the required knowledge in-house.

How have BT and NTT differed in terms of their decisions regarding the assignment problem? In the absence of available comparative information with which to answer this question we shall have to rely on circumstantial evidence. This evidence suggests that it is likely that, although both companies have accumulated knowledge through in-house research and have sourced knowledge externally, at the margin NTT is more likely than BT to have chosen the in-house option. This does seem to have been the outcome in the case of microelectronics. Thus Rudge (1990) has noted that as a result of BT's new forms of organisation analysed in this paper there has "Over the past two years ... been a continuous change in the laboratories' activities. For example, in three technical areas of major importance to BT, we have seen a progressive decline in micro-electronics, a continued high level of activity in photonics and a major growth in software systems engineering." Rudge continues, "These relative changes have not been based upon scientific interest but the ability of the laboratories to provide specialist services to the operating divisions which are in keeping with their needs *and competitive with alternative sources which are available to them*." (p. 129, emphasis added) Interestingly, a recent report on the reorganisation of Bell Laboratories stated that "Over the past year [i.e. 1991, Penzias] has... begun sliding the balance of [research] funding away from high-cost and relatively low-payoff work, such as basic physics and materials science, toward the more lucrative software and information technologies." [40]

In NTT research in areas such as microelectronics has continued, although the company's research leaders have acknowledged that they are coming under increasing pressure to demonstrate that research such as this pays. [41] Nevertheless, NTT is not without its critics regarding the wisdom of such research. In the case of microelectronics, for example, the prestigious *Nikkei Communications* has recently argued, not only that NTT's research in this area is now unnecessary as a result of the technological strength of the major Japanese manufacturing companies, but also that it may be counterproductive, interfering with the companies' research as NTT, a powerful customer, pulls research in directions that are not always productive. [42] The journal acknowledges that NTT made an important contribution to the early development of large scale integration in Japan, initiating the first major cooperative research programme in this area in 1975, a year before the much better known, but not necessarily more productive, cooperative programme initiated by MITI, the VLSI Research Project 1976–1980. [43] It argues, nevertheless, that NTT's research in this area has now outrun its usefulness. While this is

---

[40] See *Scientific American*, December 1991, p. 93.

[41] Interviews with NTT research leaders.

[42] See *Nikkei Communications*, Vol. 3, No. 4, 1990.

[43] For a detailed analysis of MITI's VLSI Research Project 1976–1980 see Fransman (1990, chapter 3). For a brief discussion of the NTT VLSI project, see p. 87.

not the place to evaluate the arguments of *Nikkei Communications* regarding NTT's microelectronics research, it is simply noted that from the NTT point of view such research is seen as being important since it facilitates the design and development of more powerful devices which are crucial determinants of the effectiveness of network elements such as switches, computers, and customer premise equipment. With this in mind, NTT researchers believe that nanometer-scale fabrication technology and quantum-device technology will open the era of new visual, intelligent, and personal telecommunications services.[44] NTT's research leaders accordingly argue that microelectronics research, and device research more generally, should continue to play an important role in the company's overall strategy.

10. Conclusion

It is neither possible nor desirable to attempt to summarise the conclusions that may be drawn from the present complex comparison of AT&T, BT and NTT. Suffice it to say here that what we have analysed in this paper are three very different 'beasts', which have reorganised themselves in the aftermath of divestiture, privatisation, and part-privatisation, and which are preparing themselves to do battle in increasingly integrated and liberalised global telecommunications markets. Although these beasts are operating in an increasingly similar global selection environment, their visions, strategies, and chosen competences are, as we have shown, significantly different. Whether these differences will be functional, in the sense that they will aid their growth in an increasingly difficult operating environment, or will be the source of significant shortcomings, is a crucial question, the answer to which will only emerge with time.

Acknowledgements

I would like to acknowledge financial support on which much of this paper was based from the Institute for Japanese–European Technology Studies, University of Edinburgh, and from the PICT Programme of the United Kingdom's Economic and Social Research Council. I would also like to thank the many senior officials in NTT and BT for the time and information that they generously gave. In particular, appreciation is due to Dr. Iwao Toda of NTT who has greatly improved my understanding of research and

---

[44] Communication from Dr. Iwao Toda, NTT.

development in NTT, inadequate though this understanding undoubtedly still remains. It is from Dr. Toda that the term 'network elements' comes and Fig. 1 in which this term is embodied. None of these people, of course, is in any way responsible for the information, analysis, and conclusions in the present paper.

Additional references

Coase, R.H., 1937. The nature of the firm. *Economics N.S.* 4, 386–405.

Fransman, M.J., 1990. *The Market and Beyond: Cooperation and Competition in Information Technology in the Japanese System* (Cambridge University Press, Cambridge).

Fransman, M.J., 1991. Explaining The Performance of The Japanese Large Company. *University of Edinburgh, JETS Paper* No. 6.

Fransman, M.J., 1992a. Japanese Failure In A High-Tech Industry? The Case Of Central Office Telecommunications Switches. *Telecommunications Policy*, April, pp. 259–276.

Fransman, M.J., 1992b. Controlled Competition In The Japanese Telecommunications Equipment Industry: The Case of Central Office Switches. In: C. Antonelli (ed.), *The Economics of Information Networks*, pp. 253–276 (North-Holland, Amsterdam).

Fransman, M.J., forthcoming. *NTT and The Globalisation of Japanese Telecommunications.*

Grupp, H. and Schnoring, T., 1992. Research and Development in Telecommunications: National Systems Under Pressure. *Telecommunications Policy*, 16 (1), 46–66.

Hounshell, D.A. and Smith, J.K., 1988. *Science and Corporate Strategy: Du Pont R&D, 1902–1980* (Cambridge University Press, Cambridge).

Molina, A.H., 1990. Building Technological Capabilities in Telecommunications Technologies. Development and Strategies in Public Digital Switching Systems. *Paper presented to the International Telecommunications Society Conference*, Venice, March, 1990.

Noll, A.M., 1991. The Future of AT&T Bell Labs and Telecommunications Research. *Telecommunications Policy* 15 (2), 101–105.

Reich, L.S., 1985. *The Making of American Industrial Research. Science and Business at GE and Bell, 1876–1926* (Cambridge University Press, Cambridge).

Richardson, G.B., 1972. The Organisation of Industry. *Economic Journal* 82 (September), 883–96.

Rudge, A.W., 1990. Clifford Paterson Lecture: Organization and Management of R&D in a Privatized British Telecom. *Sci. Publ. Affairs* 4, 115–129.

Vallance, I.D.T., 1990. The Competitive Challenge in World Communications Markets. *British Telecommunications Engineering* 9 (July), pp. 84–87.

Williamson, O.E., 1985. *The Economic Institutions of Capitalism* (The Free Press, New York).

Global Telecommunications Strategies
and Technological Changes
Edited by G. Pogorel

Chapter 15

Globalization of telecommunications operators under conditions of asymmetric national regulation

Johannes M. Bauer

Department of Telecommunication, Michigan State University, East Lansing, MI, USA

1. *Introduction*

Until about a decade ago, telecommunications presented itself as a nationally structured business. Equipped with relatively comprehensive monopoly privileges, national telecommunications operators provided domestic and, jointly with other national carriers, international telecommunications services[1]. To facilitate this joint provision of services, bilateral and multilateral agreements were established in the areas of technical standards (for example, through the Coordinating Committees of the International Telecommunication Union) as well as in the area of international tariffs through a system of accounting rates[2]. Beginning in the early 1980's, the industry is currently involved in a profound reorganization and globalization of operations[3]. Telecommunications operators form new international alliances and invest increasingly in

[1] Of course there were exceptions to this general picture. Operators such as Cable and Wireless have been operating at an international level throughout the decades. Before the strongly national structures were established, national companies such as the Bell companies or ITT maintained a diversity of subsidiaries in other countries.

[2] Drake (1994) discusses the 'ancien regime' of joint service provision in more detail.

[3] Noam and Kramer (1994) discuss the various emerging strategies. In this paper we use the term globalization to describe the expansion of a telecommunications operator beyond national boundaries in the form of foreign direct investment, the formation of alliances or joint ventures.

foreign countries[4]. This development is both a result and a driver of the global regulatory and technological changes in the telecommunications sector. In spite of this industry trend, the existing regulatory regimes are still predominantly of a national nature and exhibit a great variety of differing approaches, creating complex conditions for the internationalizing industry.

This paper investigates the globalization strategies of telecommunications operators under such asymmetric regulatory conditions. We are interested in the motives for the present internationalization of telecommunications operators, the incentives created by the heterogenous regulatory framework for the investment and internal pricing decisions of telecommunications operators and the implications of these developments for regulatory policy. For this purpose, we take the existing institutional and regulatory framework of telecommunications as our point of departure. The next section briefly reviews this framework of telecommunications as well as the emerging globalization strategies of service providers. Section three of the paper analyzes optimal strategies of globalization under asymmetric regulatory conditions. Section four discusses welfare consequences and possible implications for future regulatory approaches.

2. *The regulatory framework and emerging globalization strategies of telecommunications operators*

2.1. *The regulatory framework*

The coincidence of technological, economic, and political change has transformed the regulatory framework of telecommunications considerably. On a global scale, common trends of re-regulation are visible. Most countries have undertaken measures to liberalize the customer premises equipment market subject to minimal required standards. Similar liberalization strategies are observable in the value added networks and services segment, and the area of private networks (e.g., based on VSATs). Traditionally closed national telecommunications procurement policies are increasingly replaced by international bidding mechanisms. Price levels and structures for telecommu-

---

[4] Reliable data on foreign direct investment of telecommunications operators is difficult to generate. According to our calculations, the yearly figures for foreign direct investment of telecommunications operators in the OECD countries vary substantially between 2% and 8% of total capital investment. Two marked developments are visible: on one hand, a movement of telecommunications operators to cross-penetrate markets within the OECD area and, on the other hand, a flow of investment from the OECD countries to developing countries as well as former Communist countries in transition.

nications services are gradually adjusted to reflect cost (although considerable differences in the degree of cost-orientation remain between countries). A growing number of countries separate regulatory and operational functions as a safeguard against anticompetitive behavior in a more liberal market environment. In all, the concept of the public network, either in its PTT-version or the regulatory model of the U.S., is undergoing a fundamental transition.

It would be a mistake, however, to assume a full convergence of telecommunications regulation to a unitary new framework. Indeed, the common elements of re-regulation are equally met by persistent and new differences in the national and regional approaches. Such differences continue to exist in the extent of monopoly privileges, provisions limiting service diversification, as well as the specific regulatory instruments employed.

Over the past two decades, the concept of natural monopoly as an analytical tool to shape regulatory policy in telecommunications has gradually been abandoned or at least critically re-examined by policy-makers. Developments within the telecommunications industry, theoretical developments such as the theory of contestable markets, problems with implementing regulation in a technologically dynamic multi-product industry, and a general distrust in monopoly privileges have jointly contributed to this development (Bauer and Steinfield, 1994). Very few countries, however, have entirely eliminated monopoly privileges in the industry and the emerging boundary lines between liberalized and monopolistic segments of the industry differ in a cross-national comparison. For example, a majority of U.S. states still consider the local exchange (more precisely, local access and transport areas or LATAs) as an area in which effective competition in basic services would be impeded by the 'bottleneck power' of the local exchange carriers [5]. The European Community (EC) currently establishes a model that distinguishes between 'reserved' and 'non-reserved' (or 'competitive') segments of telecommunications. While the EC has designed an overall blueprint for a liberalization of the formerly rather closed European telecommunications markets in its Green Paper on telecommunications, it left it to the discretion of its member states to maintain monopoly privileges for the provision of the telecommunications (network) infrastructure and reserved services [6]. Yet another solution was adopted by Japan. In its telecommunications reform of the mid-1980's, facilities-based

---

[5] The monopoly position of local exchange carriers is not based in a legal but rather in a de facto monopoly. Liberalized market access conditions exist in non-basic areas such as cellular services. An increasing number of state regulatory commissions is embarking on a path of gradual deregulation (Ralls et al., 1991).

[6] In general, only voice telephony is considered such a reserved service. Case-by-case exceptions in the areas of packet switched data transmission and telex are possible. Bauer and Steinfield (1994) provide a more detailed analysis.

(type I) and non facilities-based (type II) carriers were differentiated and are subject to differing regulatory stringency.

This variation in the delineation between protected and liberalized market segments is paralleled by differing approaches to the vertical and horizontal diversification of telecommunications operators. Until 1991, the Regional Bell Operating Companies (RBOCs) in the U.S. were barred from offering a variety of information services by the Modified Final Judgement of 1982 that ended the antitrust suit against AT&T and led to the divestiture of the Bell system [7]. Other restrictions on the vertical diversification of RBOCs (eg, into the manufacturing of telecommunications equipment) are still intact. European telecommunications operators in general do not face such restrictions (although not all are vertically and horizontally diversified) [8]. The Japanese regulatory environment has established some restrictions to horizontal diversification. For instance, presently Nippon Telegraph and Telephone (NTT) is by statute prohibited from offering international services. Regulatory differences also exist with respect to the cross-provision of audio and video services by telephone companies as well as the provision of telecommunications services by cable and broadcasting companies (OECD, 1992). Finally, the ability of telecommunications operators to diversify internationally is not only impeded by restrictive market entry regulations but also various ownership restrictions imposed by the home or host countries.

In addition to these disparities, national regulatory authorities employ a multitude of regulatory instruments with quite differing degrees of regulatory intensity and effectiveness. Countries like the U.S. or the U.K. have introduced price-cap regulation to monitor the price strategies of the dominant telecommunications providers but employ differing parameters to account for inflation and estimated productivity gains. Germany, France, or Japan follow more overall approaches of cost control without the establishment of clear and transparent regulatory mechanisms. Similar international differences are visible in the priority given to antitrust supervision and control. The situation becomes even more complex in developing countries and in the former communist countries in transition. Most of these countries have no institutional framework nor tradition to cope with problems of market power and to provide safeguards against unfair competition [9]. As a result of those dif-

[7] Currently, this lifting of the line-of-business restrictions is in appeal and Congress has drafted legislation (House Bill H.R. 3515 and Senate Bill S.2112) to re-impose such limits.

[8] Some exceptions to this general rule exist. For instance, the Italian telecommunications system is segmented along different lines of business with Italcable responsible for intercontinental services, SIP for national services, ASST for long distance and international services etc. (Noam, 1992).

[9] In fact, to succeed in the privatization of telecommunications operators and to attract foreign

ferences in market entry and conduct regulation, international differentials in the explicitly or implicitly allowed rate of return on the invested capital emerge.

2.2. Globalization strategies of telecommunications operators

These specific regulatory and policy conditions in the home and host countries together with economic and technological developments set important parameters for the strategies of globalization of the telecommunications industry. They determine to a large extent the opportunities to globalize and the set of feasible globalization strategies. Globalization takes place in two interrelated basic forms, as an expansion of the provision of existing services into new geographic markets and as a process of diversification into new lines of business. These strategies are facilitated by the pattern of technological change in telecommunications which generates new options for sequential entry strategies into the provision of telecommunications services.

For instance, technological developments have blurred the boundaries between terminal equipment and switching equipment. Traffic control intelligence can be located at different points of the network, including customer premises equipment. An entry strategy into the provision of telecommunications services, therefore, might be pursued through the equipment market. Similarly, satellite-based and mobile services have opened a variety of new specialized market entry opportunities that can be expanded into the provision of more traditional telecommunications services. Market entry can also be pursued through the construction of dedicated facilities such as private networks or specialized transmission facilities such as teleports. Last, but not least, access to the services market can be accomplished through the lease of transmission capacity from the incumbent providers.

Major companies in the telecommunications equipment industry, such as AT&T (Western Electric), Alcatel (former ITT), Siemens, Philips, or L.M. Ericsson have operated at an international level for a considerable time. Bohlin and Granstrand (1991) estimate the degree of internationalization from about 5% for AT&T to almost 95% for Philips. By the same token, the loss of a protected home market for many manufacturers has resulted in a global wave of mergers, acquisitions, joint ventures and alliances. These include the cooperation between AT&T, Philips and STET, the fusion of CGE and ITT Europe into Alcatel NV; the merger of GEC/Siemens and Plessey, the cooperation between Fujitsu and GE, the purchase of Rolm by

capital, some of these countries deliberately create (or are forced to create) weak regulatory frameworks. See the discussion in Nulty (1991) for the situation in Central and Eastern Europe. Similar criticisms were made for the Latin American and British privatization policies.

Siemens, or the partial takeover of STC by Northern Telecom. Several of the large manufacturers have recently attempted to diversify into the provision of (specialized) telecommunications services. Ericsson, Alcatel, and Siemens provide VANS and Motorola is part of one of the U.K. consortia to build up personal communications services. By pursuing such strategies of vertical forward integration into niche markets, manufacturers and service providers build future entry opportunities for other, related service markets.

Unlike the manufacturers of telecommunications equipment, the telecommunications service providers were until recently basically organized at the national level. During the past decade, major telecommunications operators have launched strategies of internationalization and diversification, gradually developing into global operators. British Telecom (BT) has made inroads into foreign markets through the acquisition of Tymnet in 1989 or the offering of network integration services through its subsidiary Syncordia. France Télécom (through its subsidiary France Cable and Radio) has invested into telephone companies in Mexico and Argentina and is involved in several joint ventures in Eastern Europe. The American Regional Bell Operating Companies have acquired major shareholdings in telephone companies in Latin America (Mexico) and the Pacific region (New Zealand, Malaysia). They are engaged in a growing number of joint ventures in cellular communications concepts (Latin America, Central, Eastern, and Western Europe), the provision of international services (Eastern Europe), and in various cable franchises in the U.K. Long distance carriers such as AT&T or MCI have adopted strategies of internationalization through foreign direct investment and alliances. NTT, legally prohibited to provide international services, nevertheless has opened local offices in the U.S. and Europe. Carriers across national borders and continents engage in alliances. For instance, many PTTs and MCI cooperate in Infonet to provide worldwide value added services; Televerket of Sweden and PTT Telecom Nederland provide long distance services in the Republic of Ukraine (Noam and Kramer, 1994).

In all, various strategies of globalization and diversification emerged over the past decade, including foreign direct investment into existing or new facilities, joint ventures and strategic alliances, as well as strategies to directly sell telecommunications related services like consulting services.

3. Asymmetric regulation and globalization behavior

Telecommunications operators globalize and diversify their activities for several reasons. Among others, globalization can be a corporate response to changing demand patterns of telecommunications users, an attempt to increase profitability through the expansion in markets that yield high returns on investment, or it can be motivated as a way to escape the restrictions of national regulation. In general, all these factors will play a role in the particular strategy of telecommunications operators. In this section we develop a stylized model of globalization behavior that enables us to study the impact of asymmetric regulation on the behavior of profit maximizing firms.

3.1. A stylized model of globalization

Whereas the process of internationalization is limited to the export of goods and services from one country, globalization involves operations in several countries. In addition to its external transactions with suppliers and customers a multinational firm typically engages in vertical and horizontal intra-firm transactions across national boundaries. For the purposes of our analysis we assume that the multinational firm attempts to maximize its *global net profits* [10].

In addition to variables such as the price and quality strategies, the globalized firm possesses two additional, interrelated sets of choice variables. These are the location, level, and structure of investment as well as the pricing strategies for transactions between the different locations. In making these decisions, a profit maximizing firm has to take into account existing differences in regulatory policy, but also tariff barriers, differences in corporate tax rates, as well as the specific risks of international activities such as exchange rate fluctuations or 'cultural' risks of operating in less familiar business environments. Hence, while regulatory differences matter for the decisions of the firm, they have to be analyzed in a more general context.

In our simplified example we assume that firm 1 is the parent firm located in home country 1, and firms 2 and 3 are subsidiaries located in host countries 2 and 3 (Eden, 1985; Horst, 1971; Copithorne, 1971). Firm 3 produces inputs (goods or services) for sale to the two other firms (vertically integrated internal trade). Firms 1 and 2 sell an identical or at least overlapping bundle of goods and services in different locations and may engage in intra-firm horizontal trade. In the case of telecommunications operators a vertical trade relationship can be the supply of switching, transmission or terminal equipment by firm 3. This subsidiary could also provide information like R&D results or consulting expertise to the other firms. Due to the character of telecommunications as a service which is frequently dependent on local facilities, pure horizontal trade is at present a less frequent phenomenon.

---

[10] Other possible motives, such as establishing a position in markets with insecure but possibly high future growth opportunities (as in the case of Central and Eastern European telecommunications) or 'experimental' investments such as the stakes of the American RBOCs in British cable systems can usually be linked to the goal of profit maximization.

However, with the advancement of technology it is possible that unbundled telecommunications service elements such as switching or traffic control are provided in form of horizontal trade between subsidiaries. A similar case is the provision of database and information services for simple resale by an affiliated firm.

Given the present structures of international telecommunications we are able to make use of some assumptions that greatly simplify our analysis. First, it seems justified to assume that firms 2 and 3 possess some price flexibility (in other words, they do not operate in perfectly competitive markets)[11]. However, this flexibility might be reduced or even eliminated by regulatory oversight. In addition, we assume that the firms are able to price discriminate between their markets. For the sake of simplicity, we assume that effective regulatory oversight (if it exists at all) imposes an upper boundary on the prices the firms can charge and, hence, has effects similar to a tax on the unregulated enterprise. The global net profits of the firm are the sums of the profits of the individual firms (a more formal derivation of the conditions for a global profit maximum and their major implications for our discussion is presented in Appendix I below).

The decision of the firm can be envisioned as an interrelated two-stage process. In a first step, the production capacity for each location is determined and in a second step optimal policies to allocate production between those locations as well as policies for the internal transfer prices of transactions are implemented. Both decisions are dependent on the specific market and production conditions (represented by the cost and revenue functions in our model), the regulatory and tax conditions (in our model described by the factor T, the proportion of profit left for the firm after the effects of regulation and taxes are taken into account), the exchange rates between the different countries involved, tariffs levied on transactions between locations in different countries, as well as the possible transfer prices between different locations.

This model can be modified to accommodate a number of different situations (eg, only horizontal or only vertical trade; no regulatory or tax asymmetries; and a case without tariffs). By totally differentiating the first order conditions, important comparative static results on the impact of marginal changes in the exogenous variables can be derived. For our problem particularly interesting are the effects of asymmetric regulatory policies on the price and output decisions of the multinational telecommunications operator. In the absence of regulatory entry barriers (and other things equal) a relaxation

[11] We do not consider the limit case of a natural monopoly with marginal cost declining over the entire relevant output range. In this case it would be optimal to either produce all output in one country or the demand of each country in one organization.

of the regulatory conditions in a host country (or a tightening of regulatory control in the home country) will provide incentives for a profit maximizing telecommunications operators to increase investment in the country with more relaxed regulatory oversight and vice versa. In the case of regulatory entry barriers the globalization may take the form of alliances or joint ventures to accomplish a similar effect. Likewise, regulatory differences between countries provide incentives for a globalized telecommunications operator to use internal transfer prices to shift profits to jurisdictions with looser regulation [12]. For instance, let us assume a situation in which the combined effect of regulation and taxation in the host countries is tighter than in the home country ($T_1 > T_2$ and $T_1 > T_3$). A relaxation of regulation in the host countries would result in the expected shift of production to the country with the now improved profitability conditions and vice versa [13]. An increase in the effectiveness of regulation in the home country would induce a relocation of production to other locations. In the case of changes in one or more of the exogenous variables the overall net effect is dependent on the specific circumstances. For instance, the production relocation effect of relaxed regulatory conditions in the host country can be (more than) offset by a reduction in tariffs levied on exports from the home to the host country.

3.2. Diverse national regulation and foreign direct investment

Whereas our stylized model allows us to derive some important conclusions as to the behavior of profit-maximizing globalized telecommunications providers, the model needs to be augmented to take account of some real world characteristics. For instance, to simplify matters, the model does not directly take account of the risks associated with foreign direct investment. Depending on the specific circumstances, foreign investment may carry a higher or a lower degree of risk than domestic investment. The risk of a foreign investment project depends on the specific market situation in the host country, the prospective course of future economic development in the host country, exchange rate developments, as well as cultural and political risks emanating from uncertainty related to different business practices or institutional frameworks (Terpstra and David, 1991). The overall risk of international operations, very much like a diversification of activities at the

---

[12] From the model analysis it also becomes clear that under conditions of asymmetric regulation the multinational enterprise chooses a transfer price for intra-firm trade that is either above or below the efficient shadow price, depending on the existing differential between the effective regulation, taxation and the existing tariffs.

[13] The existence of sunk costs does not change this general result. However, the relationship between the relevant variables would become discontinuous.

national level, not only depends on the risk of each individual project but also on the covariance of risks between different activities. Indeed, an international expansion of activities can contribute to the reduction of the overall entrepreneurial risks, if the developments of major market parameters in different host countries are not perfectly correlated. In this case an international portfolio of investments can reduce the risk of shareholders and at the same time possibly stabilize or even increase the rate of return on the invested capital.

One important determinant of the risk of direct foreign telecommunications investment is the market structure in the host country. Telecommunications operators have diversified into several sub-markets of the telecommunications sector including markets with intense competition but also markets with a relatively low level of rivalry. To the first category belong the 'upper end' services such as specialized data communications services for high volume users. To some extent the bidding processes for franchises for the provision of cellular services (Central and Eastern Europe, Central and Latin America), international switching (Russia), or cable TV services show relatively intense competition but after the award of the franchises most companies operate in monopoly or duopoly market environments. Frequently, no effective regulatory oversight is established. A similar argument holds for investment in foreign telephone companies like in the cases of Argentina, Chile, Mexico, New Zealand, or Australia. At least in the former cases, the government, in an attempt to attract foreign capital did not implement effective regulatory control, thus putting the new shareholders in a rather strong, uncontested market position.

Some investment projects are undertaken in areas, whose future economic prospects are at best uncertain. Upon demand of the local governments but perhaps also as a risk-reducing strategy, telecommunications operators often choose the organizational set-up of a joint venture. This particular strategy probably can reduce the overall risk of a foreign engagement. From the perspective of the host country, the form of a joint venture is attractive for several reasons. Ideally, it provides the local partner access to foreign technology and participation in the revenues of the venture. It also provides local governments and investors a remainder of influence and discretion about the development of the telecommunications infrastructure. At present and given the anticipated future growth in the targeted foreign markets, most of the present globalization projects seem to fall into the category of medium to low risk endeavors.

3.3. National regulation and intra-firm transactions

The above theoretical analysis demonstrated that international asymmetries in regulation, tax policy, and tariffs create incentives for a multinational firm to chose profit-maximizing internal transfer prices that deviate from the free trade shadow price. This way, the firm can 'relocate' earnings into areas allowing a higher net profit on revenues. The direction of such transfers depends on the differential between the combined tax/regulation rate and the tariff rate between countries. Equation (5) in Appendix I shows that a profit-maximizing firm will attempt to set transfer prices above marginal cost (of the exporting firm) if the net return to the exporter T_i is higher than $T_j(1 + r_{ij})$, the net return of the importer per dollar of profits of imports x_{ij} or vice versa. The optimal direction of profit transfers, therefore, may be either way.

Our analysis shows that due to the particularities of telecommunications as a services sector (which somewhat limits the ability of multinational providers to trade horizontally between subsidiaries) vertically integrated firms may enjoy strategic advantages to engage in profit-shifting transfer pricing. The extent to which firms engage in profit-shifting transfer pricing may also depend on differences in the performance level between home and host countries. Technologically more sophisticated firms may possess a significant cost advantage over incumbent suppliers. This may even lead to the perverse effect that a firm in a host country with a lower performance level earns part of its profits in a 'covert' (ie undeclared and undetected) way and retransfers those profits into the home country (or another host country with a higher performance level) to cross-subsidize activities there.

The incentive to engage in foreign direct investment and transfer pricing may depend to some extent on the specific ownership structure of firms. For instance, it may not be justified to assume that public telecommunications operators, especially if they are organized as a public *administration* are acting as pure profit maximizers. Public enterprises may face a limited incentive to achieve profits if the management cannot fully dispose of those profits (as is frequently the case when profits are used to supplement the general public sector budget). This way, the fact of public ownership may act as some safeguard against the unregulated expansion of enterprises with strong monopolistic positions in their home markets.

4. Implications for telecommunications regulation

From a regulatory perspective the welfare implications of the existing asymmetric approaches to telecommunications regulation are of crucial importance. Unfortunately, due to the complex factors involved it is difficult to derive general results. As the above analysis has indicated, globalization under conditions of asymmetric regulation can lead to both welfare improvements as well as welfare deteriorations. Overall, it seems easier, to evaluate welfare consequences at the national level than at the global level. In the latter case, intricate problems of welfare comparisons might arise if the welfare improvement, say, in the host country, is accompanied by a welfare reduction in the home country. Moreover, in the medium and long run regulation itself becomes endogenous and dependent on the globalization strategies of companies.

Given these caveats, a few possible problems and available regulatory safeguards deserve mentioning. A deterioration of welfare might result if companies operating in geographically separated markets and under different forms of regulation are able to shift risks from markets with a higher risk to markets with a lower risk (and, perhaps, from shareholders to customers in captive markets) [14]. Whether or not this is possible depends largely on the specific legal and regulatory framework of the operations. A necessary precondition for risk shifting activities is some form of financial relationship between the operations in the protected and operations in non-protected markets (or lines of business). This relationship can be established as a holding company structure but also be managed through vertical or horizontal intra-firm transactions. A clear-cut institutional safeguard against risk-shifting is the requirement to set-up an independent separate subsidiary without any financial or other transactions between the different firms. If such transactions take place, regulatory supervision could avoid inefficient transactions. However, such a requirement might reduce or eliminate economic advantages of international joint production or the transfer of know-how.

The presently existing national and international regulatory approaches are not providing explicit safeguards to avoid those problems. This is, for example, evident in the case of European telecommunications operators such as Telefónica of Spain, France Télécom, or German Telekom, all of which are only subject to relatively weak forms of overall price oversight. In addition, none of these companies faces independent subsidiary requirements for their internationally diversified operations (nevertheless, in most cases subsidiaries, although not financially separated, are established). The holding company structure of the U.S. RBOCs also provides only weak safeguards against such international risk-shifting activities although, in principle, the regulatory framework would be capable of addressing the issue.

---

[14] It should be noted that a firm subject to government regulation is not operating in a risk free market. Mistaken regulation can impose risks on regulated firms that are unknown to unregulated firms.

The discussion of multinational firms engaging in transfer pricing is somewhat divided between arguments that profit-maximizing transfer pricing is enhancing global efficiency by circumventing the 'market imperfections' of tax differentials, tariffs or regulatory divergence and more nationally oriented analyses which identify transfer pricing as one obstacle in maximizing national welfare (Rugman and Eden, 1985). Asymmetries in the regulatory approaches add considerable complexity to these issues. This is especially the case if regulatory policy is established to control market power (natural monopoly, natural oligopoly) of telecommunications operators at the national level (Sharkey, 1987). Globalization provides a new set of strategies for regulated companies to evade even efficient national regulation and to press for regulatory change (Noam and Pogorel, 1994).

Matters are complicated by the fact that countries are at different stages of telecommunications infrastructure build-up. Especially countries with poorly developed telecommunication systems still follow a policy of cross-subsidization to accomplish the necessary investment into universal service infrastructures. In general, the resulting price structures are vulnerable to market entry even by competitors that are even less efficient than the incumbent supplier (ie the market structure is unsustainable). Internationally expanding and diversifying firms may, therefore, have plenty of opportunity for cream-skimming activities, leaving the overall welfare consequences ambiguous [15].

In theory, regulation is able to control inefficient international activities of telecommunications operators. One relatively simple safeguard would be a global separate subsidiary requirement. As was argued above, such a requirement may prohibit the realization of desirable economies of joint production. An alternative to this strict requirement is some form of international regulatory cooperation to detect inefficient forms of globalization. Such an approach demands a common understanding of efficiency-enhancing versus efficiency-reducing policies between the countries involved. Since efficiency gains in one country may be accompanied be reductions in efficiency in others, this condition may not be fulfilled. Moreover, complex information problems need to be overcome. The regulatory task would be easier, if unregulated, arm's length prices could be observed (which will not be the case often). A final (but improbable) alternative is the establishment of a unified international regulatory regime. Some signs, such as the international trade in service negotiations or regulatory tendencies at the level of individual nations and regions, point into this direction. However, if a symmetric global regulatory environment will be

[15] A somewhat related issue is whether firms operating in protected markets should be allowed to expand into foreign markets or whether they should have a social obligation to primarily upgrade the infrastructure within their protected markets.

achieved at all, it can be expected that a significant transition phase will have to be passed.

A last point deserves brief discussion. The existing differences in the regulation of telecommunications may have the beneficial aspect that they provide an opportunity for both companies as well as society in general to 'experiment' and, therefore, to learn about the advantages and disadvantages of different approaches. From this evolutionary perspective, a continuing diversity of approaches may be desirable even at the expense of possible welfare losses in the short run. The crucial point in a welfare evaluation is whether an optimal degree of diversity can be determined.

5. Conclusions

In this paper we established in a simplified framework how the existing international asymmetries in market entry, price, and conduct regulation influence the globalization decisions of telecommunications operators. These asymmetries may distort the incentives for international diversification and create incentives to engage in international transfer pricing and profit-shifting. On the other hand, these asymmetries also pose obstacles to efficiency increasing internationalization and diversification strategies. Existing entry barriers into the provision of telecommunications services, might prohibit strategies to utilize economies of (international) joint production or limit market access for more efficient suppliers. The effects of globalization strategies under such asymmetric regulatory conditions need to be assessed on a case by case basis but may have both welfare-increasing as well as welfare-decreasing consequences.

The level of international diversification and foreign direct investment of telecommunications operators is still relatively limited although a steady increase may be expected throughout the coming years. For the telecommunications research and policy-making community a number of important issues need to be addressed in this context. Once more observations are available, the issues raised in this paper will be accessible for more detailed empirical study and provide more precise quantitative information of the overall effect of internationalization on the profitability, efficiency, and performance of telecommunications suppliers. There is a clear need to assess the global and local welfare effects emerging from the globalization of telecommunications operators. Last, but not least, more research as to the proper international regulatory framework which facilitates efficiency enhancing globalization while, at the same time, reduces the probability of efficiency-reducing distortions, will be an important task.

Appendix I

In our model we assume that the telecommunications operator aims at a maximization of its global net profits

$$\max \Pi = \sum_{i}^{n} \Pi_i \tag{1}$$

with Π = global net profits, Π_i = net profits of firm i in country i, n = number of firms (parent company and subsidiaries). Besides traditional variables such as price and quality policies, for a globalized firm the location (and capacity) of investment in different countries as well as the pricing strategy for intra-firm transactions become important choice variables in the pursuit of its profit maximizing strategy. As Eden (1985) has shown (although in a context that does not discuss regulatory asymmetries explicitly), once the locations of operation and the production capacity in each location are determined, the global profit relation of the multinational enterprise in our three firm example (expressed in the home country's currency) is determined by the following firm-specific net profit functions

$$\Pi_1 = T_1 [R_1 - C_1 - e_3(1 + r_{31}) p_{31} x_{31} + p_{12} x_{12}] \tag{2}$$

$$\Pi_2 = e_2 T_2 \left[R_2 - C_2 - \left(\frac{e_3}{e_2} \right) (1 + r_{32})(p_{32} x_{32} - (1 + r_{12}) \frac{p_{12} x_{12}}{e_2} \right] \tag{3}$$

$$\Pi_3 = e_3 T_3 [p_{31} x_{31} + p_{32} x_{32} - C_3] \tag{4}$$

with R_i = total revenues from domestic sales of firm i, C_i = total cost of producing q_i, q_i = total output produced by firm i, T_i proportion of profit left for firm i after taking account of the effects of regulation and taxes, e_i = exchange rate of country i in terms of the home country's currency, x_{ij} = volume of exports from firm i to firm j, p_{ij} = profit-maximizing transfer price firm i charges firm j for one unit of x_{ij}, r_{ij} = ad valorem tariff rate levied by country j on x_{ij}.

Eden (1985) has also derived the first order conditions for the global net profit maximum and the profit-maximizing transfer price for the most general case of horizontal and vertical trade which are described by

$$\frac{\partial \Pi}{\partial p_{ij}} = \left[T_i - T_j(1 + r_{ij}) \right] e_i x_{ij} \gtreqless 0 \quad \text{as} \quad T_i \gtreqless T_j(1 + r_{ij}) \tag{5}$$

$$\frac{\partial \Pi}{\partial q_1} = [e_3 T_3 (p_{31} - MC_3)] + [T_1(MR_1 - MC_1 - (1 + r_{31}) e_3 P_{31})] = 0 \tag{6}$$

$$\frac{\partial \Pi}{\partial q_2} = [e_3 T_3(p_{32} - MC_3)] + [e_2 T_2(MR_2 - MC_2 - \left(\frac{e_3}{e_2}\right)(1 + r_{32})p_{32}] = 0 \quad (7)$$

$$\frac{\partial \Pi}{\partial x_{12}} = [T_1(p_{12} - MR_1)] + [T_2(e_2 MR_2 - (1 + r_{12})p_{12})] = 0 \quad (8)$$

with MR_i = marginal revenue of firm i, MC_i = marginal cost of firm i. The first square brackets in equations (6), (7), and (8) show the marginal profit of the exporter (firm i), while the second square brackets express the marginal profit of the importer firm j in the market for imports x_{ij}.

By total differentiation of the first order conditions the comparative static effects of small changes in the variables on the decisions of the firms can be analyzed. For our paper most important are the following two conclusions. First, the firm will react to any change in regulatory policy as compared to the status quo ante with a rebalancing of its investment portfolio in favor of the less restrictive regulatory jurisdiction. Second, it will set internal transfer prices so as to 'shift' profits to the less restrictive regulatory jurisdiction.

References

Bauer, J.M. and Steinfield, C., 1994. Telecommunications Initiatives of the European Communities. In: C. Steinfield, J.M. Bauer and L. Caby (eds.), *Telecommunications in Transition*, pp. 51–70 (Sage, Thousand Oaks).

Bohlin, E. and Granstrand, O., 1991. Strategic Options for National Monopolies in Transition: The Case of Swedish Telecom. *Telecommunications Policy* 15, 464.

Copithorne, L.W., 1971. International Corporate Transfer Prices and Government Policy. *Canadian Journal of Economics* 3, 324–341.

Drake, W.J., 1994. Telecommunications Standards in the New European Context. In: C. Steinfield, J.M. Bauer and L. Caby (eds.), *Telecommunications in Transition*, pp. 71–96 (Sage, Thousand Oaks).

Eden, L., 1985. The Microeconomics of Transfer Pricing. In: A.M. Rugman and L. Eden (eds.), *Multinationals and Transfer Pricing*, pp. 13–46 (St. Martin's Press, New York).

Horst, T., 1971. The Theory of the Multinational Firm: Optimal Behavior under Different Tariff and Tax Rates. *Journal of Political Economy* 79, 1059–1072.

Noam, E.M., 1992. *Telecommunications in Europe* (Oxford University Press, New York).

Noam, E.M. and Kramer, R., 1994. Telecommunications Strategies in the Developed World: A Hundred Flowers Blooming or Old Wine in New Bottles? In: C. Steinfield, J.M. Bauer and L. Caby (eds.), *Telecommunications in Transition*, pp. 272–286 (Sage, Thousand Oaks).

Noam, E.M. and Pogorel, G. (eds.), 1994. *Asymmetric Deregulation: The Dynamics of Telecommunications Policies in Europe and the United States* (Ablex, Norwood, NJ).

Nulty, T.E., 1991. Using Privatization as a Development Tool in Telecommunications — the Central and East European Situation. *Economic Symposium, 6th World Telecom Forum* (ITU, Geneva).

OECD, 1992. *Convergence Between Communications Technologies: A Policy Review. Information Computer Communications Policy 29* (OECD, Paris).

Ralls, W.R., Muth, T.A., Rosier, L.R. and Edwards, B.H., 1991. *State Telecommunication Deregulation in the United States*, 1991 Annual Report (Public Communication Associates, Lansing, MI).

Rugman, A.M. and Eden, L. (eds.), 1985. *Multinationals and Transfer Pricing* (St. Martin's Press, New York).

Sharkey, W.W., 1987. Oligopoly in International Telecommunications. *Center for Telecommunications and Information Studies, Working Paper* No. 228 (Columbia University, New York, NY).

Sherman, R., 1989. Efficiency Aspects of Diversification by Public Utilities. In: M.A. Crew (ed.), *Deregulation and Diversification of Utilities*, pp. 43–63 (Kluwer Academic Publishers, Boston, MA).

Terpstra, V. and David, K., 1991. *The Cultural Environment of International Business* (South-Western Publishing, Cincinnati, OH).

Global Telecommunications Strategies
and Technological Changes
Edited by G. Pogorel

Chapter 16

New strategies for changing structures: the complex nature of Eastern European telecommunication systems

Alfred L. Thimm

University of Vermont, Burlington, VT, USA

1. Introduction

The revolutionary technological changes in global telecommunications have provided an excellent opportunity to observe Schumpeter's 'creative destruction' process at work. Aggressive entrepreneurial telecom enterprises emerged in Europe, Japan and North America to destroy the cozy, monopolistic national equilibria. The highly competitive global telecom market that emerged has now been challenged by the sudden collapse of the Soviet empire, and the opportunities created by the pressing need to provide a new telecom infrastructure for the liberated nations of Eastern Europe and the Soviet Union.

This essay will explore the strategies of global telecommunication enterprises to penetrate emerging markets under very special circumstances; we shall also examine the role of government agencies to channel Western investments and technology into the modernization of its infrastructure.

The enormous transfer of investment capital from West Germany into the former German Democratic Republic, as well as the public discussion associated with it, has given the transformation of the East German telecommunication system a transparency that has been lacking in other East European countries; moreover, the relatively quick progress already made gives us an opportunity to draw a lesson from the East German experience that can be applied to other East European countries and to the former Soviet Union. We shall, therefore, emphasize the East German experience as a quasi-laboratory for judging enterprise strategies and national industrial policies, but shall also refer to related events in Czechoslovakia, Hungary and, if appropriate, Poland and the former Soviet Union. These countries cover the entire range of adaptive strategies to modernize their networks and to attract foreign investments.

2. *The creative destruction of Eastern Europe's telecommunication system*

"Eastern Europe foils all but the hardiest of western investors"
(*New York Times*, March 5, 1992, p. 1)

"Joint Ventures in (Eastern Europe) face unexpected roadblocks"
(*Wall Street Journal*, March 5, 1992, p. B2)

The overthrow of the Stalinist regimes in Eastern Europe (EaEu) had created enormous interest in the former East bloc among Western managers and entrepreneurs. The struggling economies of the post-Stalinist societies seemed not only to promise new markets for the products and skills of Western enterprises, Eastern Europe appeared also to provide a romantic opportunity to combine business activities with the heroic task of aiding the fledgling democracies, abandoned at Yalta, to convert from a communist to a market economy. The abysmal state of EaEu's infrastructure, roads, rails, and especially the telecommunication system, did indeed require immediate heavy investments in transportation and, above all, communication to provide a minimum base for further economic development. Telecommunication did turn out to be the crucial infrastructure deficiency. On one hand, Eastern European and Soviet telecommunication had relied on an obsolete, poorly maintained technology that was decades behind Western standards, on the other hand, telecommunication emerged as a strategic or leading industry that defines and determines the niveau of a society's technological sophistication and economic performance. The actual Western investment necessary to meet the immediate needs of the disintegrating telecom systems met, however, financial, political and technical obstacles not realized in the initial euphoria.

In spite of historical differences and an uneven pace of progress from communist to market economy, by late 1992 the Eastern European telecommunication systems still have the following common characteristics:

(1) Technological obsolescence, made worse by bad maintenance

(2) Low density usage (penetration rates vary from 7 per 100 people in Hungary to 15 per 100 in Yugoslavia).

(3) Long waiting periods for new telephones and repairs

(4) A PTT monopoly tradition reaching back to the pre-WWI period, offset in part by an eagerness to become part of Europe and hence a willingness

to accept the EC's telecommunication liberalization policy as formulated in the June 1987 Greenpaper, *Toward a Dynamic European Economy*.

(5) Telecom (PTT) administrations still managed by old apparatchiks, operating in societies where second level political and key economic positions are still being held by former communist party members. Forty years of Stalinism can not be erased over night, especially since the totalitarian nature of the old regime placed the party cadres in every strategic economic, political, and cultural position.

The modernization of an obsolete telecommunication system has technical, objective criteria that must be met and hence is more easily accomplished, or at least visualized, than the more complex transformation of a communist society into a market economy. It is not surprising, therefore, that the modernization strategies of Eastern European PTTs have precise common goals that include the following objectives:

(1) Immediate improvements in network technology

(2) Reduction in waiting lists for telephones, and sharp increases in the penetration rates that are hoped to reach average EC telephone per population figures by 1995 or at least 1997.

(3) Infusion of capital and encouragement of technology transfer through joint ventures with Western telecommunication companies (telcos).

(4) Adaption of standards and telecom administrations policies to EC directives and CEPT (Conference of European Post and Telecom administrations) procedures.

(5) Emphasis on promoting mobile telephone systems and satellite communication as relatively inexpensive ways to improve network performance and increase availability of telephones. Joint ventures with Western telecommunication companies have emphasized cellular networks and equipment manufacture.

The common aspects of Eastern European telecom strategy can be summarized by emphasizing the intent of the national telecom administrations to 'harmonize' their regulations and network standards with the emerging, post-1992 European telecommunication system. At the same time it must be recognized, however, that the liberalization and deregulation components of the EC's telecom policy clash with the 40 years of Stalinist culture that had enveloped telecom managers and users. The creation of new structures to support the new industrial strategies is difficult but necessary.[1]

The ability to overcome the Stalinist ballast varies from country to country in Eastern Europe, in accordance with history and temperament. Yet several clear patterns emerge that must be included in Western analyses of business

---

1 For the relationship between organizational structure and strategy see Chandler (1965).

opportunities. The rehabilitation and modernization of the EaEu telecommunication systems provide enormous potential opportunities for Western telcos though the political and socio-economic settling of the post-Stalinist societies create equally enormous obstacles for market penetration by capitalistic enterprises. Though these obstacles have been underestimated consistently, Western telcos' strategies have continued to pursue joint ventures with establish host countries concerns and consortia with other Western enterprises. Not sufficient time has elapsed to arrive at definite conclusions concerning the success and failures of these joint ventures and consortia, but enough evidence can be assembled to arrive at preliminary evaluations of the measures necessary to make joint ventures and consortia succeed in the Eastern European telecommunications sector.

3. *East Germany (formerly DDR)* [2]

The rapid absorption of the DDR (German acronym for German Democratic Republic) into the Federal Republic of Germany that began in November 1989 might be considered to have created special conditions for the transformation of a communist economy into a market economy. Except, possibly, for the speed of the political transformation, however, East Germany's struggle to dismantle the Stalinist apparatus mirrors the efforts to reinvent a market economy throughout Eastern Europe.

The rapid rise in east German wages during 1990, 1991 and 1992 has removed most advantages eastern Germany may have had over its eastern neighbors. As a matter of fact, there is some evidence that, over the period 1992–1993, eastern Germany may encounter greater transformation difficulties than Hungary or Czechoslovakia. [3] During the summer of 1991 30% of all east Germans were still unemployed or working short-time; by winter 1992/93 part-time work had been sharply reduced but unemployment reached 17% and was supposedly headed for 50% — a very unlikely scenario, however. [4] Still, during 1991/92 wages in east Germany have been almost ten times as

---

[2] We shall capitalize East or West Germany when referring to the political entities prior to the October 1990 unification.

[3] See Herlof et al. (1991), Müller (1991), and 'Kohls Solidarpakt eine Scheinlösung', *Wirtschaftswoche*, December 12, 1992, pp. 14–17.

[4] 'Der Wendepunkt in Osten is Erreicht', *Frankfurter Allgemeine Zeitung*, June 20, 1991, p. 15; see also 'What remedy for Ostslump?', *The Economist*, May 11, 1991, p. 65; 'Pact near to avert east German industrial collapse', *Financial Times*, November 24, 1992, p. 3. 'Umsatz boom durch die deutsche Einheit' *Wissenschaftliches Institut für Kommunikationsdienste (WIK) Newsletter*, December, p. 8.

high as in Poland (during July 1990 the difference was only 800%; Akerlof et al., 1991) and are scheduled to reach parity with west Germany by 1994/1995 according to collective bargaining agreements obtained by key unions, such as the powerful I.G. Metall.

A 1991 OECD economic survey of Germany seemed to epitomize the prevailing view that the too rapid rise in east German wages was leading to a sharp decline in its competitiveness and, possibly, to a collapse of the entire economy of the former DDR.[5] A comparison of investment allocations — both foreign and domestic — in east Germany and the other former East Block countries have led this author to conclude, however, that the higher wages in east Germany may in three to five years help east Germany to attain more quickly west German productivity standards, without the retention of obsolete industries kept alive by state subsidies. Though the east German wages are still only 50% to 60% of West Germany's, that wage level can not be supported by even those 'glamorous' production facilities that, by East Block standards, had been the most successful. Most of these factories will be closed, therefore, but will be replaced with modern facilities, as our case studies will show.[6] Czechoslovakia and, especially, Poland on the other hand are still saddled with Stalinist dinosaurs in the steel, agricultural implements, electronics and coal mining industries that can be operated, albeit with subsidies, *because* the low wage load makes their products competitive in the former East Bloc. Since the obsolete capital equipment of these relics precludes privatization or foreign investments, the ministries that used to run these industries continue in some form or other to dispense subsidies and directives to huge mills and factories to maintain employment of the workforce as well as Stalinist managers and bureaucrats. On the basis of current developments this author believes, therefore, that by 1995, east Germany will be on the way to build Europe's most modern economy while Czechoslovakia and Poland will still be in the throes of dismantling both its heavy industry and its remnants of the Communist command economy. Russia and the other successors to the former Soviet Union will, of course, be lagging behind Poland and the CSFR by several decades.

A thorough discussion of the East European transformation, except for telecommunication, is beyond the scope of this paper. We ought to emphasize, however, that the slow rate of privatizing and modernizing the state owned sector, especially the capital goods industry, is only part of the story. What has been overlooked, frequently, is the emergence of a fast growing private sector in east Germany, Hungary, Poland, and, to a lesser degree,

---

[5] 'Germany is warned over wage threat to economy', *Financial Times*, July 26, 1991, p. 1.

[6] See 'Ziemlich Weit', *Wirtschaftswoche*, June 28, 1991, pp. 42–46.

Czechoslovakia. This emerging private sector is most noticeable in the retail and service industries — the black hole of Stalinist economies — but exists also in construction and small scale manufacturing.[7] The growth of the private sector may give some indication of the economic climate that will determine the success of Western joint ventures and greenfield investments in the telecommunication/information technology sector.

In east Germany the rapid rise in wages has illuminated the technical backwardness of even the supposed world-class enterprises in Eastern Europe's most successful command economy; the collapse of the DDR champions has been a strong disincentive to Western investment in eastern Germany. The massive government investment into the former DDR's crumbling infrastructure, especially in the telecommunication areas, has provided, however, an opportunity for Western enterprises. The strategic response of key players in the global telecommunication industry — Siemens, Alcatel/S.E. Lorenz, British Telecom, Philips, various Regional Bell Holding companies (RBHCs), and especially Deutsche Bundespost (DBP) Telekom — gives us not only an unique opportunity to examine their entrepreneurial visions, but also illuminates the obstacles met in the transition period from command to market economy *all* formerly Communist countries experience. The DBP's investment in East Germany's telecommunication infrastructure provides also a model for possible EC and US strategies for government aid and investment in Eastern Europe. The transparency of transaction in East Germany's telecommunication sector creates an excellent opportunity to examine fundamental issues and predict subsequent technological, economic, and political developments in the telecommunication systems of other European countries.

East Germany's role as a preview of coming events in Czechoslovakia, Hungary and Poland applies especially to telecommunication. The telecom administrations of the Eastern/Central European countries have not only 45 years of Stalinism in common, but also a long history of PTT monopolies that preceded the Communist takeover during the immediate post-World War II period. This common history of mercantilistic and Marxist state monopolies presents a formidable obstacle to liberalization, but is, however, confronted by the objective requirements of technological modernization and the subjective desire to adopt (Western) European structures and objectives. This may mean for North American and Western European enterprises that market access agreements on the highest political levels may be fairly easily obtained, but that real, frustrating difficulties will occur on the operational or implementation levels. Before we shall take a closer look at the national telecommunica-

[7] For an excellent description of Stalinist dinosaurs see 'A Tale of Two Factories', *The Economist*, August 3, 1991, p. 51; 'Central Planning in Ukraine', *The Economist* October 16, 1993, p. 59.

tion regimes, we shall examine four mini-cases that illustrate the Kafkaesque peculiarities of the transformation of both the DDR and Eastern Europe. The first case will provide background information, the other three will present telecommunication investment strategies.

3.1. Case 1: The treuhandanstalt (liquidating agency)

Every Eastern European post-communist government wishes to privatize the established state monopolies; in each case the lack of private capital and managerial talent with market experience, and above all the quiet resistance of the old communist cadres in the ministries and factories, have made the restructuring of the economy a slow, frustrating and disappointing process that has passed in each country through almost identical phases. In Poland, the first EaEu country to attempt to transform quickly and completely a centrally planned economy, opportunistic Communist directors of state enterprises imitated American management takeovers and emerged as the new owner-managers of 'privatized' companies or co-owners of joint ventures with frequently shady foreign investors. [8] In order to safeguard the economic interest of the society, the governments of Poland and, subsequently, Czechoslovakia, East Germany and Hungary established agencies to supervise and manage the privatization of state industries. The scrutiny of the West German press and, after unification, the subsequent role of West German officials in the former DDR's privatization process have provided Western observers with a clear picture of the enormous difficulties involved, under the best conditions, in dismantling the Communist apparatus. The performance of the DDR's and now Germany's Treuhandanstalt charged with dismantling of the state owned Kombinate (vertically integrated monopolies) will provide telecommunication managers with a perspective of the difficulties faced in operating in Eastern Europe.

The first free elections in March 1990 gave the DDR a broad anti-communist government that was determined to establish a market economy. In view of the recent Polish experience, the Treuhandanstalt was organized and charged with the careful privatization of the state enterprises. Quite typically, senior bureaucrats in the various economic ministries — each major industry in Eastern Europe was, and in many cases still is, administered by a

[8] Hungary had successfully introduced market oriented economic reforms as early as January 1, 1968, and had attained a relatively successful adaption of a command economy to market requirements. The Hungarian liberalization effort, called Kadarism after the Hungarian leader, 1956–1988, was apolitical and introduced to maintain the rule of the Communist party. Poland's economic reform was part of an economic and political revolution. For a splendid discussion of earlier economic reforms see Selucky (1972).

separate 'ministry' — were posted to the Treuhandanstalt and a well known Communist apparatchik Manfred Flegel was appointed as its first director. Flegel had a reputation as an effective administrator, but was at the same time a loyal Communist who had not even belonged to the reform wing of the party. A 'sweetheart' deal, that permitted the below market value acquisition of the DDR's Interhotel chain by the West German Steigenberger Hotel and Real Estate Group but left the Interhotel chain managers in their jobs, led to the replacement of Flegel by the successful West German Railway executive Rainer-Maria Gohlke. Gohlke arrived in East-Berlin without any trustworthy West German aids and had to recruit his staff from the available bureaucracy. Isolated by his own staff he was unable to carry out a single major privatization, and resigned after 30 days.

The next and subsequently murdered Treuhand boss Detlev K. Rohwedder, previously CEO of the West German multinational steel company Hoesch, had been known for his toughness and had been careful in selecting his aides; his efforts had been helped moreover, by the accelerated unification process. The August 24, 1990 decision by the DDR's parliament to unite with the Federal Republic of Germany (FRG) on October 4th undoubtedly strengthened Rohweller's hand. Only a small number of DDR bureaucrats were to be integrated in the appropriate German ministries, a powerful incentive even for the faithful followers of the old order not to call further attention to themselves by sabotaging privatization offers. Still, even without actively preventing privatization efforts, the Treuhand's bureaucracy's skepticism towards a market economy, coupled with the ingrained habit of Stalinist economies to move up every decision to the highest possible level of the hierarchy, have prevented the closing of unprofitable operations and had generated less than two dozen privatizations by mid-September 1990. [9]

After unification in October 1990, the pace of privatization did increase and by the end of July 1992, 9000 concerns had been privatized, employing 2 millon workers, and generating about DM 200 bn in investment. Most of the large Kombinate, especially in the Chemical industry still remain money losing fossils under Treuhand administration and by December 1992, the German government seemed ready to guarantee, reluctantly, the existence of key enterprises among the 3000 companies that the Treuhand could not sell to private investors. [10] The guarantee is supposed to expire by 1994.

---

[9] See 'Der Spuk Geht Weiter' (The farce continues), *Wirtschaftswoche*, August 31, 1990, pp. 21ff. The article discloses the continued power of old SED (Communist) cadres to 'block and obstruct' the transformation of the economy.

[10] 'East Germans on Short Shifts', *The Wall Street Journal*, August 9, 1991, p. A6. See also, 'One fifth of east German land to be sold', *Financial Times*, September 14, 1991, pp. 21–24; 'Treuhand sells half of state firms', *Financial Times*, March 3, 1992, p. 2.

Mrs. Birgit Breuel, current Treuhand chief and previously deputy to Rohwedder, however, seems to have broken the resistance of the senior civil servants within her own agency and among the top-level managers of the Kombinate; yet privatization efforts are still stymied by lower level bureaucrats in the Treuhand and by management in the Kombinate. This passive resistance has expressed itself by answering the telephone only after the twentieth ring, by responding to written inquiries after a four week waiting period, or by treating interested Western visitors with the arrogance and sarcasm of a 19th century imperial 'Hofrat' (Court counselor). The most pernicious aspect of Treuhand privatization are the so-called 'Seilschaften', a mountaineering term that refers to a group of climbers that are tied to each other by one rope. These Seilschaften — as widespread in Poland, Czechoslovakia and Hungary as in East Germany — had enabled Treuhand apparatchiks to approve 'management buyouts' of enterprises governed by old party comrades, and to persuade Western companies to keep the old (ex-communist) management of newly acquired east German enterprises. "The new old chiefs have today (April 1992) more power than before; because the decide not only about promotion or demotion of their employees, but who should become unemployed."[11] If bureaucratic inertia and the old Communist boys' network can sabotage privatization in east Germany, with the consequences of unification hanging like Damocles' sword over the bureaucracy, we can appreciate easily the difficulties facing free market ministers in Poland or Czechoslovakia.

3.2. Case 2: Telecommunication technology in East Germany and Eastern Europe

The technological backwardness of east European telecommunication systems has been widely reported in the Western newspapers. Frequently, comments are made that certain network equipment or manufacturing plants are 10 to 30 years behind Western standards. There are certain factories that are only 10–15 years behind Western standards — e.g. the switching factory in Arnstadt, east Germany, acquired by Alcatel/SEL — and there are certain subsystems that are merely 30 years behind Western technology — such as the Stasi (Staats Sicherheit, DDR's secret police) special network and central office exchange in Erfurt — but the overall technological and efficiency level of East European telecommunication is even far worse than the above comments indicate. The former DDR's transmission and switching equipment had the reputation of operating Eastern Europe's most advanced telecommunication network. For instance in 1989, when the author discussed comparatively

[11] 'Seilschaften', *Wirtschaftswoche*, February 21, 1992, p. 126.

'modern features' of their respective national networks — such as modified cross-bar switches — with Czech or Hungarian telecom managers, these senior administrators would point out proudly that 'even the DDR had no more advanced equipment'. Even though the DDR had indeed EaEu's most sophisticated telecommunication system, an example will illuminate the task confronting the DBP Telekom in bringing the east German networks up to west German standards.[12] Similar tasks await Western telcos in selling their high-tech equipment in Eastern Europe.

During the silent revolution of 1989/90, the citizens committee of Erfurt, East Germany, discovered in the lowest basement of the Stasi headquarters an enormous secret telephone exchange that connected all Stasi units and offices throughout the DDR. The exchange and the entire special network was turned over to the German Post (DP), the DDR's PTT, that had been merged since with the German Federal Post (DBP). When inspecting the newly acquired exchange, the DP officials, however, were truly amazed at the 'modern facilities' that the Stasi had installed shortly before the east German revolution. Though the 'modern' Stasi exchange technology was already thirty years old it was still three to four generations younger than the mechanical relay (Strowger) switches built in 1923 that up to 1991/1992 still operated in the Leipzig and Dresden central office exchanges. Since the east German telecom system had been hopelessly overburdened, the Stasi's special network was immediately appropriated by the DP but could not be easily integrated into its network because west German equipment manufacturers no longer manufactured the necessary components while east German suppliers had never reached that level of sophistication. It is therefore, not surprising that "10 to 20 years waiting time for household telephone connections were not unusual, because the capacity in local area nets... were insufficient... and could not be provided in reasonable time." (Günther and Uhlig, 1991, p. 90).

3.3. Case 3: Alcatel/SEL's joint telecommunication venture in the DDR/East Germany

Germany's second largest telecom enterprise Standard Electric Lorenz (SEL) has been treated as a German company throughout its history, even though ITT had acquired majority ownership from the Lorenz family in the inter-war period; subsequently, SEL was sold by ITT to CGE-Alcatel in 1986, along with ITT's other European telecommunication business; as a consequence

---

[12] This mini-case is based upon the following sources: 'Der Sprung von Hebdrehwähler ins Digitale Zeitalter' (The jump from mechanical rotary switch into the digital age), *Frankfurter Allgemeine Zeitung*, June 20, 1991, p. 5, also Ricke (1990, pp. 8ff.), and 'Zur Vereinigung des Fernmeldewesens der FRG u. der DDR', unsigned *DBP memorandum*, Bonn, May 1990.

of ITT's policy to emphasize local management of its subsidiaries, virtually all ITT companies in Europe have enjoyed 'national treatment' from their respective PTTs. In spite of the fact that Alcatel N.V. has attempted to pose as a European company and is registered in Holland with such European luminaries on its supervisory board as former Prime Minister Heath, former EC commissioner Baron Davignon, and the late German statesman Franz Josef Strauss, it is very much a French enterprise. One of the questions that the 1986 acquisition of the ITT empire had raised was how the hierarchial French top-management in Paris would get along with the free-wheeling, self-assured SEL managers in Stuttgart?

After some initial difficulties, and a few management changes at SEL, Alcatel decided that both the German image of SEL and its relatively new, digital, ISDN compatible switch S-12 would be valuable in penetrating the East European market.[13] Moreover, SEL's position as a German company would be a particular asset since the DDR/east Germany should become the basis for an East-European strategy. SEL's CEO Gerhard Zeidler's strategic vision coincided fully with Alcatel's boss Pierre Sudreau. Gaining market-share in the former DDR and Eastern Europe would not only benefit Alcatel, but also strengthen the position of SEL management and switches within Alcatel. The long run strategies of SEL was to establish a position in EaEu ahead of its international competitors Siemens, AT&T, Northern Telecom, Ericsson and to "introduce [SEL's] digital switching system, through coordinated cooperation within the ALCATEL group, in the entire (former) East bloc."[14] SEL currently contributes over 20% to total Alcatel sales; by 1995 according to Zeidler, success in Eastern Europe could increase this percentage by 50%.

While SEL's German competitors, Bosch, DeTeWe, Philips/PKI and Siemens proceeded slowly and carefully in exploring joint ventures, SEL moved audaciously to establish a 6300 employee joint venture with two key manufacturing plants of the former information technology Kombinat VEB Ernst Thählmann (VEB was the DDR-German acronym for a people-owned enterprise). SEL paid DM 65 million as its 50% share of the joint venture, RFI-SEL, but provided also, as an advance, the entire working capital and top management. The management of the two DDR factories favored SEL over the offer of the more cautious Siemens management, because SEL promised to employ the entire workforce, including management, in the new joint ven-

---

[13] Alcatel has finally embarked on an expensive project to integrate its E-10 and System 12 switches in a new broadband asynchronous transfer mode (ATM) switch, designated Alcatel 1000. The claim that this switch will be available by 1994 seems wildly optimistic. See 'Alcatel unveils new strategy', *Telephony*, June 10, 1991, p. 8.

[14] SEL's CEO G. Zeidler, quoted in 'SEL: Risiko bewuszt gewählt' (Risk, knowingly chosen), *Wirtschaftswoche*, August 24, 1990, p. 153, my translation.

ture while Siemens only had been willing to promise employment for 900. Since the DDR Kombinat Ernst Thählmann had three times as many employees as needed in all its factories, the usual state of affairs in the DDR, SEL will have to expand the production of the joint venture drastically to become competitive.

In the short run the RFI-SEL joint venture depended for 60% of its capacity on existing contracts with the Soviet Union which had relied on the DDR for 20% of its telecom supply. Since January 1991, however, the Soviet Union had to pay in convertible currency for the ex-DDR telecommunications deliveries and by 1992 it had become doubtful that the successor republics — with the possible exception of the Ukraine — will abide by the original agreements. It was uncertain in 1990, however, whether the Soviets would either have the hard currency or the willingness to spend it on old analogue equipment if modern, digital equipment could be purchased on the world market for almost the same price.[15] Though east German and Eastern Europe manufacturing industries have been devastated during 1990/91 by the sudden termination of Soviet orders for their products, SEL's joint venture had been doing better than most. The backwardness of the Soviet telecommunication system had made it necessary to continue at least part of its acquisition of analogue equipment, albeit at old pre-unification prices; moreover SEL's joint venture agreement assured that the DDR/German telecommunication administration would continue to purchase its entire analogue equipment for 1990/91 from the RFI-SEL factories.

Lastly, among the risks incurred in the SEL investment were the substantial wage increases that had been granted to the workforce after the monetary reform that exchanged one DDR mark for one DM mark. The higher wages and the promise of no lay-offs until August 1991 had made it certain that SEL's joint venture would be an expensive, unprofitable operation for many years.

SEL's ambitious five year capital investment program, and the acknowledged technical competence of the East German work-force, will gradually eliminate differences between SEL's operation in Stuttgart and Arnstadt (East Germany). The changeover from traditional analogue (mechanical relais and cross-bar switches) to capital intensive digital products, such as the S-12 switch, will, however, lead to a much lower employment of labor. Labor redundancies in the unified FRG are, however, almost as difficult to obtain as in the former DDR, and very much more expensive.

The risks incurred in SEL's joint venture certainly had been known to its management. The originally projected break-even date of December 1992

[15] *ibid*, pp. 152–154.

had been highly optimistic, even though the goal of DM 400–800 million sales for 1991 turned out to have been fairly realistic; but the joint venture will give SEL a major opportunity to obtain a 50% share of the 55 billion dollar investment planned by the DBP to bring east Germany telecommunications up to Western standards by 1995. Moreover, there can be little doubt that the joint venture has opened doors to the entire East-European/Soviet market. Most of the telecom administrators in Eastern-Europe/Soviet Union retained their positions and their established ties to east Germany helps. In the long run, five to seven years from now, the joint venture may very well turn out to be a success for SEL and Alcatel; however, how many North American companies are willing or able to wait five years for a return on their investments?

3.4. Case 4: Siemens' East German strategy [16]

Siemens' East German and Eastern European strategies reflect its recognition of telecommunications as a Schumperterian leading industry that will both dominate a country's economic growth over a fifty year (Kondratieff) cycle and, simultaneously, determine its geopolitical position in the concert of Nations. [17] Alcatel/SEL acquired east German manufacturing capacity to penetrate Eastern Europe's telecom market and to supply the immediate infrastructure needs of east Germany and the Soviet Union. Siemens, by relying on established ties, however, chose to adapt itself to the DBP Telekom's efforts to rebuild the east German telecom net and bring it up to west German standards by 1997 (Siemens and Alcatel strategies were not mutually exclusive).

Siemens' capability as a 'one-stop turn-key network builder and equipment supplier' was expected to provide a competitive edge in dealing with Eastern Europe and Soviet telecommunication administrations. Still existing relationships between the former DDR's telecom equipment manufacturer KNE (Kombinat Nachrichtenelektronik) and EaEu/Soviet telecommunication technocrats could be successfully exploited by Siemens if it could project its role as the dominant force in modernizing the east German telephone system and the heir to KNE's position as EaEu's leading telecom equipment manufacturer and exporter. KNE had employed 37,000 people in 19 plants

---

[16] The main sources for this case material are a lengthy interview with: R. Frensch, executive director, Siemens, in July 1991, as well as Rosenblatt (1991)

[17] Compare Josef Schumpeter, *Business Cycles*, Vol. I passim, 1938, with Siemens' Joachim Rosenblatt. "In our industrialized world telecommunications are an indispensable resource for a country's economy... Moreover the telecommunication industry is an economic *sector in its own right* with high rates of innovation and growth. In the light of telecommunications' *pivotal dual role* Siemens is now investing heavily in the development of the telecommunications industry and network in eastern Germany" (Rosenblatt, 1991, p. 6).

with 4,500 scientists and technicians in research and development. As late as 1989 KNE had produced 500,000 line units, but allocated only 60,000 to the DDR's public telephone network and exported the rest to East Bloc neighbors for DDR Mark 3 bn. Clearly if Siemens could project itself as a reliable but technologically advanced successor to KNE it would establish a very special competitive advantage.

Siemens was founded as a telegraph manufacturing enterprise in Berlin in 1847 and had quickly developed into Germany's leading electrical engineering and telecom company with manufacturing and sales sites throughout Eastern Europe. Prior to World War I Siemens Austria, for instance, had employed over 30,000 people and was the largest manufacturer in the Habsburg monarchy. Siemens' close relationship with the German PTT, the Deutsche Bundespost (DBP), has been legendary but the company had also managed to maintain ties to the telecom administrations of Hungary and Czechoslovakia during the inter-war period. Quite amazingly some of these historical associations survived World War II. When KNE was given the lead role in developing digital switching systems for the East Bloc it first looked to Siemens for technical aid and subsequently the DDR's telecom administration Deutsche Post (DP) tried to order a packet switching system from Siemens as late as 1988. In both cases NATO's Cocom (Coordinating Committee for Multilateral Export Control) prevented the transfer of high technology equipment, but the fact that Siemens was the choice of DDR-East Bloc planners illustrates that at least part of its traditional position in Eastern Europe had been retained.

The development of corporate strategy is influenced deeply by a company's culture. Although Siemens' former CEO K.H. Kaske had been attempting, largely successfully, to strip Siemens of its excessive traditional and bureaucratic trappings, the enterprise's top management is, in my opinion, still more influenced by patriotic and historical sentiments than any other major company in contemporary Germany.

Siemens' east German strategy consisted of three strands:

(1) *Reestablishment of a full area-coverage sales and service organization* that would, in a relatively short period, integrate east German operations into Siemens' domestic structure.

(2) *Expansion of its production capacity* by acquiring smaller east German companies and by entering into joint ventures. Siemens' reemergence as a major east German electronics–telecommunication manufacturers was designed "to hasten improvements in the infrastructure and adapt and strengthen (east German) industry." [18]

---

[18] Rosenblatt (1991, p. 7) and 'Die Wege werden kürzer', *Zeitschrift für Post und Telekommunikation (ZPT)*, April 23, 1991, pp. 4–9.

(3) *Association with and support of* the DBP Telekom's short and long run strategies to first patch up existing networks and then build a leading-edge telecom system to meet and even surpass west German and EC objectives for the year 2000.

The reestablishment of a sales and service organization to cover fully east Germany was accomplished by recruiting and training east German personnel to market telecommunication systems and equipment. Interestingly, Siemens created a new corporate entity, Siemens Private Communication System Sales Gambit, to operate a broad marketing organization that transcends divisional lines. Presumably, this desirable organizational innovation has been difficult to establish in west German, where the Siemens corporation has just gone through a major restructuring and was not ready for further changes.[19] However, the new Siemens marketing organization in east Germany illustrates how the changing structures in Eastern Europe encourage organizational innovation in Western companies that penetrate the former command economies.

Siemens expanded its manufacturing capacity by investing in several joint ventures, one green field plant and one acquisition. DM 20 million were spend on modernizing Nachrichtenelektronikwerk Greifswald (NEG)as a joint venture investment; NEG is currently developing and manufacturing digital transmission systems. Since digitalization of the east German as well as the Czechoslovakian and Hungarian networks is a major objective of their respective telecommunication administrations, NEG should flourish since it still has close ties with these PTTs.

In Leipzig, Siemens acquired RFT Nachrichtenelektronik for DM 70 million and renamed it Siemens Kommunikationstechnik Gmbh (SKL). SKL will operate as an independent though wholly owned subsidiary producing public (EWSD) and private (AICOM) digital, ISDN capable switches in brand-new, state of the art facilities. For 1991 100,000 line units of public EWSD switches were scheduled, 300,000 for 1992. Note that compared with Alcatel/SEL Siemens' investments were smaller, the new organizations transparent, and the production process fully rebuilt without inheritance of excess labor; moreover, there was little exposure to future personnel or past pollution risks.

The green field factory in Schwerin started production in the summer of 1991 and produces communications cables. Since it is an entirely new production unit it is fully integrated into the Siemens structure, though it can be assumed that its location in east Germany will give Siemens an advantage in obtaining orders from the DBP Telekom for its east German network expansion.

Siemens' close involvement in the DBP Telekom's project to expand and modernize the east German telecommunications system promises consider-

---

[19] For a description of Siemens' reorganization see Franz (1990, pp. 26ff.).

able immediate and relatively riskless profits (Alcatel/SEL has also benefitted from the DBP investment in the east German infrastructure but has played a less aggressive role in the planning stage). Since both the short and long run deficiencies of Eastern European telecom systems are very similar to east Germany's, Siemens' contributions to the DBP Telekom project illustrates the opportunities the necessary modernization of the EaEu/Soviet systems provide. The short run remedial DBP Telekom plan had been to add 1.2 million digital lines to the German network by summer 1992; its intermediate goals has been to reach west German levels by 1997. Table 1 illustrates the gap that existed in 1989 between West and East German systems, and Table 2 the required additions to attain 1997 objectives.

We shall describe in the next case study the actual steps taken by the DBP Telekom to meet its 1997 objectives. Table 1 and Table 2 do indicate, however, the enormous opportunities the DM 55 bn Telekom project provided for telcos and the competitive edge Siemens and its Public Switching Systems division acquired by closely associating itself with the DBP Telekom's 'project 2000.'

During the period 1990/91 Siemens showed considerable flexibility in supporting DBP Telekom's effort to develop an inexpensive overlay network that

Table 1
Comparison of East and West German Telecommunications in 1989

	West Germany	East Germany
Population in millions	62	10.5
Telephones in millions	43	4
Main lines in millions	29	1.8 (2.4 in 1991)
Telephones per 100 households	93	16
Telephone calls per year in millions	51,710	2,400
International calls per year in millions	694	19

Source: Joachim Rosenblatt, Siemens.

Table 2
Difference between east German system in 1990 and 1997 target

East German Telecom System	1990	1997
Main lines in millions	1.8	7.2
Public Telephones	39,900	68,000
Fax machines	2,500	360,000
Data-Packet subscribers	–	5,000
BTX (Videotel) subscribers	–	380,000
Mobile Radio subscribers	–	300,000
Cable TV subscribers, millions		2.2

Source: Joachim Rosenblatt, Siemens.

would expand immediately the capacity of the former DDR/east German telecom system by 1.2 million digital lines. An overlay network uses existing fixed facilities but supplements it with software supported mobile, satellite and new digital lines, as well as additional exchanges. The Leipzig Fair in 1990, a major economic event in the DDR, gave Siemens the opportunity to set up its first mobile telephone exchange in a container in order to create additional, temporary capacity. Subsequently additional mobile radio stations have been introduced along the Autobahn leading into Berlin and in eight east German cities; by January 1991 — 3 months after the unification — 806 million mobile radio channels had been set up in containers. Similarly two/thirds of all digital exchanges provided by Siemens for the DBP Telekom were housed either in containers or special prefabricated buildings. Radio links between Berlin and the Baltic seaport Rostock supplemented by fiber-optic cables were another example of the flexible Siemens contribution to the over-lay network.

The DBP Telekom is well known for the exacting standards it ordinarily required from its suppliers, and Siemens in turn has been known, as a result of its 150 years association with the DBP, to absorb the purchasing requirements into its production process. The flexibility shown by both Siemens and the DBP is therefore noteworthy; as we shall see it was the first indication that developing the east German network would affect significantly the entire German, and perhaps even the EC telecom structure.

Along with Siemens' heavy commitment to the short-run overlay-network, we must also mention its initiative to anticipate the needs of the DBP's project last stage, the plan to introduce between 1997 and 2000 advanced telecom technologies to give east Germany the most modern network by 2000. These leading-edge, generally untested, technologies include:

(1) Fiber to the home (i.e. the extension of the fiber-optic cables to the customer's terminal equipment, 'the last mile' in telco language).

(2) Broad-band ISDN and optical transmission systems with coherent detection (i.e. wave length division multiplexing to provide additional capacity for laser beams, satellite links and to reduce transmission obstructions).

(3) Asynchronous Transfer Mode (ATM) platforms for broad-band ISDN (B-ISDN) networks. (i.e. transmission that is not related to a specific frequency or timing of the transmission facility)

Siemens has been working on a 'fiber to the home' trial project in Leipzig during 1991/92, and has been testing a transmission system with coherent detection since 1990. Both projects fit into the last phase of DBP's Telekom 'project 2000', but simultaneously fit into Siemens development programs (see Derr et al., 1990, pp. 30ff.).

From a long run view it is interesting to observe how the Siemens efforts to prepare for the advanced objectives of the DBP 2000 project also fit into the ambitious but still ill-defined EC objective to develop, as part of the EC concept 2000, ‘electronic glass fiber highways’, under the project code name METRAN (Managed European Transmission Network). This is not the place to discuss the various technologies joined in the METRAN project, but it seems that Alcatel/SEL and especially Alcatel Network Service Deutschland (ANSD) have also made major advances in developing a Managed Data Network Service (MDNS) that applies several METRAN building blocs. Siemens R&D efforts *included in its east-German strategy* will give it now operational opportunities to compete with two emerging leading technologies, Alcatel's MDNS and AT&T's SONET (synchronous optical network), an American network architecture that is compatible with B-ISDN.[20]

The four brief case studies illustrated the opportunities as well as the institutional and technical obstacles that Western telcos will encounter in their penetration of East European markets. We shall now consider in a further case study the political and economic forces that have affected the implementation of the DBP Telekom strategy to modernize the former DDR's telecommunication regime. The lessons of that major capital investment program have far reaching implications for the East European strategies of Western network operators (AT&T, British Telecom, Nynex, Pacific Telesis, U.S. West, etc.) and equipment manufacturers (Alcatel, AT&T, Bull, Ericsson, Fujitsu-ICI, Northern Telecom, Siemens, etc.)

3.5. Case 5: DBP telekom's East German strategy

Immediately after the 1989 revolution, it became obvious that East and West German telecommunication systems would have to be integrated even if, as most experts predicted then, the actual political unification of Germany was still five to ten years away. Although the poor conditions of the East German networks were not fully known either in Bonn or even in West Berlin, the accepted technological superiority of the DBP network would require massive investments in the East German telecom infrastructure in order to permit the integration of the two systems. The goal of East–West German telecom integration could be broken down into three short-run objectives:

(1) Immediate construction of new East–West telephone lines (including use of mobile and satellite networks) to provide the infrastructure for Western business activity in East Germany *and* to facilitate telephone conversations between east and west Germany.

(2) Immediate improvement of the technological performance of the East

---

[20] SONET is a hybrid architecture that contains both synchronous and a synchronous transfer modes (ATM) ATM is a requirement for B-ISDN. See ‘CEPT To Sign Pact on Broadband Net’, *Communications Week International,* March 2, 1991, p. 1 and ‘Communication networks of the future’, *Siemens: International Telecom Report,* December 1992, pp. 5–8.

German telecommunication system, with a long run goal to bring it up to West German-level by the mid or late nineties.

(3) Improvement in the provision of telephone service for east Germans, with sharp reduction in the ten year waiting period for new telephones, and an immediate introduction of consumer oriented maintenance and repair service.

The DBP had to choose among several strategies to accomplish these three clear short run objectives and to integrate them into its long run vision.

Strategy A: DBP Telekom could rely on its own resources and staff to extend to East Germans the West German telecommunication regime, defined by the Postreform law (Poststrukturgesetz) of July 1989. The East German telecom administration, the Deutsche Post (DP), would be gradually absorbed by DBP Telekom, but no significant departures from established technical and administrative DBP procedures would take place. Once the short run objective had been reached, the long run goal (Telekom 2000) could be adopted to provide East Germany by 2000 with Europe's most advanced network.

Strategy B: Short run pressures could prompt DBP Telekom to relax temporarily its monopoly position in basic (voice) telephone service and transmission network ownership. Private mobile cellular or satellite based networks might be permitted to operate in East Germany for limited periods, and large multinational corporations could be encouraged to expand their private communication systems in East Germany and make excess capacity available to third parties. The lessons of further liberalization of the system would affect the long run Telecom 2000 vision.

Strategy C: DBP Telekom could emphasize the short run needs of the East German people and their economy by installing digital radio and satellite based networks that have a much shorter construction time and lower initial costs. Though at that time about 90% of mobile telephone conversations originated or ended in fixed networks, combined radio-satellite systems could utilize available private networks and public networks with excess capacity in West Germany or Denmark. This strategy would also have the advantage of both testing technological innovations in the digital radio/cellular technology, and catching up with British and Scandinavian advances in expanding and improving mobile telecommunication networks.[21]

As it turned out DBP Telekom management chose strategy A and adopted the code name 'Telekom 2000' for a supposedly integrated short-run and

[21] For a lively but opinionated survey of mobil telecommunication technology in Germany and Europe see the 'Telekommunikation Spezial' in *Wirtschaftswoche*, June 21, 1991, pp. 49–79. Note also Lange (1990). Some of the technological advances in mobil-radio technology are 'bundled radio frequencies' that are widely used in the U.K. and Ericsson's 'Mobiltex'-system that transmit mobil packet-switched data in cellular networks. See Berntson (1989).

long-run strategy. The DBP was forced, however, by short run pressures, enlightened intervention by the Post and Telecommunication minister Christian Schwarz-Schilling, and strategy-implementation shortcomings to adopt, in an *ad hoc* manner, aspects of the two other strategy alternatives. Furthermore neither the DBP nor the German government anticipated the enormous capital investment that the economic unification process would require. Consequently the increasing, and again *ad hoc*, diversions of DBP cash flow to plug the gaping holes in the Finance minister's budget forced continuous revisions in Telekom's East German strategy. In September 1991 the DBP Telecom announced finally that it planned to invest DM 60 bn (about $35 bn at December 1992 exchange rates) in the East German telecom infrastructure during the period 1991–1997. We may assume that the actual expenditures will exceed these targets, especially in view of the increasing costs of the rapid technological changes (Goodhoat, 1991).

It is important to recognize that underestimating investment costs is a generic characteristic of Western telcos' Eastern European projects. The technological–political environment within which major capital projects take place abounds in uncertainties and furthermore deteriorates rapidly as one moves east from (East) Berlin to Warsaw to Moscow to Vladivostok. For example, the proposed mega joint-venture in 1989/90 to build a trans-Siberian terrestrial fiber-network, that would have involved numerous Western telcos, from Alcatel to U.S. West, was a potential disaster that would have cost many times its ridiculously low $500 million budget, in the unlikely case that the project ever would have been finished. Fortunately, for all concerned, the Coordinating Committee for Export Controls (Cocom) refused to release the necessary high-speed fiber technology and thus scuttled the project. DBP Telekom's experience during 1990/91 in carrying out its Telekom 2000 strategy under relatively benign conditions should be carefully studied by ambitious Western telcos before they commit themselves to broad, open-ended projects. On the other hand, the joint 1991 venture of Nokia (Finland) and the Moscow regional telephone administration to build a mobil telephone network for Moscow is an example of a well defined, manageable project; even in this case substantial budget overruns can be expected.[22] The DBP's Telekom 2000 project, however, has been a good example of an open-ended project in which the intermediate targets were moved backward and forward in time, and the desired technological goals constantly redefined, at great expense.

During 1991 the somewhat fuzzy long run objectives of the Telekom 2000 strategy began to take shape. By 1997 east and west Germany telecommunication systems were to be 'harmonized' at the advanced technological level

[22] 'Will Mobile Mania Grip the East?', *Global Telephony*, February 4, 1991, pp. 26–32.

west Germany's telecommunication was to have reached by then, according to DBP Telekom plans. This meant primarily further replacement of analog by digital ISDN narrowband networks, the rapid introduction of ISDN broad band network, and the expansion of the two West German digital mobile cellular networks. Once east Germany had reached the projected level of west Germany's telecom system, further state of the art technologies were to be introduced.

The Telekom 2000 strategy also provided the opportunity to reduce DBP Telekom's voice and network monopoly and to increase competition in the entire German telecommunication system. Post and Telekom minister Schwarz-Schilling and his advisory commission believed that by inviting privately owned mobile radio/satellite systems and leased line networks to supplement the existing fixed network monopoly and the mobile radio duopoly, the basis for further network competition could be established.

IBM, for instance, took this opportunity to put its internal private network, at the disposal of east German enterprises, and included even messenger services for new businesses without telephone in its effort to provide alternatives to the monopoly network. Private satellite services to connect east and west German computers already had been permitted by Schwarz-Schilling, over the objections of Telekom CEO Helmut Ricke, in order to make up for the slow DBP efforts to provide satellite services to big cities such as Dresden or Leipzig. Similarly, the relatively small entrepreneurial German telecom equipment supplier ANT Nachrichten and an even smaller competitor Neef Elektrotechnik received permission from the Post and Telekom Ministry (BMPT) to provide voice service via private satellite nets for east German enterprises and municipalities. The government of the Land Brandenburg has been using the General Electric's Information Service (GEIS) Mark III to gain access to the large and efficient computer center of the Nordrhein-Westfalen provincial government.

The above mentioned *ad hoc* utilization of private networks comprise violations of the DBP network and voice service monopoly and therefore represent a significant liberalization of the BMPT ministry's leased-line regulations. The 1989 post reform legislation separated regulatory authority from the managerial responsibilities of DBP Telekom's top management (Vorstand). Schwarz-Schilling has, in effect, used his regulatory authority to relax in east Germany the DBP's long standing limitations on the use of private leased line networks, over the objections of Telekom's management. The wider use of private radio-satellite and leased line networks in east Germany has provided a level of network competition that might be more permanent than Schwarz-Schilling and his advisors had been willing to admit. The reduction in the DBP's international call charges and leased lines tariffs during January 1992 can be considered to be an end product of east German *ad hoc* liberalization, similarly, the current government efforts to privatize Telecom, was greatly helped by the *de facto* liberalization in east Germany.[23]

The lessons of adapting DBP regulatory policies imaginatively to east German requirements will have long run implications for west German and the European telecom regimes. We know that successful business mergers affect the culture and structure of both enterprises. The flexible integration of the former East German DP into the DBP may reemphasize this point. Applying the preliminary lessons of the integration of the East German telecommunication network into the established West German system, we can expect that successful East–West joint ventures will require considerable flexibility and innovative adaption from Western managers and engineers.

4. Hungary: A brief survey

Hungarian 'economic reform' has groped toward a market system since the late-1960's. The two-step forward, one step back liberalization gave Hungary's economic managers a reasonably good understanding of global market forces, as well as active trading experience with the West. The quiet revolution of 1989/90 added a democratically elected government and anti-communist ministers to privatize state institutions and to lay the foundations for a free-market economy. The efforts to modernize the telecommunications system and adapt the PTT to the EC's liberalization directives provide a unique focus to examine the opportunities and obstacles Western enterprise face in Eastern/Central Europe.

(1) 1987: Magyar Telekom accepts an offer from *France Telecom* for technical assistance and approves a joint venture to provide Minitel service in Hungary.

(2) 1988: Alcatel receives a procurement contract for transmission equipment; the foundation is laid for joint venture between Alcatel/SEL and Videoton to build digital S-12 switches.

(3) 1989/90: The US company Contel formed a joint venture 'Contel Hungarian' with the PTT, Hungarian (Magyar) Telecom (HTC), to build and operate a national cellular system. Contel Hungarian was to compete with Hungary's second cellular system, the joint venture between U.S. West and Magyar Telekom. The license was withdrawn, however, in late 1990, and Contel is currently suing in international courts to regain its license.

(4) 1990: Alcatel Austria forms joint venture 'AHT Communication' with HIRANDESTCHNKIK SZOVETKEZET and the PTT subsidiary COMEX

[23] 'Deutsche Telekom to Reduce Long Distance Call Charges', *Financial Times*, January 7, 1992, p. 14.

to build PBXs and terminal equipment to increase Hungarian penetration rate from 7% to 20% by 2000.

(5) 1990/1991: The State owned Magyar Telekom (HTC) formed a joint venture, Westel Radiotelefon with U.S. West to bring cellular systems to the eagerly waiting public. In Budapest, a city wide cellular system went on line in October 1990 to become a second, though expensive, network for the capital. Westel Radiotelefon Kft. had been under great pressure to accommodate the far greater than expected demand. Westel manager Ron Sanders acknowledges that building an analog cellular system operating at 450 MHz is an expensive proposition because there are no economies of scale. The large demand for cellular phones, that come primarily from Western businesses and mangers, has made it possible to charge tariffs somewhat above Western Europe's rates. Still, Weste's initial $1250 connection fee is $250 less than that of Hungary's PTT wireline.

Westel had more than 5000 customers in late 1991, over 33,000 by October 1993 and it hopes to support 50,000 by 1995. Since wireless phones are used as substitutes for the public network, customers talk longer and from one place instead of moving around. (The old rule of thumb, accepted by DEP-Telekom and other fixed line operators, that 90% cellular traffic originates or ends in fixed networks, seems no longer valid. The analog capacity of Westel's 45 MHz transmission will be exhausted soon, therefore. The Hungarian communication ministry licensed, therefore, two competing 900 MHz cellular operators, one to provide analog the other digital GSM service. Unless Westel obtains the digital GSM license, U.S. West's investment will be threatened.

As part of the HTC modernization strategy, Northern Telecom supplied 22,000 digital lines to support the cellular network that generates a surprisingly large number of cellular to cellular calls. Alcatel-SEL formed a joint venture with the privatized Hungarian company Videoton to build digital switches for Hungary's overlay digital net. Similarly, Siemens A.G. has formed a joint venture with Terta, a local manufacturer of transmission equipment; Northern Telecom Ltd. and its Austrian associate, Austrian Telecom, have jointed with BHG, a newly privatized Hungarian company, to produce transmission and switching equipment.

All of the above manufacturers hope to get a substantial share of the $6bn the PTT plans to spend on network modernization over the next five to seven years. The Western telcos also expect to penetrate other former socialist countries, especially Croatia and Poland, that have long historical ties to Hungary. Alcatel-SEL began manufacturing its System-12 switch during the summer of 1991, and has high expectations since its S-12 switch has been designed especially for medium sized central offices.

(6) 1991: Budapest's vastly improved communication system is tied into the European network through Ericsson international gateway switches. HTC

has been careful to avoid relying on one Western telco and has given U.S. West, Ericsson, Northern Telecom and Siemens various parts of its overall investment project. HTC's strategy has been to improve first the Budapest telecom network and then to modernize the country's system.

(7) 1991/1992: Installation of a digital overly network, similar to east Germany's is the next step in the HTC's modernization program, that now also includes plans for privatization. The HTC intends to retain its monopoly on the overlay network but will allow local services being offered by competing enterprises.

Among the Western telcos engaged in supporting the modernization of the Hungarian system, U.S. West seems to have made the largest capital investment. So far U.S. West has invested over $10 million in the joint venture Westel, to be followed by $15–30 million in subsequent years. As long as the Florint is not convertible the risks for U.S. West are serious; A base of 30,000 subscribers and a convertible Florint made the U.S. West joint venture a financial success, though Hungary's possible overemphasis on mobil networks might create additional competition.

Typical for East European telecom strategies, the U.S. West venture is supported by an alliance with Ericsson and Motorola; both telcos promised to supply the telephones and switches for the special 450 MHz frequencies. Ericsson, Motorola and U.S. West have been aggressively penetrating Eastern European markets, with noteworthy successes in Hungary and Czechoslovakia. Each enterprise has been effective in using its technical competence in specific areas, such as mobile telephones, switches, or network management as its competitive edge. Ericsson has also been able to draw on its historical ties in Central/Eastern Europe as an additional competitive advantage that has made it a major player in Hungary, Czechoslovakia, Poland and the former Soviet Union. A quick examination of Ericsson's performance in Hungary should be of considerable interest to all Western telcos.

4.1. Case 6: Ericsson in Hungary

The Swedish company, L.M. Ericsson was founded at the turn of the century as a telecommunications equipment company. Its small home market forced it from its very beginning to rely on international sales *and* foreign manufacturing sites for the necessary economies of scale and scope. By 1911 Ericsson had a state of the art plant in Hungary, and a very strong market position in the Austria–Hungarian empire. The depression of the 1930's and the protectionist policies of that period hit Ericsson very hard, and forced it, among other emergency measures, to sell its Hungarian plant to Germany's S.E. Lorenz (now Alcatel/SEL); SEL in turn lost its site after Soviet/Hungarian nationalization of foreign and domestic enterprises. The state-owned enterprise BHG that took over the former Ericsson plant, renewed the ties with the factory's founder in 1968 by licensing its cross-bar switch and exporting it to the east block countries. Sweden's neutrality was, of course, a great advantage for Ericsson's relations with Hungary though the enterprise did, generally, abide by COCOM restrictions.

In 1988/89 Hungary requested proposals for Western participation in an ambitious project to provide a digital network for its rural population. Western telcos had to have a Hungarian partner, however, to be considered. Competing against Alcatel/SEL and Northern Telecom, which were allied with state owned companies, Ericsson allied itself in 1989 with a small privately-owned (under an obscure 1988 law) but highly efficient enterprise Muzertechnika, whose 400 employees outperformed BHG's work force of 8,000.[24] Ericsson was confident that it could submit the low cost bid, but was skeptical whether the Communist authorities would abide by the results.

By the time the bids were opened, however, the revolution had occurred and the switching contract really went to the lowest bidders. Ericsson obtained 65% of the contract, whose worth is still not fully determined, and Siemens (Austria) allied with another privately owned enterprise Telefongyar won the other 35%. Neither Alcatel (SEL) nor Northern Telecom who had allied themselves with Stalinist dinosaurs had come even close. Ericsson's managerial skills, reputation, historical connection, and political sophistication were rewarded.

4.2. Recent developments

The flexibility of Western telcos, their combination of cooperation and competition, have startled the Hungarians. HTC plans to license two more cellular carriers but members of parliament wonder if there is not too much competition already. The HTC supports a revision of the Communications Act that would give it authority to increase network competition, and solicit additional Western telcos' interest. Parliament finally approved the liberalization of the telecommunication system in October 1992, but the implementation has been left up to the HTC which no longer reports to the Ministry of Transport and Communication. No rapid changes in existing conditions should be expected. The PTT hopes that U.S. West will show interest in building regional digital overlay networks. A $6 bn telecom market is expected with a yearly demand for 300,000 additional lines. The telecom administration will accept further liberalization in order to be a part of Europe. A proposal for partial privati-

---

[24] 'Liberalisation Express Powers through Hungary', *Telephony*, June 3, 1991, p. 40; see also 'Eastern Europe: Starting from Scratch', *International Telecommunications Survey*, *Financial Times*, October 18, 1993, p. 6

zation of the PTT has been discussed in parliament since 1990. NYNEX, Bell Atlantic, and Cable and Wireless have expressed interest in buying the 30% share supposedly reserved for foreign ownership; little further progress has been made, however, and one must not forget that European conservatives, such as the Antell government, have no objections to government ownership of the infrastructure.

Hungary presents right now the best potential market for Western telcos in Eastern/Central Europe. The HTC's ambitious modernization program of the country's network simply can't be realized without Western aid. HTC wants to add two million lines by the year 2000 (currently it has 800,000 lines), and hopes to cut its present 10–12 year wanting lists in half by 1993. In order to generate up to 65% of the $6 bn required, it must raise its tariffs from the current very low levels (a telephone connection costs $1.50 per month and 3¢ per call). Higher tariffs will make HTC a more desirable partner for the Western telcos that will have to invest the $2.10 bn if the HTC project is to be realized. The HTC will ultimately accept further liberalization and even partial privatization in order to be part of Europe. If the Austrian PTT, until five years ago one of the most mercantilistic in Europe, could adapt itself to the EC deregulation policies, so can the Hungarians. The precipitous fall in domestic demand that led to cutbacks in planned investments by General Electric, General Motors, Ford and Julius Meinl (an upscale Austrian grocery chain) does not seem to have affected the pent up demand for telephones and telecommunication services, nor the eagerness of Western telcos to gain a foothold in Hungary.

5. *The Czech and Slovak Republics*

Prior to the 'velvet revolution' in November 1989, Czechoslovakia (now the Czech and Slovak Republics) had Eastern Europe's most stolid Stalinist government. Unlike Hungary or Poland all Czechs and Slovaks were employees of the state or state sponsored organizations, thus making privatization most difficult. For thousand years an integral part of the Central European economy, Czechoslovakia had become a supplier of sturdy but unsophisticated capital goods for the Soviet Union with a weak and primitive consumer goods and services, especially telephone services-industry. The new government, based upon a loose coalition of classical liberals and social democrats, was slow in committing itself fully to a market economy; strong element to of the new elites, including the charismatic president Vaclav Havel, kept searching for a 'third way'; in consequence privatization in the CSFR had lagged far behind Hungary, East Germany and Poland during 1990/1991. The telecommunication ministry, however, had accepted an immediate modernization

Table 3
Comparison of networks in Eastern Europe, 1989

	Czechoslovakia	Hungary	Poland
Main lines	2.2. million	307,000	3.1 million
Percentage connected to manual exchanges	0.2	8.1	8.4
Main lines per 1000 people	14	8.6	8.1
Telex lines	12,000	13,000	35,000
PBX lines	1.9 million	742,000	2.1 million

strategy that would rely on privatization to gain Western telcos' support; the actual implementation of the plan was controversial, however, and the telecommunication ministry has had five different ministers between 1989 and 1992.

U.S. West and Bell Atlantic have gained the leading role in the CSFR's modernization program. Their strategy to work with the national PTT and to provide the network operation skills to support the modernization plans have been amazingly successful. In July 1990 the two telcos obtained exclusive rights to build a national analog cellular network for $60 million, and to operate later this system jointly with the Czechoslovak PTT. In addition the two carriers will participate in the modernization of the existing system which, by Eastern European standards, was far superior to Poland's and Hungary's. Modernization of the entire network is scheduled to take 10 years and $4 bn. Eurotel, a joint venture of Bell Atlantic, U.S. West, and the CSFR's PTT, completed the first step of this long run strategy by opening cellular service between Prague and Washington in September 1991. Eurotel has currently about 5,000 customers in Prague, Brnó and Bratislava, and expects to serve 12,000 customers by the end of 1992.[25] Equipment for the system has been provided by Finland's Nokia and Denmark's Dancall.

Network modernization will also include terrestrial microwave links, an optical fiber overlay system, and upgrading the crossbar switches to support a 40% digital system by 2000. Lastly, the Bell company will supply data services for business users in Prague, Bratislava, and Brnó through a joint 49 : 51 venture with the PTT. In 1992, during the finally launched privatization drive under finance minister Klaus, further plans have emerged to license several cellular network operators. Consortia headed by BT, Preussen-Elektra (a German Utility), and McCaw Cellular Communications have been bidding for licenses, but the strong position of US West/Bell Atlantic has not been

[25] 'Czechs Turn up Cellular Service', *Telephony*, September 16, 1991, p. 3. See also 'Grosse Welle', *Wirtschaftswoche*, January 1, 1992, pp. 18–19, and 'Upgrade in Czechoslovakia', *Communicationsweek International*, July 2, 1990, p. 11.

threatened. Critics have pointed out that regional telephone monopolies such as the BOCs have an advantage in dealing with national telecom monopolies.

Total foreign investment in the CSR has amounted to $1 bn with two-thirds coming from Germany yet American Bell companies have dominated Western telecommunication investments in the CSFR often in alliance with Scandinavian equipment manufactures; other Western telcos, especially Alcatel/SEL have also gained significant footholds. Alcatel/SEL's joint manufacturing venture with the Czech enterprise Tesla Liptovski Hradok was chosen as a vendor of S-12 switches by the PTT ministry in September 1991.[26] It is important to note that SEL was able to secure this contract over tough competition because it was willing to manufacture, jointly with a Czech enterprise (SEL owns 60% of the venture); joint production capacity in the CSFR is planned around 250K units/year, with an initial investment of DM 20 million. Production began in the fall of 1991.

Of considerable interest is IBM's joint venture with the government owned equipment manufacturer Telsa and the Czech and Slovak PTTs. In a quite unorthodox manner IBM has developed and installed a telecommunication system that added digital channels to radio wave transmission used primarily for TV. Regulations would prohibit such a network in the US or Germany. Supported by IBM computers, the TV channels have been used to build a private digital telecom system that provides voice, data and facsimile services and also gives customers access to IBM's international networks that supply information and electronic mail. In this imaginative joint venture IBM supplies the technology, the PTT, the network links, and Telsa routine transmission components. We must notice again how the pressing needs of EaEu can provide innovation opportunities.

One must not believe that the joint-ventures listed above have already made significant changes in the CSFR's telcom system. Though Western managers and several important native businesses have obtained improved services, especially through new cellular mobile and packet networks, the average Czech is still facing the obsolete system the command economy created.[27] The PTT's ambitious long-run goals include telephones for 85% of households, and 165,000 fax machines at a cost of about $6 bn. The opportunities for Western telcos are obvious, provided they are willing to transfer technology through joint ventures, commit themselves to local manufacturing, and are willing to face the risks of a system in transition.

---

[26] 'CSFR entscheidet Sich Für System-12', *ISDN Report*, September 1991, p. 8.

[27] 'Starting With a Blank Page', *Telephony*, July 1, 1991, pp. 22–25; 'Eastern Promise', *Financial Times*, February 27, 1992, p. 16; 'Starting from Scratch', *Financial Times*, October 18, 1993, p. 6

6. Summary

Our discussion and case study of Eastern Europe's political and economic transformation can be summarized in four statements:

(1) Transition from a communist command economy to a free market society is far more difficult than expected in 1989. Privatization has been expanding at a snails pace throughout EaEu. Management at all levels has still not adapted itself to the demands of a market economy, and work ethic and attitudes have not yet changed significantly.

(2) Telecommunications' role as a vital infrastructure for a modern market society has provided a friendly and more reliable climate for joint ventures than can be found in other industries. Even in the telecommunication field, however, we find that agreements made at the highest level can be sabotaged by middle and low level apparatchiks, no one has yet made a profit [28].

(3) Enormous infusion of capital into east Germany, and in particular into east German telecommunication, has accelerated the transformation of the archaic infrastructure and can serve as a model to predict developments in Czechoslovakia, Hungary and Poland for the years 1992–1995.

(4) Significant risks, but also enormous opportunities are associated with capital investments in Eastern Europe, though the probability of success in the telecommunications section is probably higher than elsewhere, especially if joint ventures have well defined objectives and managerial structures.

6.1. Recommendations

(1) Uncertainties and risks involved require thorough knowledge of the political, social and economic conditions in a EaEu country *before* investment and joint-venture decisions are considered.

(2) The political economic consequences of dismantling Stalinist dinosaurs will be felt in the period 1992–1924, in all former East Bloc countries, including east Germany. Investments and joint ventures, even in telecommunications, must be able to experience three to five profitless years; moreover, exit-barriers will be high even after five years.

(3) Eastern/Central Europe presents, however, the most rapidly growing telecommunication market for the 1990s and beyond. Moreover, EaEu is the platform for penetrating the former Soviet Union. No telco that wishes to be a global player can afford to disregard Eastern Europe.

---

[28] Ingo Vogelzang, 'Zur Privatisierung von Telefongesellschaften', *WIK*, November, 1992; Hans Bauer, 'Catching Up in Communications', *Siemens: International Telecom Report*, July–August, 1993, pp. 14–18.

(4) Success in EaEu requires establishment of manufacturing and research facilities. Reliance on the U.S. Department of Commerce and MAFF (Market Access Fact Finding) talks is not a substitute for being firmly, and for the long run, represented in a EaEu country. Manufacturing operation in EaEu will be difficult to manage and will require a long learning period for both Western managers and Eastern work force.

(5) A proper technology transfer strategy must provide immediate aid for a crumbling telecom infrastructure, as well as long run support for the development of a high technology network system and equipment industry.

References

Akerlof, G.A. et al., 1991. A Solution for East Germany. *Financial Times*, May 21, 1991.

Berntson, Göran, 1989. Mobiltex — A New Network for Mobil Data Communication. *Ericsson Review* No. 1.

Chandler, Alfred D., 1965. *Strategy and Structure* (MIT Press).

Derr, F., Marko, H. and Neilinger, S., 1990. Optical Transmission Systems with Coherent Detection. *Siemens Review*, September/October, pp. 30ff.

Franz, Herman, 1990. Prepared for the Next Century: The New Structure of Siemens AG. *Siemens Review*, September/October, pp. 26ff.

Goodhoat, David, 1991. Big Boost for East German Phone Links. *Financial Times*, September 12, p. 3.

Günther, W, and Uhlig. H., 1991. Telekommunikation in der DDR. *Wissenschaftliches Institut für Kommunikationsdienste* No. 90, Vol. 1, p. 90.

Herlof, G., Hessenins, H., Rose, A. and Yellen, J., 1991. *East Germany in from the Cold.* Brookings Papers on Economic Activity.

Lange, Klaus, 1990. *Chancen and Risiken der Mobilfunktechnologien*, October (Wissenschaftliches Institut für Kommunikationsdienste, (WIK), Bad Honnef).

Müller, J., 1991. Das Fernmeldewesen in Osteuropa. *Zeitschrift für Post und Telekommunikation (ZPT)*, November, pp. 54–56,

Ricke, H., 1990. Aus Zwei Mach Eins (Make one out of two). *Zeitschrift für Post und Telekommunikation (ZPT)*, November 28.

Rosenblatt, Joachim, 1991. Telecommunications in Eastern Germany. *(Siemens) International Telecom Report*, May/June.

Schumpeter, Josef, 1938. *Business Cycles*, Vol I.

Selucky, Radoslav, 1972. *Economic Reforms in Eastern Europe* (Praeger, New York).

Subject index